FORM FACTORS IN COMPLETELY INTEGRABLE MODELS OF QUANTUM FIELD THEORY

ADVANCED SERIES IN MATHEMATICAL PHYSICS

Published

Vol. 1: Mathematical Aspects of String Theory
edited by S-T Yau

Vol. 2: Bombay Lectures on Highest Weight Representations of Infinite Dimensional Lie Algebras
by V G Kac and A K Raina

Vol. 3: Kac-Moody and Virasoro Algebras: A Reprint Volume for Physicists
edited by P Goddard and D Olive

Vol. 4: Harmonic Mappings, Twistors and σ-Models
edited by P Gauduchon

Vol. 5: Geometric Phases in Physics
edited by A Shapere and F Wilczek

Vol. 7: Infinite Dimensional Lie Algebras and Groups
edited by V Kac

Vol. 8: Introduction to String Field Theory
by W Siegel

Vol. 9: Braid Group, Knot Theory and Statistical Mechanics
edited by C N Yang and M L Ge

Vol. 10: Yang-Baxter Equations in Integrable Systems
edited by M Jimbo

Vol. 11: New Developments in the Theory of Knots
edited by T Kohno

Vol. 12: Soliton Equations and Hamiltonian Systems
by L A Dickey

Vol. 15: Non-Perturbative Quantum Field Theory – Mathematical Aspects and Applications
Selected Papers of Jürg Fröhlich

Vol. 16: Infinite Analysis – Proceedings of the RIMS Research Project 1991
edited by A Tsuchiya, T Eguchi and M Jimbo

Advanced Series in Mathematical Physics – Vol. 14

FORM FACTORS IN COMPLETELY INTEGRABLE MODELS OF QUANTUM FIELD THEORY

F. A. SMIRNOV

St. Petersburg Branch of Steklov Mathematical Institute

World Scientific

Singapore • New Jersey • London • Hong Kong

Published by

World Scientific Publishing Co. Pte. Ltd.

P. O. Box 128, Farrer Road, Singapore 9128

U.S.A. office: Suite 1B, 1060 Main Street, River Edge, NJ 07661

U.K. office: 73 Lynton Mead, Totteridge, London N20 8DH

FORM FACTORS IN COMPLETELY INTEGRABLE MODELS OF QUANTUM FIELD THEORY

ISBN 981-02-0244-X
981-02-0245-8 (pbk)

Printed in Singapore by Utopia Press.

CONTENTS

CONTENTS

INTRODUCTION

During the last several years the theory of completely integrable models in (1 + 1)-dimensions has attracted much attention of the experts in quantum field theory. The possibility to obtain exact, non-perturbative results for these models is of great importance. Recent activity in the field was initiated by the papers by Faddeev and Korepin [1,2] devoted to the quasiclassical quantization of solitons in sine-Gordon model (SG). The connection of SG with massive Thirring model established by Coleman [3] stimulated further interest to SG model in particular and to the quantization of solitons in general. This interest was stimulated also by the analogy of (1 + 1)-dimensional nonlinear σ-model (NLσ) and (3 + 1)-dimensional Yang-Mills theory outlined by Polyakov [4].

Korepin and Faddeev [1] realized the most important feature of the completely integrable quantum models, which is the factorizability of scattering. They predicted the exact S-matrix for SG model with the special values of coupling constant corresponding to the absence of the reflection of solitons. This hypothesis was used by Zamolodchikov as the basis for phenomenological calculation of S-matrix for SG model [5]. Further development of the phenomenological (bootstrap) approach based on factorizability, unitarity and crossing-symmetry led to the exact calculation of S-matrices for many completely integrable models (see the review by A.B. and Al. B. Zamolodchikovs [6] and the papers by Karowski and collaborators [7-9]).

The quantum inverse scattering method (QISM) formulated by Faddeev, Sklyanin and Takhtajan [10-12] allows us to unify the theories of the completely integrable quantum models. QISM makes it possible to pass from local fields to the physical creation–annihilation operators. Many

S-matrices predicted phenomenologically have been confirmed dynamically in the framework of QISM (the paper by Korepin [13] was important in this respect). The achievements of QISM are reflected in several reviews [14-16].

In the context of this book, the paper by Thacker and collaborators [17] is worth special attention. In this paper the quantum version of the Gelfand-Levitan-Marchenko (GLM) equations was formulated. These equations allow us to get the inverse transform from the creation–annihilation operators of physical particles to the local fields. Further development of GLM method was given in the papers by Göckeler, Kulish and the author [18-20]. Let us note, however, that all the papers mentioned above deal with models whose physical vacuum coincides with the Fock vacuum of the local fields. These models are rather trivial (in particular relativistic models cannot be of this kind). That is why their consideration is mostly of methodological character. The first model with nontrivial ground state considered in GLM method was XXZ ferromagnet with domain wall [21].

Let us describe the situation in QISM up to 1983. Up to that time the method had been applied to many models and had allowed one to calculate exact S-matrices. The theory of factorizable S-matrices was further developed (the paper by Kulish, Reshetikhin and Sklyanin [22] is of particular importance). However, no remarkable out-of-shell results had been obtained. Thus the calculation of the Green functions and the objects connected with them became the most important problem.

In 1984 Korepin proposed a method of presenting the Green functions for Bose gas with nonzero density in the form of special series [23]. We cannot describe this method in more detail (see review [24]) than saying that its greatest advantage is the calculation of the long distance asymptotics of the Green functions for several interesting models.

The development of GLM method was also continued. In the paper [25] the author obtained GLM equations for SG model. These equations allowed one to calculate the matrix elements of the local field (form factors) corresponding to the physical states which contain SG bosons (breathers) [26]. The spectrum of SG contains, besides breathers, two-component particules–solitons. According to Ref. [27] the GLM equation when applied to solitons lead to a set of matrix Riemann problems. At first sight there is no hope of solving these Riemann problems. Here, however, two circumstances help. First, for certain values of the coupling constant, the reflection of solitons disappears. In that case the Riemann problems become rather simple and soliton form factors for the reflectionless case have been calculated in the papers [28, 29]. Second, GLM equations provide some additional information. This additional information, known soliton form factors in reflectionless case and known breather form factors, provided the author with the heuristic base necessary for the calculation of soliton form factors [27, 30] (the details are given in [31]). The very possibility of obtaining exact formulae for form factors reflects the remarkable selfconsistency and beauty of completely integrable models.

It is necessary to mention the papers by Kirillov and Khamitov [32, 33]. In these papers the following programme proposed by E.K. Sklyanin was realized. It is known that the sinh-Gordon model, possessing phenomenologically very simple spectrum which involves only one boson, cannot be investigated in the usual framework of QISM. However, one can try to write down the hypothetical GLM equations for the model by analogy with SG. The proposed method of verification of the equations is very natural: one has to make sure that the local fields defined by the equations do satisfy local commutativity.

Analyzing the results of the papers [32, 33] and having in mind possible applications to more complicated models, the author realizes that the GLM equations are not themselves essential for the proof of the local commutativity and all the necessary requirements can be formulated directly in terms of form factors. It should be mentioned that attempts to formulate the full set of requirements on form factors had been made by Karowski and collaborators [34], but their set was not complete and possessed a number of drawbacks because the factorizability was not taken into account in all its aspects.

In the paper [35, 36] Kirillov and the author, considering the particular example of SU(2)-invariant Thirring model (ITM), realized the following programme. The requirements on form factors were formulated which were shown to ensure local commutativity. These requirements had the form of a system of equations, which were solved explicitly. This last step was the most complicated one because we had to solve a matrix Riemann problem. The SG experience was of great importance.

In the paper [37] Kirillov and the author realized a similar programme for the O(3) nonlinear σ-model (NLS). The importance of this model was mentioned above. In spite of the fact that the spectrum and the S-matrix for this model were predicted several years ago, for a long time the model could not be treated by QISM because the usual formalism of the method was hardly applicable to it.

It should be mentioned that the structure of the form factors described in Refs. [30, 35, 37] is very general. It was demonstrated in the paper [38] where the formulae for the form factors in SU(N)-invariant Thirring model (chiral Gross-Neveu model) were presented.

In the present work we summarize the results of the papers [27, 30, 31, 35-38]. We follow the axiomatic approach outlined in the papers [35-38]. The main goals of the present work are: first, to describe the general scheme of the calculation of form factors; second, to present complete proofs and reasonings which had been omitted in the original papers because of their limited volume.

Let us describe briefly the organization of the work. It consists of eleven sections and appendices.

Section 0 contains the necessary general information about completely integrable models of quantum field theory. The spectra and S-matrices of SG, ITM, NLS are presented.

In Sec. 1 we formulate a set of requirements on form factors (axioms). The origination of these axioms is clarified by considering the well-known example of two-particle form factors. A sketch of possible phenomenological deduction of the axioms is given in Appendix 2.

Section 2 is the ideological base of our programme. It contains the local commutativity theorem which is formulated as follows. If the form factors of some operator satisfy the axioms formulated in Sec. 1 then the operator is local.

In Sec. 3 we present the explicit formulae for SG soliton form factors.

In Sec. 4 we show that the SG form factors defined in Sec. 3 do satisfy the axioms.

In Sec. 5 we discuss the SG form factors corresponding to both solitons and breathers.

In Sec. 6 some additional properties of SG form factors are established which allow us to identify the operators defined by the form factors with natural local operators occurring in the model.

In Sec. 7 the form factors in ITM are obtained.

In Sec. 8 a procedure is described of obtaining NLS form factors from ITM ones. The explicit formulae for NLS form factors are presented and their properties studied.

Section 9 deals with the asymptotics of the form factors for large momentum transfer.

In Sec. 10 the singularities of the commutators at the origin of coordinates are studied. The asymptotics obtained in Sec. 9 are essential here.

Appendix 1 considers the SU(N)-invariant Thirring model.

The content of Appendix 2 has already been mentioned.

I am grateful to L.D. Faddeev for his interest in my work and continued support.

I also gratefully acknowledge my thanks for many useful discussions with my colleagues at the Leningrad Branch of Steklov Mathematical Institute: A.G. Izergin, V.E. Korepin, P.P. Kulish, N. Yu. Reshetikin, M.A. Semenov-Tian-Shansky, E.K. Sklyanin and L.A. Takhtajan. Special thanks are due to A.N. Kirillov for fruitful collaboration.

FORM FACTORS IN COMPLETELY INTEGRABLE MODELS OF QUANTUM FIELD THEORY

0

COMPLETELY INTEGRABLE MODELS OF QUANTUM FIELD THEORY

In this section we present briefly the necessary facts concerning the completely integrable models of quantum field theory with massive spectra. We parametrize the massive particles by their rapidities β ($p_0 = m \text{ ch } \beta, p_1 = m \text{ sh } \beta, m$ being the mass of the particle), and the isotropic index ε if the particle possesses internal degree of freedom. The scattering in these models is factorized: the many-particle scattering process is reduced to two-particle processes, momenta of particles are not subjected to any changes. The S-matrix of two particles with rapidities β_1, β_2 will be denoted by $S_{\varepsilon_1,\varepsilon_2}^{\varepsilon'_1,\varepsilon'_2}(\beta)$, where $\beta = |\beta_1 - \beta_2|$, $\varepsilon_1, \varepsilon_2$ are characteristic of the internal degrees of freedom of "in", $\varepsilon'_1, \varepsilon'_2$ of "out" particles. The factorization of scattering is equivalent to the validity of the equation of triangles [6] (Yang-Baxter equation [38, 40]):

$$
\begin{aligned}
&S_{\varepsilon'_1,\varepsilon'_2}^{\varepsilon_1,\varepsilon_2}(\beta_1 - \beta_2) S_{\varepsilon''_1,\varepsilon'_3}^{\varepsilon'_1,\varepsilon_3}(\beta_1 - \beta_3) S_{\varepsilon''_2,\varepsilon''_3}^{\varepsilon'_2,\varepsilon'_3}(\beta_2 - \beta_3) \\
&\quad = S_{\varepsilon'_2,\varepsilon'_3}^{\varepsilon_2,\varepsilon_3}(\beta_2 - \beta_3) S_{\varepsilon'_1,\varepsilon''_3}^{\varepsilon_1,\varepsilon'_3}(\beta_1 - \beta_3) S_{\varepsilon''_1,\varepsilon''_2}^{\varepsilon_1,\varepsilon'_2}(\beta_1 - \beta_2) \; .
\end{aligned}
$$

Note that summation over repeated indices is always implied. The S-matrix should also satisfy the requirements of unitarity and crossing symmetry:

$$
\begin{gathered}
S_{\varepsilon''_1,\varepsilon''_2}^{\varepsilon_1,\varepsilon_2}(\beta) S_{\varepsilon'_1,\varepsilon'_2}^{\varepsilon''_1,\varepsilon''_2}(-\beta) = \delta_{\varepsilon'_1}^{\varepsilon_1} \delta_{\varepsilon'_2}^{\varepsilon_2} \; , \\
S_{\varepsilon_1,\varepsilon_2}^{\varepsilon'_1,\varepsilon'_2}(\pi i - \beta) = C_{\varepsilon_1 \varepsilon''_1} S_{\varepsilon'''_1,\varepsilon_2}^{\varepsilon''_1,\varepsilon'_2}(\beta) C^{\varepsilon'''_1 \varepsilon'_1}
\end{gathered}
\tag{1}
$$

where C is a matrix satisfying two requirements: $C^t = C, C^2 = I$. The value of C will be important.

In the basic part of the book three models are considered which are the sine-Gordon model (SG), SU(2)-invariant Thirring model (ITM) and O(3) nonlinear σ-model (NLS).

SG model is described by the Lagrangian

$$\mathcal{L}_{\text{SG}} = \int ((\partial_\mu u)^2 + \frac{m^2}{\beta^2}(\cos(\beta u) - 1))dx ,$$

where u is a Bose-field. Great attention has been paid to the quantization of SG model [1,2,5,12,13,41-45]. Let us summarize the necessary results. We use the following quantity as the coupling constant,

$$\xi = \frac{\pi\beta^2}{8\pi - \beta^2} , \quad 0 < \xi < \infty .$$

SG model is equivalent [3] to the neutral sector of massive Thirring model (MTM) whose Lagrangian is

$$\mathcal{L}_{\text{MTM}} = \int \left(i\overline{\psi}\gamma_\mu\partial_\mu\psi - \frac{g}{2}(\overline{\psi}\gamma_\mu\psi)^2 - m\overline{\psi}\psi \right) dx ,$$

where $\psi = \{\psi_i\}_{i=1,2}$ is the Fermi field, $\gamma_0 = \sigma^1, \gamma_1 = i\sigma^2, \sigma^i (i = 1,2,3)$ are the Pauli matrices, $g = \frac{\pi}{2}\left(\frac{\pi-\xi}{\xi}\right)$.

The correspondence between local operators was established by Coleman [3]:

$$\begin{aligned} \varepsilon^{\mu\nu}\partial_\nu u &= -\frac{2\pi}{\beta} j^\mu , \\ \cos\beta u &= -\overline{\psi}\psi , \end{aligned} \tag{2}$$

where j^μ is the fermion current $\overline{\psi}\gamma^\mu\psi$.

The spectrum of SG model contains solitons possessing an internal degree of freedom (soliton-antisoliton). The possible values of the corresponding index ε are chosen as $\pm\frac{1}{2}$. Solitons correspond to fermions in MTM. Two ranges should be distinguished. The range $\pi < \xi < \infty$ is the repulsive one, there are no bound states for these values of ξ. The point $\xi = \pi$ corresponds to the free fermion case, interaction disappears and the model is equivalent to the massive free Fermi field. The range $0 < \xi < \pi$ is an attractive one; here bound states (breathers) exist. These are $\left[\frac{\pi}{\xi}\right]$ breathers. The mass of m-breather $\left(m = 1, \dots, \left[\frac{\pi}{\xi}\right]\right)$ is equal to $2M \sin \frac{m\xi}{2}$ (M is the soliton mass). For the sake of uniformity we allow ourselves an inaccuracy by prescribing to m-breather the index $\varepsilon = m$. Strictly speaking, m does not correspond to any internal degree of freedom. However, this prescription will be very useful.

Soliton-soliton scattering matrix is given by [5]:

$$
\begin{aligned}
&S_{\frac{1}{2},\frac{1}{2}}^{\frac{1}{2},\frac{1}{2}}(\beta) = S_{-\frac{1}{2},-\frac{1}{2}}^{-\frac{1}{2},-\frac{1}{2}}(\beta) = S_0(\beta) \ , \\
&S_{\frac{1}{2},-\frac{1}{2}}^{\frac{1}{2},-\frac{1}{2}}(\beta) = S_{-\frac{1}{2},\frac{1}{2}}^{-\frac{1}{2},\frac{1}{2}}(\beta) = -\frac{\mathrm{sh}\, \frac{\pi}{\xi}\beta}{\mathrm{sh}\, \frac{\pi}{\xi}(\beta - \pi i)} S_0(\beta) \ , \\
&S_{-\frac{1}{2},\frac{1}{2}}^{\frac{1}{2},-\frac{1}{2}}(\beta) = S_{\frac{1}{2},-\frac{1}{2}}^{-\frac{1}{2},\frac{1}{2}}(\beta) = \frac{\mathrm{sh}\, \frac{\pi^2 i}{\xi}}{\mathrm{sh}\, \frac{\pi}{\xi}(\beta - \pi i)} S_0(\beta) \ , \\
&S_{\varepsilon_1,\varepsilon_2}^{\varepsilon_1',\varepsilon_2'}(\beta) = 0 \quad \varepsilon_1' + \varepsilon_2' \neq \varepsilon_1 + \varepsilon_2 \ , \\
&\qquad S_0(\beta) = -\exp\left\{ -i \int_0^\infty \frac{\sin(\kappa\beta)\, \mathrm{sh}\, \left(\frac{\pi - \xi}{3}\kappa\right)}{\kappa\, \mathrm{ch}\, \left(\frac{\pi\kappa}{2}\right) \mathrm{sh}\, \left(\frac{\xi\kappa}{2}\right)} d\kappa \right\} \ .
\end{aligned}
\tag{3}
$$

Soliton-breather scattering matrix is given by

$$
\begin{aligned}
&S_{m,\frac{1}{2}}^{\varepsilon_1,\varepsilon_2}(\beta) = S_{\frac{1}{2},m}^{\varepsilon_2,\varepsilon_1}(\beta) = \delta_{\frac{1}{2}}^{\varepsilon_2} \delta_m^{\varepsilon_1} S_{S,m}(\beta) \ , \\
&S_{m,-\frac{1}{2}}^{\varepsilon_1,\varepsilon_2}(\beta) = S_{-\frac{1}{2},m}^{\varepsilon_2,\varepsilon_1}(\beta) = \delta_{-\frac{1}{2}}^{\varepsilon_2} \delta_m^{\varepsilon_1} S_{S,m}(\beta) \ , \\
&\qquad S_{S,m}(\beta) = \prod_{j=1}^{m} \frac{i \cos \frac{\xi}{2} + \mathrm{sh}\, \left(\beta - \frac{i\xi}{2}(m+1-2j)\right)}{i \cos \frac{\xi}{2} - \mathrm{sh}\, \left(\beta - \frac{i\xi}{2}(m+1-2j)\right)} \ .
\end{aligned}
\tag{4}
$$

Breather-breather scattering matrix is given by

$$
\begin{aligned}
S_{m,n}^{\varepsilon_1,\varepsilon_2}(\beta) &= \delta_m^{\varepsilon_1} \delta_n^{\varepsilon_2} \mathrm{cth}\, \frac{1}{2}\left(\beta - \frac{i(m+n)}{2}\xi\right) \mathrm{cth}\, \frac{1}{2}\left(\beta - i\frac{|m-n|}{2}\xi\right) \\
&\times \mathrm{th}\, \frac{1}{2}\left(\beta + i\frac{(m+n)}{2}\xi\right) \mathrm{th}\, \frac{1}{2}\left(\beta + i\frac{|m-n|}{2}\xi\right) \prod_{j=1}^{\min(m,n)-1} \\
&\times \mathrm{th}^2\, \frac{1}{2}\left(\beta + \frac{i\xi}{2}(|m-n|+2j)\right) \mathrm{cth}^2\, \frac{1}{2}\left(\beta - \frac{i\xi}{2}(|m-n|+2j)\right) \ .
\end{aligned}
\tag{5}
$$

From Eq. (3) it follows that the values of the coupling constant $\xi = \pi/\nu, \nu = 1, 2, \dots,$ are very specific because at these points $S_{\frac{1}{2},-\frac{1}{2}}^{-\frac{1}{2},\frac{1}{2}}(\beta) =$

$S^{\frac{1}{2},-\frac{1}{2}}_{-\frac{1}{2},\frac{1}{2}}(\beta) = 0$, i.e., the reflection of solitons is absent. In particular, for $\xi = \pi$ the S-matrix is equal to -1, which corresponds to the free fermion case.

Nontrivial matrix C in Eq. (1) exists only for soliton-soliton scattering: $C = \sigma^1$.

An important circumstance should be mentioned which was a subject of controversy a few years ago. In the papers [42, 43] it is established that there are phase transitions at the points $\xi = \pi\kappa, \kappa \geq 2$. The existence of the phase transitions results in changes of the spectrum of excitations at these points. In the paper [45] it is established that the spectrum of excitations is the same for all the values of coupling constant in the range $\xi > \pi$ (this is the picture we follow). The contradiction has the following solution. In papers [42, 43] and [45], different regularizations were used for the quantization of SG model which is sensitive to the regularization used for $\xi > \pi$. Thus papers [42, 43] and [45] deal with different regularized models. As is mentioned above, we follow the picture presented in [45], which is in our opinion more natural from the phenomenological point of view: there is no need for phase transitions because the S-matrix is a regular function of the coupling constant. This choice does not mean that we reject the regularization given in [42, 43]; we only consider this regularization more useful, rather, for statistical mechanics than for quantum field theory.

The second model we deal with is ITM with the Lagrangian

$$\mathcal{L}_{\mathrm{ITM}} = \int (i\overline{\psi}\gamma_\mu\partial_\mu\psi - g(\overline{\psi}\sigma^a\gamma_\mu\psi)(\overline{\psi}\sigma^a\gamma_\mu\psi))dx \ ,$$

where $\psi = \{\psi_i^\alpha\}$ is the Fermi field, $i = 1,2$ being the spinor index, $\alpha = 1,2$ being the isotopic index, σ^a are the Pauli matrices acting on the isotopic indices. The model possesses the property of asymptotic freedom. Mass is generated dynamically via the mechanism of dimensional transmutation [46]. Physical excitations are two-component massive particles (kinks) which will be parametrized by their rapidities and isotopic $\varepsilon = 1,2$. Mass of the kink will be denoted by M. The S-matrix is given by

$$S^{\varepsilon_1',\varepsilon_2'}_{\varepsilon_1,\varepsilon_2}(\beta) = S_0(\beta)\frac{1}{\beta - \pi i}(\beta\delta^{\varepsilon_1'}_{\varepsilon_1}\delta^{\varepsilon_2'}_{\varepsilon_2} - \pi i\delta^{\varepsilon_2'}_{\varepsilon_1}\delta^{\varepsilon_1'}_{\varepsilon_2}) \ ,$$
$$S_0(\beta) = \frac{\Gamma\left(\frac{1}{2} + \frac{\beta}{2\pi i}\right)\Gamma\left(-\frac{\beta}{2\pi i}\right)}{\Gamma\left(\frac{1}{2} - \frac{\beta}{2\pi i}\right)\Gamma\left(\frac{\beta}{2\pi i}\right)} \ . \tag{6}$$

Kinks do not create any bound states. The S-matrix satisfies (1) with $C = i\sigma^2$. Also the S-matrix is SU(2)-invariant, which means that

$$S^{\varepsilon_1 \varepsilon_2}_{\varepsilon''_1 \varepsilon''_2}(\beta) g^{\varepsilon''_1}_{\varepsilon'_1} g^{\varepsilon''_2}_{\varepsilon'_2} = g^{\varepsilon_1}_{\varepsilon''_1} g^{\varepsilon_2}_{\varepsilon''_2} S^{\varepsilon''_1 \varepsilon''_2}_{\varepsilon'_1 \varepsilon'_2}(\beta) \tag{7}$$

for arbitrary $g \in \mathrm{SU}(2)$.

Notice that the charge conjugation matrix C for ITM does not satisfy the condition $C^t = C$ (actually $C^t = -C$). This fact reflects the unusual property of ITM kinks: they are spin-$\frac{1}{4}$ particles. However it is possible to modify the base in the space of states and to introduce auxiliary particles which possess the usual statistics. The details are given in Sec. 7. The connection between ITM and SG is also discussed in Sec. 7.

NLS model exhibits the greatest difficulty for quantization [47, 49], although phenomenologically the spectrum of excitations has been predicted several years ago [50-54]. The model is described by the Lagrangian

$$\mathcal{L}_{\mathrm{NLS}} = \frac{1}{g} \int (\partial_\mu n^a \partial_\mu n^a) dx \ ,$$

where $a = 1, 2, 3$, and the constraint $(n^1)^2 + (n^2)^2 + (n^3)^2 = 1$ applies. The model possesses asymptotic freedom. Physical excitations are massive (due to dimensional transmutation) vector particles. We parametrize them by the rapidity β and isotopic index $\varepsilon = 1, 2, 3$ and denote the mass of the particle by M. The S-matrix is given by

$$\begin{aligned} S^{\varepsilon'_1, \varepsilon'_2}_{\varepsilon_1, \varepsilon_2}(\beta) &= \frac{1}{(\beta + \pi i)(\beta - 2\pi i)} \\ &\times \left\{ \beta(\beta - \pi i) \delta^{\varepsilon'_1}_{\varepsilon_1} \delta^{\varepsilon'_2}_{\varepsilon_2} - 2\pi i (\beta - \pi i) \delta^{\varepsilon'_2}_{\varepsilon_1} \delta^{\varepsilon'_1}_{\varepsilon_2} + 2\pi i \delta_{\varepsilon_1, \varepsilon_2} \delta^{\varepsilon'_1, \varepsilon'_2} \right\} . \end{aligned} \tag{8}$$

The S-matrix is O(3)-invariant, i.e., satisfies (7) with $g \in \mathrm{O}(3)$. Matrix C in Eq. (1) is equal to the unit matrix. Bound states are absent.

Formally, this is all the necessary preliminary information about the completely integrable models we need. If the reader should think this information is too brief we refer him to the original papers cited in this section.

1
THE SPACE OF PHYSICAL STATES. THE NECESSARY PROPERTIES OF FORM FACTORS.

For the description of the space of physical states we use the Zamolodchikov-Faddeev operators. The existence of these operators is an axiom in our approach. Their concrete realizations in the framework of QISM for some models can be found in [25, 55]. Generally speaking, we could avoid using these operators but in that case the methodology would become more complicated and bulky. The Zamolodchikov-Faddeev operators $Z^{\varepsilon}(\beta), Z^*_{\varepsilon}(\beta)$ create the following algebra:

$$\begin{aligned}
&Z^{\varepsilon_1}(\beta_1)Z^{\varepsilon_2}(\beta_2) = S^{\varepsilon_1,\varepsilon_2}_{\varepsilon'_1,\varepsilon'_2}(\beta_1-\beta_2)Z^{\varepsilon'_2}(\beta_2)Z^{\varepsilon'_1}(\beta_1) \ , \\
&Z^*_{\varepsilon_1}(\beta_1)Z^*_{\varepsilon_2}(\beta_2) = Z^*_{\varepsilon'_2}(\beta_2)Z^*_{\varepsilon'_1}(\beta_1)S^{\varepsilon'_1,\varepsilon'_2}_{\varepsilon_1,\varepsilon_2}(\beta_1-\beta_2) \ , \\
&Z^{\varepsilon_1}(\beta_1)Z^*_{\varepsilon_2}(\beta_2) = Z^*_{\varepsilon'_2}(\beta_2)S^{\varepsilon'_2\varepsilon_1}_{\varepsilon_2\varepsilon'_1}(\beta_2-\beta_1)Z^{\varepsilon'_1}(\beta_1) + \delta^{\varepsilon_1}_{\varepsilon_2}\delta(\beta_1-\beta_2) \ .
\end{aligned} \tag{9}$$

There is the physical vacuum $|ph\rangle$, and the operator $Z^{\varepsilon}(\beta)$ annihilates the vacuum. Operator $Z^*_{\varepsilon}(\beta)$ creates a particle with rapidity β and internal degree of freedom (for breathers in SG – with the number) ε.

The space of states is generated by vectors

$$|\beta_n,\dots,\beta_1\rangle_{\varepsilon_n,\dots,\varepsilon_1} \equiv Z^*_{\varepsilon_n}(\beta_n)\dots Z^*_{\varepsilon_1}(\beta_1)|ph\rangle \ ,$$

and the dual space

$$^{\varepsilon_1,\dots,\varepsilon_n}\langle\beta_1,\dots,\beta_n| = \langle ph|Z^{\varepsilon_1}(\beta_1)\dots Z^{\varepsilon_n}(\beta_n) \ .$$

Let us order the rapidities $\beta_1, \ldots, \beta_n$ in accordance with increasing values, $\beta_{i_1} < \beta_{i_2} \cdots < \beta_{i_n}$. "In" and "out" states are given by the formulae

$$|\beta_1, \ldots, \beta_n\rangle^{\text{in}}_{\varepsilon_1, \ldots, \varepsilon_n} = |\beta_{i_n}, \ldots, \beta_{i_1}\rangle_{\varepsilon_{i_n}, \ldots, \varepsilon_{i_1}} ,$$
$$|\beta_1, \ldots, \beta_n\rangle^{\text{out}}_{\varepsilon_1, \ldots, \varepsilon_n} = |\beta_{i_1}, \ldots, \beta_{i_n}\rangle_{\varepsilon_{i_1}, \ldots, \varepsilon_{i_n}} .$$

"In" and "out" states form the complete base in the space of state. It is clear from Eq. (9) that "in" states transform to "out" states via the total S-matrix.

Consider any local operator $O(x_0, x_1)$ and construct the matrix element

$$\begin{aligned} {}_{\varepsilon'_1, \ldots, \varepsilon'_m}\langle \beta'_1, \ldots, \beta'_m | O(x_0, x_1) | \beta_n, \ldots, \beta_1\rangle_{\varepsilon_n, \ldots, \varepsilon_1} \\ \equiv \exp(i\Sigma p^\mu(\beta'_j)x_\mu - i\Sigma p^\mu(\beta_j)x_\mu) \\ \times f^{\varepsilon'_m, \ldots, \varepsilon'_1}_{\varepsilon_1, \ldots, \varepsilon_n}(\beta'_m, \ldots, \beta'_1 | \beta_1, \ldots, \beta_n) . \end{aligned}$$

The function $f^{\varepsilon'_m, \ldots, \varepsilon'_1}_{\varepsilon_1, \ldots, \varepsilon_n}(\beta'_m, \ldots, \beta'_1 | \beta_1, \ldots, \beta_n)$ will be called "form factor". Usually, in the Quantum Field Theory the term "form factor" is used for the matrix element calculated between "in" or "out" states (frequently this term is used only for two-particle matrix elements). The term "form factor" in our meaning is a special generalization of the usual one for the theory with factorized scattering. If the sets β and β' are ordered we obtain the matrix element taken between "in" or "out" states. For arbitrary sets β and β' we have some "mixed", neither "in" nor "out" states.

In the present section the requirements which form factors should satisfy are formulated. To give an intuitive idea about these requirements we would like to consider, following [34], the two-particle form factors whose properties are well-known from the general course of Quantum Field Theory (see for example [56, 57]). Consider for example the two-fermion form factor of current j_μ in MTM (two-soliton form factor of $\varepsilon_{\mu\nu}\frac{\beta}{2\pi}\partial_\nu u$ in SG):

$$\langle ph | j_\mu(0) | \beta_1, \beta_2\rangle^{\text{in}}_{\varepsilon_1, \varepsilon_2} \equiv F(\beta_1, \beta_2)_{\varepsilon_1, \varepsilon_2} \left(\sum_{j=1}^{2} (e^{\beta_j} - (-1)^\mu e^{-\beta_j}) \right) .$$

From manuals on Quantum Field Theory it is known that F depends really on the variable $S = (p(\beta_1) + p(\beta_2))^2 = 2M^2(1 + \text{ch}\,(\beta_1 - \beta_2))$, being the boundary value of an analytical function of S defined on the complex plane with a cut from $4M^2$ to ∞ (Fig. 1). The possible poles are situated on the

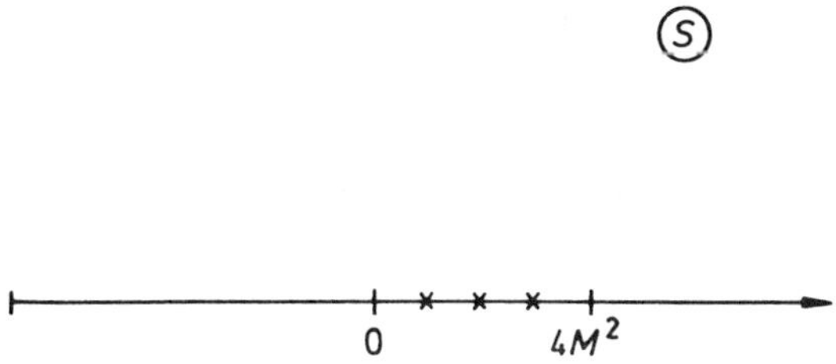

Fig. 1.

segment $[0, 4M^2]$. These poles correspond to bound states. The value of the analytical function at the point $S - i0 (S \in \mathbb{R})$ coincides with

$$\langle ph | j_\mu(0) | \beta_1, \beta_2 \rangle^{\text{out}}_{\varepsilon_1, \varepsilon_2} .$$

Evidently the plane with the cut transforms into the half-strip $0 < \operatorname{Im} \beta < 2\pi, \operatorname{Re} \beta > 0$, in terms of the variable $\beta = |\beta_1 - \beta_2|$. The line $\operatorname{Im} \beta = 0$ corresponds to the upper, while $\operatorname{Im} \beta = 2\pi$ to the lower shore of the cut. The segment $[0, 4M^2]$ is a pre-image of the segment $[0, 2\pi i]$, being covered twice by inverse transformation. Consider for definiteness case $\beta_2 > \beta_1$ and return to the function $f(\beta_1, \beta_2)_{\varepsilon_1, \varepsilon_2}$. It is clear from the very definition that

$$f(\beta_1, \beta_2)_{\varepsilon_1, \varepsilon_2} S^{\varepsilon_1, \varepsilon_2}_{\varepsilon'_1, \varepsilon'_2}(\beta_1 - \beta_2) = f(\beta_2, \beta_1)_{\varepsilon'_2, \varepsilon'_1} . \tag{10}$$

The general properties described above result in the following relation for $\beta_2 > \beta_1$:

$$f(\beta_1, \beta_2 + 2\pi i)_{\varepsilon_1, \varepsilon_2} = f(\beta_2, \beta_1)_{\varepsilon_2, \varepsilon_1} . \tag{11}$$

Equation (11) applies to all the values of β_1, β_2 due to continuity.

Combining (10) and (11) one obtains the following equation:

$$f(\beta_1, \beta_2 + 2\pi i)_{\varepsilon_1, \varepsilon_2} = f(\beta_1, \beta_2)_{\varepsilon'_1 \varepsilon'_2} S^{\varepsilon'_1, \varepsilon'_2}_{\varepsilon_1, \varepsilon_2}(\beta_1 - \beta_2) , \tag{12}$$

which is the Riemann problem for the functions $f(\beta_1, \beta_2)$, being considered as functions of β_2. Function $f(\beta_1, \beta_2)$ has simple poles on the segment $\operatorname{Re} \beta_2 = \beta_1$, which corresponds to bound states. Two simple poles at the points $\beta_2 = \beta_1 + i\alpha, \beta_2 = 2\pi i + \beta_1 - i\alpha$ correspond to the bound state with the mass $2M \cos \frac{\alpha}{2}$. These facts taken together provide us with the possibility of explicit calculation of $f(\beta_1, \beta_2)_{\varepsilon_1, \varepsilon_2}$. The Riemann problem in this case is solvable trivially because the S-matrix can be diagonalized via a transformation which does not depend on β_1, β_2.

Let us return to the general case. Consider the form factors $f(\beta_1, \dots, \beta_n)_{\varepsilon_1, \dots, \varepsilon_n}$. It will become clear from what follows that other form factors can be re-constructed from these. It follows from the very definition that

$$f(\beta_1, \dots, \beta_i, \beta_{i+1}, \dots, \beta_n)_{\varepsilon_1, \dots, \varepsilon_i, \varepsilon_{i+1}, \dots, \varepsilon_n} S^{\varepsilon_i, \varepsilon_{i+1}}_{\varepsilon'_i, \varepsilon'_{i+1}}(\beta_i - \beta_{i+1})$$
$$= f(\beta_1, \dots, \beta_{i+1}, \beta_i, \dots, \beta_n)_{\varepsilon_1, \dots, \varepsilon'_{i+1}, \varepsilon'_i, \dots, \varepsilon_n} . \quad (13)$$

Equation (13) will be called the symmetry property.

How can one generalize Eq. (12) to the case of arbitrary n? In the paper [34] some attempt was made in the framework of the general field theoretical approach. However, the meaning of the equations presented in [34] is not very clear. Form factors which really depend on $n-1$ real variables enter these equations as functions of all possible S-variables, i.e. of $\frac{n(n-1)}{2}$ variables.

Using GLM equations, the author came in papers [27, 30] to the simple and fruitful generalization of Eq. (12):

$$f(\beta_1, \dots, \beta_n + 2\pi i)_{\varepsilon_1, \dots \varepsilon_n} = f(\beta_n, \beta_1, \dots, \beta_{n-1})_{\varepsilon_n, \varepsilon_1, \dots, \varepsilon_{n-1}} . \quad (14)$$

In this equation the specific features of factorized scattering are taken into account. We will not present here any derivation of Eq. (14). Its validity will be proven *a posteriori* in the next section.

Using Eq. (13), Eq. (14) can be rewritten in the form

$$f(\beta_1, \dots, \beta_n + 2\pi i)_{\varepsilon_1, \dots, \varepsilon_n} = f(\beta_1, \dots, \beta_n)_{\varepsilon'_1, \dots, \varepsilon'_n}$$
$$\times S^{\varepsilon'_{n-1}, \varepsilon'_n}_{\varepsilon_{n-1}, \tau_1}(\beta_{n-1} - \beta_n) S^{\varepsilon'_{n-2}, \tau_1}_{\varepsilon_{n-2}, \tau_2}(\beta_{n-2} - \beta_n) \dots S^{\varepsilon'_1, \tau_{n-2}}_{\varepsilon_1, \varepsilon_n}(\beta_1 - \beta_n) . \quad (15)$$

Equation (14) is an axiom in our approach. The equivalent Eq. (15) is a Riemann problem for the tensor-valued function. At first sight there is no hope of solving this problem. However, the miraculous fact is that for all the models considered this problem can be solved explicitly. In our opinion this is a consequence of some hidden symmetry of the problem.

What should be said about $f(\beta_1, \dots \beta_n)_{\varepsilon_1, \dots, \varepsilon_n}$, considered as a function of β_n? We require it to be an analytical function in the strip $0 \leq \text{Im } \beta_n \leq 2\pi$

whose only singularities are simple poles. In contrast to the two-particle form factor, the n-particle one has annihilation poles at the points $\beta_n = \beta_j + \pi i$, $j = 1, \dots, n-1$, with residues

$$
\begin{aligned}
&2\pi i \operatorname{res} f(\beta_1, \dots, \beta_n)_{\varepsilon_1, \dots, \varepsilon_n} \\
&\quad = f(\beta_1, \dots, \hat{\beta}_j, \dots, \beta_{n-1})_{\varepsilon'_1, \dots, \hat{\varepsilon}'_j, \dots, \varepsilon'_{n-1}} C_{\varepsilon_n, \varepsilon'_j} \\
&\quad \times \Big\{ \delta^{\varepsilon'_1}_{\varepsilon_1} \dots \delta^{\varepsilon'_{j-1}}_{\varepsilon_{j-1}} S^{\varepsilon'_{n-1}, \varepsilon'_j}_{\varepsilon_{n-1}, \tau_1}(\beta_{n-1} - \beta_j) S^{\varepsilon'_{n-2}, \tau_1}_{\varepsilon_{n-2}, \tau_2}(\beta_{n-2} - \beta_j) \dots \\
&\quad \times S^{\varepsilon'_{j+1}, \tau_{n-j-2}}_{\varepsilon_{j+1}, \varepsilon_j}(\beta_{j+1} - \beta_j) - S^{\varepsilon'_j, \varepsilon'_1}_{\tau_1, \varepsilon_1}(\beta_j - \beta_1) \dots \\
&\quad \times S^{\tau_{j-3}, \varepsilon'_{j-2}}_{\tau_{j-2}, \varepsilon_{j-2}}(\beta_j - \beta_{j-2}) S^{\tau_{j-2}, \varepsilon'_{j-1}}_{\varepsilon_j, \varepsilon_{j-1}}(\beta_j - \beta_{j-1}) \delta^{\varepsilon'_{j+1}}_{\varepsilon_{j+1}} \dots \delta^{\varepsilon'_{n-1}}_{\varepsilon_{n-1}} \Big\} , \\
&\qquad\qquad\qquad\qquad\qquad\qquad \beta_n = \beta_j + \pi i \qquad (16)
\end{aligned}
$$

where C is the matrix involved in Eq. (1), and for SG breathers $C_{\varepsilon_1 \varepsilon_2} = \delta_{\varepsilon_1, \varepsilon_2}$. These singularities are the only singularities of the form factors for models containing no bound states. Eq. (16) is an axiom in our approach.

For models with bound states (SG $\xi < \pi$) there are additional poles. It is sufficient to indicate these poles in the strip $0 < \operatorname{Im} \beta_n < \pi$ because, using Eqs. (13) and (14) one can obtain from them the poles situated in the strip $\pi < \operatorname{Im} \beta_n < 2\pi$. The SG model possesses a complicated hierarchy of bound states for $\xi < \pi$: breather is the bound state of two solitons, soliton is the bound state of soliton and breather, higher breather is the bound state of lower ones. These possibilities correspond to the following singularities of $f(\beta_1, \dots, \beta_{n-1}, \beta_n)_{\varepsilon_1, \dots, \varepsilon_{n-1}, \varepsilon_n}$ (for the sake of simplicity we consider only the poles at the points $\beta_n = \beta_{n-1} + i\alpha$; the poles at the points $\beta_n = \beta_j + i\alpha, j < n-1$ can be considered via the symmetry):

1. If $\varepsilon_n = \pm\frac{1}{2}, \varepsilon_{n-1} = \pm\frac{1}{2}$, there are simple poles at the points $\beta_n = \beta_{n-1} + \pi i - i\xi m$ with residues

$$
\begin{aligned}
&\operatorname{res} f(\beta_1, \dots, \beta_{n-1}, \beta_n)_{\varepsilon_1, \dots, \varepsilon_{n-1}, \varepsilon_n} = a_m (-1)^{\frac{2\varepsilon_n + 1}{2} m} C_{\varepsilon_{n-1}, \varepsilon_n} \\
&\quad \times f\left(\beta_1, \dots, \beta_{n-2}, \beta_{n-1} + \frac{\pi i}{2} - \frac{i\xi m}{2}\right)_{\varepsilon_1, \dots, \varepsilon_{n-2}, m} , \\
&a_m = \left(\frac{1}{2\pi} S_0(\pi i - i\xi m) \frac{\xi}{m} \sin\frac{\pi^2}{\xi}\right)^{\frac{1}{2}} . \qquad (17)
\end{aligned}
$$

2. If $\varepsilon_n = m, \varepsilon_{n-1} = \pm\frac{1}{2}$, there are simple poles at the points $\beta_n =$

$\beta_{n-1} + \frac{\pi i}{2} + \frac{i\xi}{2}(m-2p), p = 0, \ldots, m-1$, with the residues

$$\begin{aligned}
&\text{res } f(\beta_1, \ldots, \beta_{n-2}, \beta_{n-1}, \beta_n)_{\varepsilon_1 \ldots \varepsilon_{n-2}, \pm\frac{1}{2}, m} \\
&\quad = a_m^p f\Bigg(\beta_1, \ldots, \beta_{n-2}, \beta_{n-1} \\
&\quad + i\xi(m-p), \beta_{n-1} + \frac{\pi i - i\xi p}{2}\Bigg)_{\varepsilon_1, \ldots, \varepsilon_{n-2}, \pm\frac{1}{2}, p}, \\
&a_m^p = 2\pi a_{m-p}^0 a_{m-p,p}, p > 0\,, \\
&a_m^0 = \left(\frac{1}{2\pi i}\text{res } S_m(\beta)\right)^{\frac{1}{2}}, \quad \beta = \frac{1}{2}(\pi + \xi m)i
\end{aligned} \tag{18}$$

where a_{m_1,m_2} will be defined later.

We set

$$f(\beta_1, \ldots, \beta_{\kappa-1}, \beta_\kappa)_{\varepsilon_1, \ldots, \varepsilon_1, 0} \equiv f(\beta_1, \ldots, \beta_{\kappa-1})_{\varepsilon_1, \ldots, \varepsilon_{\kappa-1}}\,.$$

3. If $\varepsilon_\kappa = \pm\frac{1}{2}, \varepsilon_{n-1} = m$, there are simple poles at the points $\beta_n = \beta_{n-1} + \frac{\pi i}{2} + \frac{i\xi}{2}(m-2p), p = 0, \ldots, m-1$, with the residues

$$\begin{aligned}
&\text{res } f(\beta_1, \ldots, \beta_{n-2}, \beta_{n-1}, \beta_n)_{\varepsilon_1, \ldots, \varepsilon_{n-2}, m, \pm\frac{1}{2}} \\
&\quad = a_m^p f\Big(\beta_1, \ldots, \beta_{n-2}, \beta_{n-1} \\
&\quad + \frac{i\xi}{2}(m-p), \beta_{n-1} + \frac{\pi i - i\xi m}{2}\Big)_{\varepsilon_1, \ldots, \varepsilon_{n-2}, p, \pm\frac{1}{2}}\,.
\end{aligned} \tag{19}$$

4. If $\varepsilon_n = m_1, \varepsilon_{n-1} = m_2$, there are simple poles at the points $\beta_n = \beta_{n-1} + \frac{i\xi}{2}(m_1 + m_2 - 2p), p = 0, \ldots, \min(m_1, m_2) - 1$ and $\beta_n = \beta_{n-1} + \pi i - \frac{i\xi}{2}(m_1 + m_2 - 2p), p = 1, \ldots, \min(m_1, m_2)$, (for $m_1 = m_2$ the value $p = m_1$ is omitted) with residues

$$\begin{aligned}
&\text{res } f(\beta_1, \ldots, \beta_{n-2}, \beta_{n-1}, \beta_n)_{\varepsilon_1, \ldots, \varepsilon_{n-2} m_1, m_2} = a_{m_1,m_2}^p \\
&\quad \times f\left(\beta_1, \ldots, \beta_{n-2}, \beta_{n-1} + \frac{i\xi}{2}(m_2 - p), \beta_{n-1}\right. \\
&\quad \left. + \frac{i\xi}{2}(m_1 - p)\right)_{\varepsilon_1, \ldots, \varepsilon_{n-2}, m_1 + m_2 - p, p},
\end{aligned}$$

$$\text{res } f(\beta_1, \dots, \beta_{n-2}, \beta_{n-1}, \beta_n)_{\varepsilon_1,\dots,\varepsilon_{n-2} m_1 m_2} = \tilde{a}^p_{m_1,m_2}$$
$$\times f\left(\beta_1, \dots, \beta_{n-2}, \beta_{n-1} + \frac{i\xi p}{2}, \beta_{n-1} + \pi i - \frac{i\xi}{2}(m_1 + m_2 - 2p)\right)_{\varepsilon_1,\dots,\varepsilon_{n-2}, m_1-p, m_2-p} ,$$
$$a^p_{m,m_2} = 2\pi a_{m_1,m_2-p} a_{m_2-p,p} , \quad p > 0 ,$$
$$a_{m,m_2} = \left(\frac{1}{2\pi i} \text{ res } S_{m_1,m_2}(\beta)\right)^{\frac{1}{2}} , \quad \beta = i\xi \frac{m_1 + m_2}{2} .$$
$$\tilde{a}^p_{m_1,m_2} = 2\pi a_{m_1-p,p} a_{m_2-p,p} . \tag{20}$$

This system of relations is axiomized. However one should not be afraid of the complexity of the system: form factors corresponding to states which contains breathers are secondary objects. As it follows from (17) they can be obtained as residue of soliton form factors. So the properties (18), (19), (20) should be indirectly contained in the soliton form factors.

Once more about the analytical properties. It follows from (15) that $f(\beta_1, \dots, \beta_n)_{\varepsilon_1,\dots,\varepsilon_n}$ can be continued analytically to all the complex planes of the variable β_n although its structure of singularities in the plane is rather complicated. From the symmetry property (13) it is clear that $f(\beta_1, \dots, \beta_n)_{\varepsilon_1,\dots,\varepsilon_n}$ is an analytical function of all its arguments. In what follows we shall be interested in the case where the points are moved from the real axis, for example,

$$f(\beta_1, \dots, \beta_m, \beta_{m+1} + \sigma, \dots, \beta_n + \sigma)_{\varepsilon_1,\dots,\varepsilon_n} . \tag{21}$$

What are the singularities of (21) in the strip $0 \leq \text{Im } \sigma \leq \pi$? It is clear that there are only simple poles on the segments $\text{Re } \sigma = \beta_i - \beta_j$, the position of which can be found from (17)–(20). In the absence of bound states the function (21) is regular up to the line $\text{Im } \sigma = \pi$.

Let us turn to form factors of the general kind $f(\beta'_m, \dots, \beta'_1 | \beta_1, \dots, \beta_n)^{\varepsilon'_m,\dots,\varepsilon'_1}_{\varepsilon_1,\dots,\varepsilon_n}$. Suppose the set β' is separated from the set β, which means that there exists such a constant δ that $|\beta_i - \beta'_j| > \delta \ \forall i, j$. Under these conditions we postulate the formula

$$f(\beta'_m, \dots, \beta'_1 | \beta_1, \dots, \beta_n)^{\varepsilon'_m,\dots,\varepsilon'_1}_{\varepsilon_1,\dots,\varepsilon_n} = \prod_{j=1}^{m} C^{\varepsilon'_j \varepsilon''_j}$$
$$\times f(\beta'_m - \pi i, \dots, \beta'_1 - \pi i, \beta_1, \dots, \beta_n)_{\varepsilon''_m,\dots,\varepsilon''_1,\varepsilon_1,\dots,\varepsilon_n} , \tag{22}$$

which is equivalent (14) to the following:

$$f(\beta'_m,\dots,\beta'_1|\beta_1,\dots,\beta_n)^{\varepsilon'_m,\dots,\varepsilon'_1}_{\varepsilon_1,\dots,\varepsilon_n}=\prod_{j=1}^{m}C^{\varepsilon'_j\varepsilon''_j}$$

$$\times f(\beta_1,\dots,\beta_n,\beta'_m+\pi i,\dots,\beta'_1+\pi i)_{\varepsilon_1,\dots,\varepsilon_n,\varepsilon''_m,\dots,\varepsilon''_1}\,.$$

Why did we require the sets β and β' to be separated? From (22) and (16) it follows that $f(\beta'_m,\dots,\beta_1|\beta_1,\dots,\beta_n)$ as defined by (22) has simple poles at the points $\beta_i=\beta'_j$. The understanding of these singularities should be clarified. That will be done in the next section.

Let us formulate explicitly our axioms.

Axiom 1. Form factor $f(\beta_1,\dots,\beta_n)_{\varepsilon_1,\dots,\varepsilon_n}$ possesses the symmetry property (13).

Axiom 2. Form factor $f(\beta_1,\dots,\beta_n)_{\varepsilon_1,\dots,\varepsilon_n}$ satisfies Eq. (14).

Axiom 3. Form factor $f(\beta_1,\dots,\beta_n)_{\varepsilon_1,\dots,\varepsilon_n}$ has annihilation poles with the residues (16).

Axiom 4. If the spectrum of the model contains different kinds of particles and bound states, the form factors have additional singularities. The positions of these singularities can be found from general physical reasonings. For example, for SG Eqs. (17-20) hold.

For separated sets $\{\beta'_m,\dots,\beta'_1\},\{\beta_1,\dots,\beta_n\}$, form factor $f(\beta'_m,\dots,\beta'_1|\beta_1,\dots,\beta_n)^{\varepsilon'_m,\dots,\varepsilon'_1}_{\varepsilon_1,\dots,\varepsilon_n}$ is defined by Eq. (22). A refinement of the last statement suitable for the case of $\beta'_m,\dots\beta'_1$ and $\beta_1,\dots,\beta_n$ not being separated will be given in the next section.

In the next section, from these axioms the local commutativity theorem will be deduced which is an ideological foundation of the book.

In SG model we shall deal with, besides others, the operators $\exp\left(\pm\frac{i\beta u}{2}\right)$. Axioms 1, 2 are to be slightly changed when dealing with these operators which should be called quasi-local. The fact is that solitons are not very common for the Bose field theory excitations. Specific properties of the operators $\exp\left(\pm\frac{i\beta u}{2}\right)$ may be more easily understood in terms of equivalent MTM, in which the operators are equal to $\exp(\pm\pi i\times\int_{-\infty}^{x}j_0(x)dx')$. It is clear that local operators exist which do not commute with $\exp(\pm\pi i\int_{-\infty}^{x}j_0(x')dx')$ on the space-like interval. This specific character of the operators $\exp\left(\pm\frac{i\beta u}{2}\right)$ stipulates some modifications in Eqs. (14, 16), namely, in the RHS of Eq. (14) the multiplier $(-1)^{\frac{2\varepsilon_n+1}{2}}$ appears. The same multiplier appears in the second term of the RHS of Eq. (16). Besides these, operators $\exp\left(\pm\frac{i\beta u}{2}\right)$ have nonzero vacuum averages, and the corresponding two-soliton form factors have an annihilation pole with residue

$$2\pi i \text{ res } f(\beta_1, \beta_2)_{\varepsilon_1,\varepsilon_2} = C_{\varepsilon_1\varepsilon_2}\ , \quad \beta_2 = \beta_1 + \pi i\ .$$

We put the vacuum averages of $\exp\left(\pm\frac{i\beta u}{2}\right)$ to equal unity.

Although the programme Axioms 1-4 $\mapsto$ local commutativity, which will follow below in this monograph, seems to be most beautiful, the realization of the programme of paper [45], i.e. the obtaining of Axioms 1-4 from general axioms of quantum field theory, is also of interest. In Appendix 2 we outline the latter programme heuristically.

2

THE LOCAL COMMUTATIVITY THEOREM

In the present section, the theorem is proven which states that two local operators commute on a space-like interval if their form factors satisfy the system of axioms formulated in Sec. 1.

First, let us introduce a system of brief notations. Consider the set of rapidities $\beta_1, \ldots, \beta_n$. Let us connect a space h_i with every β_i. The spaces h_i are isomorphic to $\mathbb{C}^2$ for ITM and SG with $\xi > \pi$, to $\mathbb{C}^3$ for NLS, to $\mathbb{C}^2 \oplus \mathbb{C}^{[\pi/\xi]}$ for SG with $\xi < \pi$. The base in the space h_i is enumerated by the index ε. S-matrix $S(\beta_i - \beta_j)$ is considered as an operator in the space $h_i \otimes h_j$, its matrix elements being given by Eqs. (3, 4, 5, 6, 8). The unitarity condition, crossing-symmetry condition and Yang-Baxter equations in these notations look as follows:

$$\begin{aligned} &S(\beta_1 - \beta_2)S(\beta_2 - \beta_1) = I \ , \\ &S(\beta_1 - \beta_2 + \pi i) = C(\beta_1)S^{t(\beta_1)}(\beta_2 - \beta_1)C(\beta_1) \ , \\ &S(\beta_1 - \beta_2)S(\beta_1 - \beta_3)S(\beta_2 - \beta_3) \\ &\quad = S(\beta_2 - \beta_3)S(\beta_1 - \beta_3)S(\beta_1 - \beta_2) \ , \end{aligned} \tag{23}$$

where $C(\beta_i)$ is the matrix C acting on $h_i, t(\beta_i)$ is the transposition with respect to h_i, in SG model with $\xi < \pi$ the matrix C is equal to $\sigma^1 \oplus I_{[\pi/\xi]}$. Matrix C satisfies the relations $C^t = C, C^2 = I$.

Besides, we suppose that the following relation holds:

$$C(\beta_1)C(\beta_2)S(\beta_1 - \beta_2) = S(\beta_1 - \beta_2)C(\beta_1)C(\beta_2) \ .$$

Form factors $f(\alpha_m,\dots,\alpha_1|\beta_1,\dots,\beta_n)^{\varepsilon'_m\dots\varepsilon'_1}_{\varepsilon_1\dots\varepsilon_n}$ create a (m,n) tensor. Let us assume β_i to be always connected with h_i, which means that when β_i and β_j change places the corresponding spaces h_i and h_j also change places. The meaning of the last, nebulous phrase will become clear after rewriting Eqs. (13, 14, 16) in new notations:

$$
\begin{aligned}
&f(\beta_1\dots\beta_i,\beta_{i+1},\dots,\beta_n)S(\beta_i-\beta_{i+1}) = f(\beta_1,\dots,\beta_{i+1},\beta_i,\dots,\beta_n)\ ,\\
&f(\beta_1,\dots,\beta_{n-1},\beta_n+2\pi i) = f(\beta_n,\beta_1,\dots,\beta_{n-1})\ ,\\
&\text{res}\ f(\beta_1,\dots,\beta_n) = \frac{1}{2\pi i}f(\beta_1,\dots,\hat{\beta}_j,\dots,\beta_{n-1})e(\beta_j,\beta_n)\\
&\quad\times[S(\beta_{n-1}-\beta_j)\dots S(\beta_{j+1}-\beta_j) - S(\beta_j-\beta_1)\dots S(\beta_j-\beta_{j-1})]\ ,\\
&\hfill \beta_n = \beta_j+\pi i\ ,
\end{aligned}
$$

where $e(\beta_1,\beta_2) = C^{\varepsilon,\varepsilon'}e_{1,\varepsilon}\otimes e_{2,\varepsilon'}$, with $e_{i,\varepsilon}$ forming the base in h_i.

Equation (22) rewrites as follows:

$$
\begin{aligned}
&f(\alpha_m,\dots,\alpha_1|\beta_1,\dots,\beta_n)\\
&\quad = C(\alpha_1,\dots,\alpha_m)(f(\alpha_m-\pi i,\dots,\alpha_1-\pi i,\beta_1,\dots,\beta_n))^{t(\alpha_1,\dots,\alpha_m)}\ ,
\end{aligned}
$$

where $C(\alpha_1,\dots,\alpha_m) = C(\alpha_1)\dots C(\alpha_m)$, $t(\alpha_1,\dots,\alpha_m)$ denotes transposition with respect to the spaces connected with $\alpha_1,\dots,\alpha_m$. It has been mentioned above that Eq. (22) gives the correct formula for form factors if the sets $\alpha_1,\dots\alpha_m$ and $\beta_1,\dots,\beta_n$ are separated.

From Eqs. (13, 14, 16) it follows that $f(\alpha_\kappa,\dots,\alpha_1|\beta_1,\dots,\beta_m)$ possesses the properties:

1. $f(\alpha_\kappa,\dots\alpha_1|\beta_1,\dots,\beta_i,\beta_{i+1},\dots,\beta_m)S(\beta_i-\beta_{i+1})$
   $= f(\alpha_\kappa,\dots,\alpha_1|\beta_1,\dots,\beta_{i+1}\beta_i,\dots,\beta_m)\ ,$

2. $S(\alpha_{i+1}-\alpha_i)f(\alpha_\kappa,\dots,\alpha_{i+1},\alpha_i,\dots,\alpha_1|\beta_1,\dots,\beta_m)$
   $= f(\alpha_\kappa,\dots,\alpha_i,\alpha_{i+1},\dots\alpha_1|\beta_1,\dots,\beta_m)\ ,$

3. $f(\alpha_\kappa-\pi i,\alpha_{\kappa-1},\dots,\alpha_1|\beta_1,\dots,\beta_m)$
   $= f^{t(\alpha_\kappa)}(\alpha_{\kappa-1},\dots\alpha_1|\beta_1,\dots,\beta_m,\alpha_\kappa)C(\alpha_\kappa)\ ,$ (24)

4. $f(\alpha_\kappa,\dots,\alpha_1|\beta_1,\dots,\beta_{m-1},\beta_m+\pi i)$
   $= C(\beta_m)f^{t(\beta_m)}(\beta_m,\alpha_\kappa,\dots,\alpha_1|\beta_1,\dots,\beta_{m-1})\ .$ (25)

5. There are simple poles at the points $\alpha_i = \beta_j$ with residue

$$\begin{aligned}
&2\pi i \text{ res } f(\alpha_\kappa, \ldots, \alpha_1|\beta_1, \ldots, \beta_m) \\
&\quad = -S(\alpha_i - \alpha_{i+1}) \ldots S(\alpha_i - \alpha_\kappa) \\
&\quad \times f(\alpha_\kappa, \ldots, \hat{\alpha}_i, \ldots, \alpha_1|\beta_1, \ldots, \hat{\beta}_j, \ldots, \beta_m) \\
&\quad \times I(\alpha_i, \beta_j) S(\beta_m - \beta_j) \ldots S(\beta_{j+1} - \beta_j) \\
&\quad + S(\alpha_{i-1} - \alpha_i) \ldots S(\alpha_1 - \alpha_i) I(\alpha_i, \beta_j) \\
&\quad \times f(\alpha_\kappa, \ldots, \hat{\alpha}_i, \ldots, \alpha_1|\beta_1, \ldots, \hat{\beta}_j, \ldots, \beta_m) \\
&\quad \times S(\beta_j - \beta_1) \ldots S(\beta_j - \beta_{j-1}) \qquad (26)
\end{aligned}$$

where $I(\alpha_i, \beta_j)$ is the tensor identifying the spaces connected with α_i and β_j, its notation in components being $\delta_{\varepsilon_j}^{\varepsilon_i'}$.

Let us further shorten our notations. Namely, let us denote the set of the rapidities by capital Latin letters, for example, $A = \{\alpha_j\}_{j=1}^{\kappa}, B = \{\beta_j\}_{j=1}^{m}$. We are still associating a space h with every rapidity. The number of elements in A will be denoted by $n(A)$. The set of rapidities ordered in accordance with increasing (decreasing) values will be denoted by $\overrightarrow{A}(\overleftarrow{A})$. Suppose $\alpha_1 < \alpha_2 < \ldots < \alpha_\kappa$, i.e. $\overrightarrow{A} = \{\alpha_1, \ldots, \alpha_\kappa\}$, the product $Z(\alpha_1) \ldots Z(\alpha_\kappa)$ will be denoted by $Z(\overrightarrow{A})$ while the product $Z^*(\alpha_\kappa) \ldots Z^*(\alpha_1)$ by $Z^*(\overleftarrow{A})$. The last definition needs some refinement. We understand $Z(\alpha)(Z^*(\alpha))$ to be a vector (covector) in the space connected with α. The products of these operators are always understood as tensor ones with respect to this internal structure.

Let the set A be divided into two subsets $A = A_1 \cup A_2$, and consider the operators $S(\overrightarrow{A}|\overrightarrow{A}_1), S(\overleftarrow{A}_1|\overleftarrow{A})$ defined by

$$Z(\overrightarrow{A}) = S(\overrightarrow{A}|\overrightarrow{A}_1) Z(\overrightarrow{A}_2) Z(\overrightarrow{A}_1) ,$$

$$Z^*(\overleftarrow{A}) = Z^*(\overleftarrow{A}_1) Z^*(\overleftarrow{A}_2) S(\overleftarrow{A}_1|\overleftarrow{A}) .$$

It can be easily shown that

$$S(\overleftarrow{A}_1|\overrightarrow{A}) S(\overrightarrow{A}|\overrightarrow{A}_1) = I . \qquad (27)$$

Supposing that $\alpha_\kappa > \alpha_{\kappa-1} \ldots > \alpha_1, \beta_m > \beta_{m-1} > \ldots \beta_1$, we define "$\delta$-function" $\Delta(A|B)$ by the formula

$$\Delta(A|B) = \delta_{\kappa m} \prod_{i=1}^{\kappa} \delta(\alpha_i - \beta_i) I(\alpha_i, \beta_i) ,$$

where $I(\alpha_i, \beta_i)$ identifies the spaces associated with α_i and β_i. The full base of "in"-states consists of the vectors

$$|\overleftarrow{B}\rangle = Z^*(\overleftarrow{B})|ph\rangle ,$$

the dual base being

$$\langle\overrightarrow{A}| = \langle ph|Z(\overrightarrow{A}) .$$

In what follows we shall deal with several sets of this type. The writings $\overrightarrow{A}_1 \cup \overrightarrow{A}_2$ and $\overrightarrow{A_1 \cup A_2}$ will be distinguished; the first one means that the ordered set A_2 is added to A_1 while the second means that $A_1 \cup A_2$ is ordered.

Now we are in a position to refine the definition of the matrix elements of the general type. Let A and B be arbitrary sets of rapidities. We postulate the following form of the matrix element:

$$\langle\overrightarrow{A}|O(0,0)|\overleftarrow{B}\rangle = \sum_{\substack{A=A_1\cup A_2\\ B=B_1\cup B_2}} S(\overrightarrow{A}|\overrightarrow{A}_1)f(\overleftarrow{A}_1 + i0|\overrightarrow{B}_1) \times \Delta(A_2, B_2)S(\overleftarrow{B}_1|\overleftarrow{B})(-1)^{n(B_2)} . \tag{28}$$

$\overleftarrow{A}_1 + i0$ means that all the rapidities involved in A_1 are slightly moved onto the upper half-plane. It is clear that we find $f(\overleftarrow{A}|\overrightarrow{B})$ to be the RHS of (28) for separated sets A and B.

The following Lemma is very useful for the proof of local commutativity.

Lemma 1. The formula (28) is equivalent to the following:

$$\langle\overrightarrow{A}|O(0,0)|\overleftarrow{B}\rangle = \sum_{\substack{A=A_1\cup A_2\\ B=B_1\cup B_2}} S(\overrightarrow{A}|\overrightarrow{A}_2)f(\overleftarrow{A}_1 - i0|\overrightarrow{B}_1) \times \Delta(A_2, B_2)S(\overleftarrow{B}_2|\overleftarrow{B})(-1)^{n(B_2)} . \tag{29}$$

Proof. From the formula for the residues (26), it follows that

$$f(\overleftarrow{A} + i0|\overrightarrow{B}) = \sum_{\substack{A=A_1\cup A_2\cup A_3\\ B=B_1\cup B_2\cup B_3}} S(\overrightarrow{A}|\overrightarrow{A_1 \cup A_3})S(\overrightarrow{A_1 \cup A_3}|\overrightarrow{A}_3) \times f(\overleftarrow{A}_1 - i0|\overrightarrow{B}_1)\Delta(A_2, B_2)\Delta(A_3, B_3)S(\overleftarrow{B}_3|\overleftarrow{B_1 \cup B_3}) \times S(\overleftarrow{B_1 \cup B_3}|\overleftarrow{B})(-1)^{n(B_3)} . \tag{30}$$

Substituting Eq. (30) into Eq. (28) one obtains

$$\begin{aligned}\langle \overrightarrow{A}|O(0,0)|\overleftarrow{B}\rangle = \sum_{\substack{A=A_1\cup A_2\cup A_3\cup A_4\\ B=B_1\cup B_2\cup B_3\cup B_4}} & S(\overrightarrow{A}|\overrightarrow{A_1\cup A_2\cup A_3})\\ &\times S(\overrightarrow{A_1\cup A_2\cup A_3}|\overrightarrow{A_1\cup A_3})S(\overrightarrow{A_1\cup A_3}|\overrightarrow{A}_3)\\ &\times f(\overleftarrow{A}_1 - i0|\overrightarrow{B}_1)\Delta(A_2,B_2)\Delta(A_3,B_3)\Delta(A_4,B_4)\\ &\times S(\overleftarrow{B}_3|\overleftarrow{B_1\cup B_3})S(\overleftarrow{B_1\cup B_3}|\overleftarrow{B_1\cup B_2\cup B_3})\\ &\times S(\overleftarrow{B_1\cup B_2\cup B_3}|\overleftarrow{B})(-1)^{n(B_3)+n(B_4)}\,. \end{aligned}\tag{31}$$

Let us transform Eq. (31), replacing $\Delta(A_2,B_2)\Delta(A_4,B_4)$ by $[S(\overrightarrow{A_2\cup A_4}|\overrightarrow{A}_2)]^{-1}\Delta(A_2,B_2)\Delta(A_4,B_4)[S(\overleftarrow{B}_2|\overleftarrow{B_2\cup B_4})]^{-1}$, and denote $A_2\cup A_4$ by A_5, $B_2\cup B_4$ by B_5. Then Eq. (31) transforms to

$$\begin{aligned}\langle \overrightarrow{A}|O(0,0)|\overleftarrow{B}\rangle = \sum_{\substack{A=A_1\cup A_3\cup A_5\\ B=B_1\cup B_3\cup B_5\\ B_2\subset B_5}} & S(\overrightarrow{A}|\overrightarrow{A_1\cup A_3})S(\overrightarrow{A_1\cup A_3}|\overrightarrow{A}_3)\\ &\times f(\overleftarrow{A}_1 - i0|\overrightarrow{B}_1)\Delta(A_5,B_5)\Delta(A_3,B_3)S(\overleftarrow{B}_3|\overleftarrow{B_1\cup B_3})\\ &\times S(\overleftarrow{B_1\cup B_3}|\overleftarrow{B})(-1)^{n(B)+n(B_1)+n(B_2)}\,. \end{aligned}$$

Notice now that the summation over B_2 gives

$$\sum_{B_2\subset B_5}(-1)^{n(B_2)} = \begin{cases} 1\,, & B_5=\emptyset\\ 0\,, & B_5\neq\emptyset\,,\end{cases}$$

and consequently

$$\begin{aligned}\langle \overrightarrow{A}|O(0,0)|\overleftarrow{B}\rangle = \sum_{\substack{A=A_1\cup A_3\\ B=B_1\cup B_3}} & S(\overrightarrow{A}|\overrightarrow{A}_3)f(\overleftarrow{A}_1 - i0|\overrightarrow{B}_1)\\ &\times \Delta(A_3,B_3)S(\overleftarrow{B}_3|\overleftarrow{B})(-1)^{n(B_3)}\,. \end{aligned}$$

Q.E.D.

Remark. Everything is slightly different when dealing with quasi-local operators $\exp\left(\pm\frac{i\beta u}{2}\right)$. Recalling that the space h for SG model is isomorphic to $\mathbb{C}^2 \oplus \mathbb{C}^{[\frac{\pi}{\xi}]}$, let us introduce an operator d which acts in h as $(-I_2) \oplus (I_{[\frac{\pi}{\xi}]})$, where $I_2, I_{[\frac{\pi}{\xi}]}$ are unit operators in $\mathbb{C}^2$ and $\mathbb{C}^{[\frac{\pi}{\xi}]}$ respectively, i.e. the operator d reflects the soliton space and leaves unchanged the breather one. For the quasi-local operators the matrix $d(\alpha_i)$ should be put in before the second term in the RHS of (26). Eq. (29) remains valid without any modifications while in Eq. (28) the multiplier $(-1)^{n(B_2)}$ is to be replaced by $(-1)^{n(B_1)}d(B_1)$.

From Lemma 1 follows the existence of two presentations for the matrix element of the commutator of two local operators. These presentations are given by the following.

Lemma 2. Consider two operators O_1, O_2 which are constructed through the form factors $f_1(\beta_1, \ldots, \beta_n), f_2(\beta_1, \ldots, \beta_n)$ via the procedure described above. There are two possible presentations of an arbitrary matrix element of the commutator $[O_1(x), O_2(y)]$ (hereafter x, y refer to space coordinates). The first is

$$\begin{aligned}
&\langle \overrightarrow{A}|[O_1(x), O_2(y)]|\overleftarrow{B}\rangle \exp(i\kappa(A)x - i\kappa(B)y) \\
&= \sum_{\substack{A=A_1\cup A_2\cup A_3 \\ B=B_1\cup B_2\cup B_3}} S(\overrightarrow{A}|\overrightarrow{A}_1)S(\overrightarrow{A_2\cup A_3}|\overrightarrow{A}_3) \\
&\times \sum_{n(C)=0}^{\infty} \int dCG_-(x-y|C, A_1, B_1, A_3) \\
&\times \Delta(A_3, B_3)S(\overleftarrow{B}_2|\overleftarrow{B_2\cup B_3})S(\overleftarrow{B_2\cup B_3})|\overleftarrow{B}) ,
\end{aligned} \tag{32}$$

and the second is

$$\begin{aligned}
&\langle \overrightarrow{A}|[O_1(x), O_2(y)]|\overleftarrow{B}\rangle \exp(i\kappa(A)x - i\kappa(B)y) \\
&= \sum_{\substack{A=A_1\cup A_2\cup A_3 \\ B=B_1\cup B_2\cup B_3}} S(\overrightarrow{A}|\overrightarrow{A}_1)S(\overrightarrow{A_2\cup A_3}|\overrightarrow{A_3}) \\
&\times \sum_{n(C)=0}^{\infty} \int dCG_+(x-y|C, A_1, B_1, A_3) \\
&\times \Delta(A_3, B_3)S(\overleftarrow{B}_3|\overleftarrow{B_2\cup B_3})S(\overleftarrow{B}_1|\overleftarrow{B}) ,
\end{aligned} \tag{33}$$

where $\kappa(A) = \sum_{\alpha \in A} \kappa(\alpha), \kappa(\alpha) = M \text{ sh } \alpha, M$ is the mass (in SG model with $\xi < \pi, M$ is to be understood as the matrix diag $\Big(M, M, 2M \sin \frac{\xi}{2}, \ldots$ $2M \sin \left(\frac{\xi}{2}\left[\frac{\pi}{\xi}\right]\right)\Big)$ acting in the space h connected with α). Integration over C means integration over all rapidities involved in C. The function G_- is given by

$$
\begin{aligned}
G_-(x-y|C, A_1, B_1, A_3) &= (-1)^{n(B_1)+n(B_2)+n(C)} \\
&\times f_1(\overleftarrow{A}_1 + i0|\overrightarrow{C} \cup \overrightarrow{B}_2) S(\overleftarrow{C}|\overleftarrow{A}_3 \cup \overleftarrow{C}) f_2(\overleftarrow{C} \cup \overleftarrow{A}_2 - i0|\overrightarrow{B}_1) \\
&\times \exp(i(x-y)(\kappa(C) + \kappa(B_2) + \kappa(A_3) + \kappa(A_2))) \\
&- f_2(\overleftarrow{A}_2 - i0|\overrightarrow{B}_1 \cup \overrightarrow{C}) S(\overrightarrow{C} \cup \overrightarrow{A}_3|\overrightarrow{C}) f_1(\overleftarrow{A}_1 \cup \overleftarrow{C} + i0|\overrightarrow{B}_2) \\
&\times \exp(i(x-y)(-\kappa(C) + \kappa(B_2) + \kappa(A_3) + \kappa(A_2))) \,. \qquad (34)
\end{aligned}
$$

The function $G_+(x - y|C, A_1, B_1, A_3)$ can be obtained from $-G_-(x - y|C, A_1, B_1, A_3)$ by the replacement $x \leftrightarrow y, f_1 \leftrightarrow f_2$.

Proof. Consider the matrix element $\langle \overrightarrow{A}|O_1(x)O_2(y)|\overleftarrow{B}\rangle$. Let us put a full set of states between $O_1(x)$ and $O_2(y)$:

$$
\begin{aligned}
&\langle \overrightarrow{A}|O_1(x)O_2(y)|\overleftarrow{B}\rangle \\
&\quad = \sum_{n(D)=0}^{\infty} \int dDdF \langle \overrightarrow{A}|O_1(x)|\overleftarrow{D}\rangle \Delta(D, F) \langle \overrightarrow{F}|O_2(y)|\overleftarrow{B}\rangle \,.
\end{aligned}
$$

Let us use the formula (28) for $\langle \overrightarrow{A}|O_1(x)|\overleftarrow{D}\rangle$ and the formula (29) for $\langle \overrightarrow{F}|O_2(y)|\overleftarrow{B}\rangle$:

$$
\begin{aligned}
&\langle \overrightarrow{A}|O_1(x)O_2(y)|\overleftarrow{B}\rangle \\
&\quad = \sum_{A=A_1\cup A_4} \sum_{B=B_1\cup B_4} \sum_{n(D)=0}^{\infty} \int dDdF \sum_{D=D_1\cup D_2} \sum_{F=F_1\cup F_2} S(\overrightarrow{A}|\overrightarrow{A}_1) \\
&\quad \times f_1(\overleftarrow{A}_1 + i0|\overrightarrow{D}_1) \Delta(A_4, D_2) S(\overleftarrow{D}_1|\overleftarrow{D}) \Delta(D, F) \\
&\quad \times S(\overrightarrow{F}|\overrightarrow{F}_2) f_2(\overleftarrow{F}_1|\overrightarrow{B}_1 + i0) \Delta(F_2, B_4) S(\overleftarrow{B}_4|\overleftarrow{B}) \\
&\quad \times \exp(-i\kappa(A)x + i\kappa(D)x - i\kappa(F)y + i\kappa(B)y)
\end{aligned}
$$

$$= \sum_{A=A_1\cup A_2\cup A_3} \sum_{B=B_1\cup B_2\cup B_3} \sum_{n(C)=0}^{\infty} \int dC S(\overrightarrow{A}|\overrightarrow{A}_1)$$
$$\times f_1(\overleftarrow{A}_1 + i0|\overrightarrow{C \cup B_2}) S(\overleftarrow{C \cup B_2}|\overleftarrow{C \cup B_2 \cup A_3 \cup A_2})$$
$$\times \Delta(A_3, B_3) S(\overrightarrow{C \cup B_2 \cup B_3 \cup A_2}|\overrightarrow{B_2 \cup B_3})$$
$$\times f_2(\overleftarrow{A_2 \cup C}|\overrightarrow{B_1} + i0) S(\overleftarrow{B_2 \cup B_3}|\overleftarrow{B})$$
$$\times \exp(-i\kappa(A_1)x + i\kappa(B_2)x + i\kappa(C)(x-y) - i\kappa(B_1)y - i\kappa(A_2)y)\ .$$

Let us use two identities:

$$Z^*(\overleftarrow{C \cup B_2}) Z^*(\overleftarrow{A_3 \cup A_2}) S(\overleftarrow{C \cup B_2}|\overleftarrow{C \cup B_2 \cup A_2 \cup A_3})$$
$$\times S(\overrightarrow{C \cup B_2 \cup A_2 \cup A_3}|\overrightarrow{B_2 \cup A_3}) = Z^*(\overleftarrow{B_2 \cup A_3}) Z^*(\overleftarrow{C \cup A_2})$$
$$Z^*(\overleftarrow{C \cup B_2}) Z^*(\overleftarrow{A_3 \cup A_2}) S(\overrightarrow{A_2 \cup A_3}|\overrightarrow{A}_3) S(\overrightarrow{C \cup B_2}|\overrightarrow{B_2})$$
$$\times S(\overleftarrow{C}|\overleftarrow{A}_3 \cup \overleftarrow{C}) S(\overleftarrow{C}|\overleftarrow{C \cup A_2}) S(\overleftarrow{B}_2|\overleftarrow{B_2 \cup A_3})$$
$$= Z^*(\overleftarrow{B_2 \cup A_3}) Z^*(\overleftarrow{C \cup A_2})\ . \tag{35}$$

When deriving these identities Eq. (27) should be used. From Eqs. (35) it follows that

$$S(\overleftarrow{C \cup B_2}|\overleftarrow{C \cup B_2 \cup A_2 \cup A_3}) \Delta(A_3, B_3) S(\overrightarrow{C \cup B_2 \cup A_2 \cup B_3}|\overrightarrow{B_2 \cup B_3})$$
$$= S(\overrightarrow{A_2 \cup A_3}|\overrightarrow{A}_3) S(\overrightarrow{C \cup B_2}|\overrightarrow{B}_2) \Delta(A_3, B_3)$$
$$\times S(\overleftarrow{C}|\overleftarrow{B}_3 \cup \overleftarrow{C}) S(\overleftarrow{C}|\overleftarrow{C \cup A_2}) S(\overleftarrow{B}_2|\overleftarrow{B_2 \cup B_3})\ ,$$

hence

$$\langle \overrightarrow{A}|O_1(x)O_2(y)|\overleftarrow{B}\rangle = \sum_{A=A_1\cup A_2\cup A_3} \sum_{B=B_1\cup B_2\cup B_3} \sum_{n(C)=0}^{\infty} \int dC S(\overrightarrow{A}|\overrightarrow{A}_1)$$
$$\times S(\overrightarrow{A_2 \cup A_3}|\overrightarrow{A}_3) f_1(\overleftarrow{A}_1 + i0|\overrightarrow{C} \cup \overrightarrow{B}_2) S(\overleftarrow{C}|\overleftarrow{A}_3 \cup \overleftarrow{C}) \Delta(A_3, B_3)$$
$$\times f_2(\overleftarrow{C} \cup \overleftarrow{A}_2|\overrightarrow{B}_1 + i0) S(\overleftarrow{B}_2|\overleftarrow{B_2 \cup B_3})$$
$$\times S(\overleftarrow{B_2 \cup B_3}|\overleftarrow{B})(-1)^{n(B_1)+n(B_2)+n(C)}$$
$$\exp(i\kappa(C)(x-y) - i\kappa(A_1)x + i\kappa(B_2)x + i\kappa(B_1)y - i\kappa(A_2)y)\ .$$

The matrix element $\langle \overrightarrow{A}|O_2(y)O_1(x)|\overleftarrow{B}\rangle$ can be considered in a similar way using again the formulae (28), (29) for O_1, O_2 respectively. In this way we

obtain in Eq. (32). Equation (33) can be obtained using the formulae (29), (28) for O_1, O_2 respectively.
Q.E.D.

Remark. When dealing with quasi-local operators $\exp\left(\pm\frac{i\beta u}{2}\right)$ some modifications have to be made. If O_2 is local and O_1 is quasi-local, the multipliers $d(C \cup B_2)$ and $d(B_2)$ appear in the formula for O before the first and second terms respectively. If both O_1 and O_2 are quasi-local the common multiplier $(-1)^{n(B_1)+n(B_2)+n(C)}$ disappears while the multipliers $d(C \cup B_2)$ and $d(B_2)$ appear before the first and second terms respectively.

Now we are in a position to prove the main theorem.

Theorem 1. Suppose the form factors of two operators satisfy Axioms 1-5 and the additional requirement

$$f_i(\beta_1, \ldots, \beta_\kappa, \beta_{\kappa+1}+\sigma, \ldots, \beta_n+\sigma) = O(e^{S|\sigma|}) , \qquad |\sigma| \sim \infty ,$$

where S is some number common for all κ, n. Then every matrix element of the commutator $[O_1(x), O_2(y)]$ is a distribution whose support consists of one point $x = y$.

Proof. Let us consider for simplicity a model without bound states (SG $\xi > \pi$, ITM, NLS).

The matrix element of the commutator $\langle \overrightarrow{A} |[O_1(x), O_2(y)]| \overleftarrow{B} \rangle$ depends essentially only on $x - y$; that is why we can put $y = 0$. Consider any finite function from C^∞-$\varphi(x)$. This function can be presented in the form $\varphi(x) = \varphi_+(x) + \varphi_-(x), \varphi_\pm(x) = \varphi(x)\theta(\pm x)$, θ being Heaviside's function. Consider the convolution

$$\begin{aligned} &\int_{-\infty}^{\infty} (\langle \overrightarrow{A} |[O_1(x), O_2(0)]| \overleftarrow{B} \rangle e^{i\kappa(A)x)})\varphi(x)dx \\ &= \int_{-\infty}^{0} (\langle \overrightarrow{A} |[O_1(x), O_2(0)]| \overleftarrow{B} \rangle e^{i\kappa(A)x})\varphi_-(x)dx \\ &+ \int_{0}^{\infty} (\langle \overrightarrow{A} |[O_1(x), O_2(0)]| \overleftarrow{B} \rangle e^{i\kappa(A)x})\varphi_+(x)dx \ . \end{aligned} \tag{36}$$

A trick first used in Ref. [32] concerns the operation of substituting (32) for the first term in (36) and (33) for the second term in (36). For the first term one has

$$\int_{-\infty}^{0} \langle \overrightarrow{A} | [O_1(x), O_2(0)] | \overleftarrow{B} \rangle e^{i\kappa(A)x} \varphi_-(x) dx$$

$$= \sum_{\substack{A=A_1 \cup A_2 \cup A_3 \\ B=B_1 \cup B_2 \cup B_3}} S(\overrightarrow{A} | \overrightarrow{A}_1) S(\overrightarrow{A_2 \cup A_3} | \overrightarrow{A}_3)$$

$$\times \sum_{n(C)=0}^{\infty} \left\{ \int_{-\infty}^{0} dx \varphi_-(x) \int dC G_-(x|C, A_1, B_1, A_3) \right\} \Delta(A_3, B_3)$$

$$\times S(\overleftarrow{B}_2 | \overleftarrow{B_2 \cup B_3}) S(\overleftarrow{B_2 \cup B_3} | \overleftarrow{B}) .$$

Consider the expression in the brackets above:

$$\int_{-\infty}^{0} dx \varphi_-(x) \int dC G_-(x|C, A_1, B_1, A_3)$$

$$= \int dC f_1(\overleftarrow{A}_1 + i0 | \overrightarrow{C} \cup \overrightarrow{B}_2) S(\overleftarrow{C} | \overleftarrow{A}_3 \cup \overleftarrow{C}) f_2(\overleftarrow{C} \cup \overleftarrow{A}_2 | \overrightarrow{B}_1 + i0)$$

$$\times \hat{\varphi}_-(\kappa(C) + \kappa(B_2) + \kappa(A_3) + \kappa(A_4))$$

$$- \int dC f_2(\overleftarrow{A}_2 - i0 | \overrightarrow{B}_1 \cup \overleftarrow{C}) S(\overrightarrow{C} \cup \overrightarrow{A}_3 | \overrightarrow{C}) f_1(\overleftarrow{A}_1 \cup \overleftarrow{C} | \overrightarrow{B}_2 - i0)$$

$$\times \hat{\varphi}_-(-\kappa(C) + \kappa(B_2) + \kappa(A_3) + \kappa(A_4)) , \tag{37}$$

where $\hat{\varphi}_-$ is the Fourier transform of the function φ_-. Function $\hat{\varphi}_-(\kappa)$ is analytical for Im $\kappa < 0$ because $\varphi_-(x) = 0$ for $x > 0$. Suppose that the support of $\varphi(x)$ does not contain the point $x = 0$, then $\hat{\varphi}_-(\kappa)$ would decrease faster than any power of κ^{-1} for $\kappa \to \infty$. Let $C = \{\gamma_1, \ldots, \gamma_{n(C)}\}$ and consider new variables $\Delta_i = \gamma_i - \gamma_{i+1}, i = 1, \ldots, n(C)-1, \sigma = \frac{1}{n(C)} \Sigma \gamma_i$. For fixed $\Delta_1, \ldots, \Delta_{n(C)-1}$, the function $\hat{\varphi}_-(\kappa(C) + \kappa(B_2) + \kappa(A_3) + \kappa(A_4))$ decreases faster than any power of $e^{-|\sigma|}$ for $\sigma \to \pm\infty$ while the function $f_1(\overleftarrow{A}_1 | \overrightarrow{C} \cup \overrightarrow{B}_2) S(\overleftarrow{C} | \overleftarrow{A}_3 \cup \overleftarrow{C}) f_2(\overleftarrow{C} \cup \overleftarrow{A} | \overrightarrow{B}_1)$ increases no faster than a fixed power of $e^{|\sigma|}$ as it follows from the estimation presented in the statement of the theorem. From the reasoning presented at the end of the previous section it is clear that f_1 and f_2 are regular in the strip $-\pi + 0 <$ Im $\sigma < 0$ (the absence of bound states is essential here). The same can be said about the matrix $S(\overleftarrow{C} | \overleftarrow{A}_3 \cup \overleftarrow{C})$. All these facts together with an understanding of the singularities prescribed in (37) allow us to move the contour of integration of the first integral in Eq. (37) from the real axis to the line Im $\sigma = -\pi + 0$. Making this transformation we can rewrite the first integral in Eq. (37) in the form

$$\int dC f_1(\overleftarrow{A}_1 + i0|(\overleftarrow{C} - \pi i + i0) \cup \overrightarrow{B_2})$$
$$\times S(\overleftarrow{C} - \pi i|\overleftarrow{A}_3 \cup \overleftarrow{C} - \pi i) f_2((\overleftarrow{C} - \pi i + i0) \cup \overleftarrow{A}_2|\overrightarrow{B}_1 + i0)$$
$$\times \hat{\varphi}_-(-\kappa(C) + \kappa(B_2) + \kappa(A_3) + \kappa(A_4)) \,.$$

Notice now that it is possible to change the ordering of C in f_1, f_2 and S simultaneously. Using (23), (25), (26) we can write the above as

$$\int dC f_1^{t(C)}(\overleftarrow{A}_1 \cup \overleftarrow{C}|\overrightarrow{B}_2 - i0) S^{t(C)}(\overrightarrow{C} \cup \overrightarrow{A}_3|\overrightarrow{C})$$
$$\times f_2^{t(C)}(\overleftarrow{A}_2 - i0|\overrightarrow{B}_1 \cup \overrightarrow{C})\hat{\varphi}_-(-\kappa(C) + \kappa(B_2) + \kappa(A_3) + \kappa(A_4)) \,, \tag{38}$$

where $t(C)$ is the transposition with respect to all the spaces associated with the rapidities involved in C. It is clear now that (38) is cancelled by the second integral in Eq. (37) (evidently $f_1^{t(C)} S^{t(C)} f_2^{t(C)} = f_2 S f_1$).

The integral

$$\int (\langle \overrightarrow{A}|[O_1(x), O_2(0)]|\overleftarrow{B}\rangle e^{i\kappa(A)x})\varphi_+(x)dx \tag{39}$$

can be treated in a similar way. Using the presentation (33) for the matrix element we make sure that the integral (39) is equal to zero if the support of φ does not contain the point $x = 0$.

For the case with bound states (SG, $\xi < \pi$), the proof encounters serious complications when deforming the contour of integration across the poles of the integrand. As a result, we obtain a discrepancy between the first and the second integral in Eq. (37). The system of relations (17), (18), (19), (20) stipulates the vanishing of the discrepancy. The proof of this fact is very tangled and we would not present it here. Q.E.D.

Remark 1. The theorem can be generalized to take into consideration the operators $\exp\left(\pm\frac{i\beta u}{2}\right)$. The additional signs which were considered in the Remark to Lemma 2 are compensated by signs appearing during the analytical continuation $\sigma \to \sigma - \pi i + i0$ due to the changes in (14).

Remark 2. The Lorentz transformation corresponds to simultaneous shifts of the rapidities in form factors. That is why the generalization of Theorem 1 to the space-like interval is trivial.

Remark 3. In proving Theorem 1 we have ignored completely the problem of convergence. We shall return to this question in Sec. 10.

3

SOLITON FORM FACTORS IN SG MODEL

In this section formulae are presented for the soliton form factors in SG model. All the necessary properties of these form factors are established in the following sections. We shall consider the coupling constant ξ of generic position under $\xi \neq \frac{\pi}{\nu}$; the reflectionless case $\xi = \frac{\pi}{\nu}$ will be considered in Sec. 6.

The section consists of four subsections; three of them are devoted to the description of some auxiliary objects and the fourth contains the formulae for the form factors.

I. The definition of the operation "⟨ ⟩". Consider a set of rapidities $\beta_1, \ldots, \beta_{2n}$ (we deal only with operators preserving the topological charge, which is why only the even-number soliton form factors differ from zero) Associate a space $h_i \simeq \mathbb{C}^2$ with every β_i. In the space h_i^* we consider a base $e_{i,\varepsilon}$ ($\varepsilon = \pm\frac{1}{2}, e_{i,-\frac{1}{2}} \simeq (0,1), e_{i,\frac{1}{2}} \simeq (1,0)$). The natural base of the tensor product $H_n^* = \Pi \otimes h_j$ consists of the vectors $e_{\varepsilon_1 \ldots \varepsilon_{2n}} = e_{1,\varepsilon_1} \otimes \ldots \otimes e_{2n,\varepsilon_{2n}}$. The uncharged sub-space stretched on $e_{\varepsilon_1 \ldots \varepsilon_{2n}}$ with $\Sigma\varepsilon_j = 0$ will be denoted by $H_{n,0}^*$.

Consider the operator $\hat{S}_{ij}(\beta)$ acting nontrivially only in $h_i \otimes h_j$:

$$\hat{S}_{ij}(\beta) = \frac{1}{2}\,\mathrm{sh}^{-1}\left(\frac{\pi}{\xi}(\beta - \pi i)\right)\left\{I_i \otimes I_j\left(\mathrm{sh}\,\frac{\pi}{\xi}(\beta - \pi i) - \mathrm{sh}\,\frac{\pi}{\xi}\beta\right)\right.$$
$$\left. + \sigma_i^3 \otimes \sigma_j^3\left(\mathrm{sh}\,\frac{\pi}{\xi}(\beta - \pi i) + \mathrm{sh}\,\frac{\pi}{\xi}\beta\right) - \mathrm{sh}\,\frac{\pi^2 i}{\xi}(\sigma_i^1 \otimes \sigma_j^1 + \sigma_i^2 \otimes \sigma_j^2)\right\},$$

where $\sigma_i^1 \sigma_i^2, \sigma_i^3, I_i$ are the Pauli matrices and the unit matrix acting in h_i.

It is evident that $\hat{S}_{i,j}$ differs from the soliton S-matrix only by the absence of the multiplier $S_0(\beta)$. Our first purpose is to construct certain special base in $H^*_{n,0}$. To this end, let us introduce a new object. Consider an auxiliary space $h_0 \simeq \mathbb{C}^2$ and construct a matrix

$$\begin{aligned}\begin{pmatrix} A(\sigma|\beta_1,\ldots,\beta_{2n}) & B(\sigma|\beta_1,\ldots,\beta_{2n}) \\ C(\sigma|\beta_1,\ldots,\beta_{2n}) & D(\sigma|\beta_1,\ldots,\beta_{2n}) \end{pmatrix} &\equiv T(\sigma|\beta_1,\ldots,\beta_{2n}) \\ &= \hat{S}_{0,1}(\sigma-\beta_1)\hat{S}_{0,2}(\sigma-\beta_2)\ldots\hat{S}_{0,2n}(\sigma-\beta_{2n})\,. \end{aligned} \tag{40}$$

The product and division to blocks are due to the space h_0. This definition is standard in the framework of QISM [14-16]. The matrix (40) is a monodromy matrix of "high level Bethe Ansatz" [58]. The following involutions hold when $\sigma \in \mathbb{R}$:

$$\begin{aligned} A^*(\sigma|\beta_1,\ldots,\beta_{2n}) &= D(\sigma+\pi i|\beta_1,\ldots,\beta_{2n})\prod_{j=1}^{2n}\frac{\operatorname{sh}\frac{\pi}{\xi}(\sigma-\beta_j)}{\operatorname{sh}\frac{\pi}{\xi}(\sigma-\beta_j+\pi i)}\,, \\ B^*(\sigma+\pi i|\beta_1,\ldots,\beta_{2n}) &= C(\sigma|\beta_1,\ldots,\beta_{2n})\prod_{j=1}^{2n}\frac{\operatorname{sh}\frac{\pi}{\xi}(\sigma-\beta_j-\pi i)}{\operatorname{sh}\frac{\pi}{\xi}(\sigma-\beta_j)}\,. \end{aligned} \tag{41}$$

The operators $\hat{S}_{ij}$ satisfy the Yang-Baxter equation:

$$\begin{aligned} &\hat{S}_{0,i}(\sigma-\beta_i)\hat{S}_{0,i+1}(\sigma-\beta_{i+1})\hat{S}_{i,i+1}(\beta_i-\beta_{i+1}) \\ &\quad= \hat{S}_{i,i+1}(\beta_i-\beta_{i+1})\hat{S}_{0,i+1}(\sigma-\beta_{i+1})\hat{S}_{0,i}(\sigma-\beta_i)\,, \end{aligned}$$

hence

$$\begin{aligned} &\hat{S}^{-1}_{i,i+1}(\beta_i-\beta_{i+1}) \\ &\times\begin{pmatrix} A(\sigma|\beta_1,\ldots,\beta_i,\beta_{i+1},\ldots,\beta_{2n}) & B(\sigma|\beta_1,\ldots,\beta_i,\beta_{i+1},\ldots,\beta_{2n}) \\ C(\sigma|\beta_1,\ldots,\beta_i,\beta_{i+1},\ldots,\beta_{2n}) & D(\sigma|\beta_1,\ldots,\beta_i,\beta_{i+1},\ldots,\beta_{2n}) \end{pmatrix} \\ &\times\hat{S}_{i,i+1}(\beta_i-\beta_{i+1}) \\ &= \begin{pmatrix} A(\sigma|\beta_1,\ldots,\beta_{i+1},\beta_i,\ldots,\beta_{2n}) & B(\sigma|\beta_1,\ldots,\beta_{i+1},\beta_i,\ldots,\beta_{2n}) \\ C(\sigma|\beta_1,\ldots,\beta_{i+1},\beta_i,\ldots,\beta_{2n}) & D(\sigma|\beta_1,\ldots,\beta_{i+1},\beta_i,\ldots,\beta_{2n}) \end{pmatrix}. \end{aligned} \tag{42}$$

As in Sec. 2, we imply h_i to be always associated with β_i, which means that if β_i and β_{i+1} change places their spaces h_i and h_{i+1} also change places.

In what follows we shall often denote, when it does not cause confusion, the operators $A(\sigma|\beta_1,\ldots,\beta_{2n})$, etc. by $A(\sigma)$, etc. From the Yang-Baxter equations it follows that operators A, B, C, D satisfy a great number of commutation relations which can be written in the form [14]

$$R(\sigma_1-\sigma_2)T(\sigma_1)\otimes T(\sigma_2)=T(\sigma_2)\otimes T(\sigma_1)R(\sigma_1-\sigma_2)\ , \qquad (43)$$

where the tensor product is taken only with respect to the space h_0, and $R(\sigma)$ is a (4×4)-matrix acting in $\mathbb{C}^2\otimes\mathbb{C}^2$:

$$R(\sigma)=\begin{pmatrix} \mathrm{sh}\,\frac{\pi}{\xi}(\sigma-\pi i) & 0 & 0 & 0 \\ 0 & -\mathrm{sh}\,\frac{\pi^2 i}{\xi} & \mathrm{sh}\,\frac{\pi}{\xi}\sigma & 0 \\ 0 & \mathrm{sh}\,\frac{\pi}{\xi}\sigma & -\mathrm{sh}\,\frac{\pi^2 i}{\xi} & 0 \\ 0 & 0 & 0 & \mathrm{sh}\,\frac{\pi}{\xi}(\sigma-\pi i) \end{pmatrix}.$$

In particular, one finds from Eq. (43) that the operators $A(\sigma_1)$ and $A(\sigma_2)$ commute; the same can be said about the operators $B(\sigma_1)$ and $B(\sigma_2)$.

Consider a set of covectors

$$w_{\varepsilon_1,\dots,\varepsilon_{2n}}(\beta_1,\dots,\beta_{2n})=\langle 0|\prod_{p:\varepsilon_p=\frac{1}{2}}B(\beta_p|\beta_1,\dots,\beta_{2n})\ ,$$

where $\varepsilon_i=\pm\frac{1}{2}, \Sigma\varepsilon_i=0, \langle 0|=e_{-\frac{1}{2}-\frac{1}{2}\cdots-\frac{1}{2}}$. From what follows it will be clear that $w_{\varepsilon_1,\dots,\varepsilon_{2n}}(\beta_1,\dots,\beta_{2n})$ create a base in $H^*_{n,0}$. Let us list the properties of the base, formulating them as Lemmas.

Lemma 1.

$$\begin{aligned} & w_{\varepsilon_1,\dots,\varepsilon_i\varepsilon_{i+1},\dots,\varepsilon_{2n}}(\beta_1,\dots,\beta_i,\beta_{i+1},\dots,\beta_{2n})\hat{S}_{i,i+1}(\beta_i-\beta_{i+1}) \\ & = w_{\varepsilon_1,\dots,\varepsilon_{i+1},\varepsilon_i,\dots,\varepsilon_{2n}}(\beta_1,\dots,\beta_{i+1},\beta_i,\dots,\beta_{2n})\ . \end{aligned} \qquad (44)$$

Proof. The proof follows immediately from (42).

Lemma 2. The covectors $w_{\varepsilon_1,\dots,\varepsilon_{2n}}(\beta_1,\dots,\beta_{2n})$ are eigencovectors for $A(\sigma|\beta_1,\dots,\beta_{2n})$.

Proof. From (43) one gets in particular

$$\begin{aligned} & B(\sigma_1)A(\sigma_2) \\ & = -\frac{\mathrm{sh}\,\frac{\pi}{\xi}(\sigma_2-\sigma_1-\pi i)}{\mathrm{sh}\,\frac{\pi}{\xi}(\sigma_2-\sigma_1)}A(\sigma_2)B(\sigma_1)+\frac{\mathrm{sh}\,\frac{\pi^2 i}{\xi}}{\mathrm{sh}\,\frac{\pi}{\xi}(\sigma_2-\sigma_1)}A(\sigma_1)B(\sigma_2)\ . \end{aligned}$$

Using this relation and the identity

$$\langle 0|A(\sigma) = \langle 0| \prod_{j=1}^{2n} \frac{\text{sh } \frac{\pi}{\xi}(\sigma - \beta_j)}{\text{sh } \frac{\pi}{\xi}(\sigma - \beta_j - \pi i)} \,,$$

one can derive [14, 59] an identity

$$\langle 0| \prod_{j=1}^{k} B(\tau_j) A(\sigma) = \prod_{j=1}^{2n} \frac{\text{sh } \frac{\pi}{\xi}(\sigma - \beta_j)}{\text{sh } \frac{\pi}{\xi}(\sigma - \beta_j - \pi i)} \prod_{j=1}^{k} \frac{\text{sh } \frac{\pi}{\xi}(\sigma - \tau_j - \pi i)}{\text{sh } \frac{\pi}{\xi}(\sigma - \tau_j)}$$
$$\times \langle 0| \prod_{j=1}^{k} B(\tau_j) + \sum_{l=1}^{k} \langle 0| \prod_{j\neq l} B(\tau_j) B(\sigma) \frac{\text{sh } \frac{\pi^2 i}{\xi}}{\text{sh } \frac{\pi}{\xi}(\sigma - \tau_l)}$$
$$\times \prod_{j=1}^{2n} \frac{\text{sh } \frac{\pi}{\xi}(\tau_l - \beta_j)}{\text{sh } (\tau_l - \beta_j - \pi i)} \prod_{j=1}^{k} \frac{\text{sh } \frac{\pi}{\xi}(\tau_l - \tau_j)}{\text{sh } \frac{\pi}{\xi}(\tau_l - \tau_j - \pi i)} \,.$$

It is quite obvious that if $\tau_1, \ldots, \tau_k$ is a subset of $\beta_1, \ldots, \beta_{2n}$ all the terms except the first one disappear. Hence

$$w_{\varepsilon_1,\ldots,\varepsilon_{2n}}(\beta_1, \ldots, \beta_{2n}) A(\sigma)$$
$$= \prod_{p:\varepsilon_p = -\frac{1}{2}} \frac{\text{sh } \frac{\pi}{\xi}(\sigma - \beta_p)}{\text{sh } \frac{\pi}{\xi}(\sigma - \beta_p - \pi i)} w_{\varepsilon_1,\ldots,\varepsilon_{2n}}(\beta_1, \ldots, \beta_{2n}) \,. \tag{45}$$

Q.E.D.

Lemma 3. The bases $e_{\varepsilon_1,\ldots,\varepsilon_{2n}}$ and $w_{\varepsilon_1,\ldots,\varepsilon_{2n}}(\beta_1, \ldots, \beta_{2n})$ are connected via a triangular transformation, which means that in the decomposition

$$w_{\varepsilon_1,\ldots,\varepsilon_{2n}}(\beta_1, \ldots, \varepsilon_{2n}) = \sum_{\varepsilon'} C_{\varepsilon_1,\ldots,\varepsilon_{2n}}^{\varepsilon'_1,\ldots,\varepsilon'_{2n}}(\beta_1, \ldots, \beta_{2n}) e_{\varepsilon'_1,\ldots,\varepsilon'_{2n}} \,,$$

only those $C_{\varepsilon_1,\ldots,\varepsilon_{2n}}^{\varepsilon'_1,\ldots,\varepsilon'_{2n}}$ for which $(\varepsilon'_1, \ldots, \varepsilon'_{2n}) \leq (\varepsilon_1, \ldots, \varepsilon_{2n})$ differ from zero. The multi-indices are ordered as binaries, for example, $(\frac{1}{2}, \frac{1}{2}, -\frac{1}{2}, -\frac{1}{2}) < (\frac{1}{2}, -\frac{1}{2}, \frac{1}{2}, -\frac{1}{2}) < (-\frac{1}{2}, \frac{1}{2}, \frac{1}{2}, -\frac{1}{2}) < (\frac{1}{2}, -\frac{1}{2}, -\frac{1}{2}, \frac{1}{2}) < (-\frac{1}{2}, \frac{1}{2}, -\frac{1}{2}, \frac{1}{2}) < (-\frac{1}{2}, -\frac{1}{2}, \frac{1}{2}, \frac{1}{2})$. The coefficient $C_{-\frac{1}{2}\cdots-\frac{1}{2}\frac{1}{2}\cdots\frac{1}{2}}^{-\frac{1}{2}\cdots-\frac{1}{2}\frac{1}{2}\cdots\frac{1}{2}}$ is equal to

$$C_{-\frac{1}{2}\cdots-\frac{1}{2}\frac{1}{2}\cdots\frac{1}{2}}^{-\frac{1}{2}\cdots-\frac{1}{2}\frac{1}{2}\cdots\frac{1}{2}}(\beta_1, \ldots, \beta_{2n}) = \prod_{i\leq n} \prod_{j>n} \frac{\text{sh } \frac{\pi}{\xi}(\beta_i - \beta_j)}{\text{sh } \frac{\pi}{\xi}(\beta_i - \beta_j + \pi i)} \,.$$

Proof. Consider some $w_{\varepsilon_1\ldots\varepsilon_{2n}}(\beta_1,\ldots,\beta_{2n})$ and suppose all $\varepsilon_j=-\frac{1}{2}$ for $j\geq k$ (k is arbitrary). Present matrix (40) as a product of two matrices:

$$\begin{pmatrix} A_1(\sigma) & B_1(\sigma)\\ C_1(\sigma) & D_1(\sigma)\end{pmatrix} = \hat{S}_{0,1}(\sigma-\beta_1)\ldots\hat{S}_{0,k-1}(\sigma-\beta_{k-1}) ,$$

$$\begin{pmatrix} A_2(\sigma) & B_2(\sigma)\\ C_2(\sigma) & D_2(\sigma)\end{pmatrix} = \hat{S}_{0,k}(\sigma-\beta_k)\ldots\hat{S}_{0,2n}(\sigma-\beta_{2n}) .$$

Evidently the matrix elements of these matrices commute. The operator $B(\sigma)$ can be presented as follows:

$$B(\sigma)=A_1(\sigma)B_2(\sigma)+B_1(\sigma)D_2(\sigma) .$$

To obtain $w_{\varepsilon_1,\ldots,\varepsilon_{2n}}(\beta_1,\ldots,\beta_{2n})$ we shall successively apply the operators $B(\beta_p), p:\varepsilon_p=\frac{1}{2}$ to $\langle 0|$. Taking into account that the covectors $\langle 0|B_1(\beta_{j_1})\ldots B_1(\beta_{jl})$ are the eigencovectors for $A_1(\sigma)$ with the eigenvalues (45) and $\langle 0|D_2(\sigma)=\langle 0|$, one gets

$$\begin{aligned} w_{\varepsilon_1,\ldots,\varepsilon_{2n}}(\beta_1,\ldots,\beta_{2n}) &= \langle 0|\prod_{\varepsilon_p=\frac{1}{2}} B_1(\beta_p)\\ &= \left(\prod_{j=1}^{k-1}\otimes e_{j,-\frac{1}{2}}\right)\prod_{\varepsilon_p=\frac{1}{2}} B_1(\beta_p)\otimes\left(\prod_{j=k}^{2n} e_{j,-\frac{1}{2}}\right) . \end{aligned}$$

Consequently the decomposition of $w_{\varepsilon_1,\ldots,\varepsilon_{2n}}(\beta_1,\ldots,\beta_{2n})$ contains only those $e_{\varepsilon_1,\ldots,\varepsilon_{2n}}$ for which $\varepsilon_k=\ldots=\varepsilon_{2n}=-\frac{1}{2}$. The calculation of $C^{-\frac{1}{2},\ldots-\frac{1}{2},\frac{1}{2},\ldots,\frac{1}{2}}_{-\frac{1}{2},\ldots,-\frac{1}{2},\frac{1}{2},\ldots,\frac{1}{2}}$ is rather trivial.

Q.E.D.

Lemma 4. The covector $w_{\varepsilon_1,\ldots,\varepsilon_{2n}}(\beta_1,\ldots,\beta_{2n})$ has simple poles only at the points $\beta_j=\beta_i=\pi i+i\xi k, k\in \mathrm{Z}$ for $j>i,\varepsilon_j=\frac{1}{2},\varepsilon_j=-\frac{1}{2}$, the corresponding residues being

$$\begin{aligned} &\mathrm{res}\ w_{\varepsilon_1,\ldots,\varepsilon_{i-1},-\frac{1}{2},\varepsilon_{i+1},\ldots,\varepsilon_{j-1},\frac{1}{2},\varepsilon_{j+1},\ldots\varepsilon_{2n}}(\beta_1,\ldots,\beta_{2n})\\ &=\frac{\xi}{\pi}\sin\frac{\pi^2}{\xi}w_{\varepsilon_1,\ldots,\varepsilon_{i-1},\varepsilon_{i+1},\ldots,\varepsilon_{j-1},\varepsilon_{j+1},\ldots,\varepsilon_{2n}}\\ &\times(\beta_1,\ldots,\beta_{i-1},\beta_{i+1},\ldots,\beta_{j-1},\beta_{j+1},\ldots,\beta_{2n}) \end{aligned}$$

$$\otimes\left(e_{i,\frac{1}{2}}\otimes e_{j,-\frac{1}{2}}+(-1)^k e_{i,-\frac{1}{2}}\otimes e_{j,\frac{1}{2}}\right)\prod_{\substack{p:\varepsilon_p=-\frac{1}{2}\\ p\neq i}}^{2n}\frac{\operatorname{sh}\frac{\pi}{\xi}(\beta_i-\beta_p+\pi i)}{\operatorname{sh}\frac{\pi}{\xi}(\beta_i-\beta_p)}$$

$$\times\prod_{k>j}\frac{\operatorname{sh}\frac{\pi}{\xi}(\beta_k-\beta_i)}{\operatorname{sh}\frac{\pi}{\xi}(\beta_k-\beta_i-\pi i)}\hat{S}_{j-1,i}(\beta_{j-1}-\beta_i)\dots\hat{S}_{i+1,i}(\beta_{i+1}-\beta_i)\,.$$

Proof. Consider first the case $j=2n, i=2n-1$. Present matrix (40) as a product of two matrices:

$$\begin{pmatrix}A_1(\sigma) & B_1(\sigma)\\ C_1(\sigma) & D_1(\sigma)\end{pmatrix}=\hat{S}_{0,1}(\sigma-\beta_1)\dots\hat{S}_{0,2n-2}(\sigma-\beta_{2n-2})\,,$$

$$\begin{pmatrix}a(\sigma) & b(\sigma)\\ c(\sigma) & d(\sigma)\end{pmatrix}=\hat{S}_{0,2n-1}(\sigma-\beta_{2n-1})\hat{S}_{0,2n}(\sigma-\beta_{2n})\,.$$

At first glance it is clear that the pole at the point $\beta_{2n}=\beta_{2n-1}+\pi i+i\xi k$ may exist only for $\varepsilon_{2n}=\frac{1}{2}$. Let us act on $\langle 0|$ by $B(\beta_{2n})=B_1(\beta_{2n})d(\beta_{2n})+A_1(\beta_{2n})b(\beta_{2n})$:

$$\langle 0|B(\beta_{2n})=\langle 0|\Big\{B_1(\beta_{2n})+\prod_{j=1}^{2n-2}\frac{\operatorname{sh}\frac{\pi}{\xi}(\beta_{2n}-\beta_j)}{\operatorname{sh}\frac{\pi}{\xi}(\beta_{2n}-\beta_j-\pi i)}$$

$$\times\left[\frac{\operatorname{sh}\frac{\pi}{\xi}(\beta_{2n}-\beta_{2n-1})}{\operatorname{sh}\frac{\pi}{\xi}(\beta_{2n}-\beta_{2n-1}-\pi i)}\sigma_{2n}^-+\frac{\operatorname{sh}\frac{\pi^2 i}{\xi}}{\operatorname{sh}\frac{\pi}{\xi}(\beta_{2n}-\beta_{2n-1}-\pi i)}\sigma_{2n-1}^-\right]\Big\}\,,$$

hence

$$\operatorname{res}\langle 0|B(\beta_{2n})=\prod_{j=1}^{2n-2}\frac{\operatorname{sh}\frac{\pi}{\xi}(\beta_{2n-1}-\beta_j+\pi i)}{\operatorname{sh}\frac{\pi}{\xi}(\beta_{2n-1}-\beta_j)}$$

$$\times\frac{\xi}{\pi}\sin\frac{\pi^2}{\xi}\langle 0|(\sigma_{2n}^-+(-1)^k\sigma_{2n-1}^-)\,,\quad(\beta_{2n}=\beta_{2n-1}+\pi i-i\xi k)\,.$$

Now let us apply $B(\sigma)$ to $\langle 0|(\sigma_{2n}^-+(-1)^k\sigma_{2n-1}^-)$, $(\sigma\neq\beta_{2n},\beta_{2n}=\beta_{2n-1}+\pi i+i\xi k)$. Notice that

$$\langle 0|(\sigma_{2n}^- + (-1)^k\sigma_{2n-1}^-)d(\sigma) = \langle 0|\Big\{\sigma_{2n}\frac{\text{sh}\,\frac{\pi}{\xi}(\sigma-\beta_{2n-1}-\pi i)}{\text{sh}\,\frac{\pi}{\xi}(\sigma-\beta_{2n-1}-2\pi i)}$$

$$+(-1)^k\sigma_{2n-1}^-\frac{\text{sh}\,\frac{\pi}{\xi}(\sigma-\beta_{2n-1})}{\text{sh}\,\frac{\pi}{\xi}(\sigma-\beta_{2n-1}-\pi i)}$$

$$-\sigma_{2n}^-\frac{\text{sh}\,\frac{\pi^2 i}{\xi}}{\text{sh}\,\frac{\pi}{\xi}(\sigma-\beta_{2n-1}-\pi i)\text{sh}\,\frac{\pi}{\xi}(\sigma-\beta_{2n-1}-2\pi i)}\Big\}$$

$$=\frac{\text{sh}\,\frac{\pi}{\xi}(\sigma-\beta_{2n-1})}{\text{sh}\,\frac{\pi}{\xi}(\sigma-\beta_{2n-1}-\pi i)}\langle 0|(\sigma_{2n}^- + (-1)^k\sigma_{2n-1}^-)\ ,$$

$$\langle 0|(\sigma_{2n}^- + (-1)^k\sigma_{2n-1}^-)b(\sigma)$$

$$=\langle 0|\Big\{-(-1)^k\sigma_{2n}^-\sigma_{2n-1}^-\frac{\text{sh}\,\frac{\pi}{\xi}(\sigma-\beta_{2n-1}-\pi i)\text{sh}\,\frac{\pi^2 i}{\xi}}{\text{sh}\,\frac{\pi}{\xi}(\sigma-\beta_{2n-1}-2\pi i)\text{sh}\,\frac{\pi}{\xi}(\sigma-\beta_{2n-1}-\pi i)}$$

$$+(-1)^k\frac{\text{sh}\,\frac{\pi^2 i}{\xi}}{\text{sh}\,\frac{\pi}{\xi}(\sigma-\beta_{2n-1}-2\pi i)}\sigma_{2n-1}^-\sigma_{2n}^-\Big\} = 0\ ,$$

hence

$$\langle 0|(\sigma_{2n}^- + (-1)^k\sigma_{2n-1}^-)B(\sigma)$$

$$=\frac{\text{sh}\,\frac{\pi}{\xi}(\sigma-\beta_{2n-1})}{\text{sh}\,\frac{\pi}{\xi}(\sigma-\beta_{2n-1}-\pi i)}\langle 0|(\sigma_{2n}^- + (-1)^k\sigma_{2n-1}^-)B_1(\sigma)\ .$$

It is clear now that

$$\text{res}\ w_{\varepsilon_1,\dots,\varepsilon_{2n}}(\beta_1,\dots,\beta_{2n})$$

$$=\frac{\pi}{\xi}\sin\frac{\pi^2}{\xi}\delta_{\varepsilon_{2n},\frac12}\delta_{\varepsilon_{2n-1},-\frac12}\prod_{p:\varepsilon_p=-\frac12}^{2n-2}\frac{\text{sh}\,\frac{\pi}{\xi}(\beta_{2n-1}-\beta_p+\pi i)}{\text{sh}\,\frac{\pi}{\xi}(\beta_{2n-1}-\beta_p)}$$

$$\times(e_{2n-1,-\frac12}\otimes e_{2n,\frac12}+(-1)^k e_{2n-1,\frac12}\otimes e_{2n,-\frac12})$$

$$\otimes\, w_{\varepsilon_1,\dots,\varepsilon_{2n-2}}(\beta_1,\dots,\beta_{2n-2})\ ,\quad \beta_{2n}=\beta_{2n-1}+\pi i+i\xi k\ .$$

The proof for arbitrary i,j can be obtained using (44) and the formula

$$(e_{i,-\frac12}\otimes e_{j,\frac12}+(-1)^p e_{i,\frac12}\otimes e_{j,-\frac12})\hat{S}_{ki}(\sigma)\hat{S}_{kj}(\sigma-\pi i)$$

$$=\frac{\text{sh}\,\frac{\pi}{\xi}\sigma}{\text{sh}\,\frac{\pi}{\xi}(\sigma-\pi i)}(e_{i,-\frac12}\otimes e_{j,\frac12}+(-1)^p e_{i,\frac12}\otimes e_{j,-\frac12})\ .$$

Q.E.D.

In Sec. 9 we shall need a formula for the scalar products of vectors decomposed via $w_{\varepsilon_1,\dots\varepsilon_{2n}}(\beta_1,\dots,\beta_{2n})$. It is evident that the base $\{w_{\varepsilon_1,\dots,\varepsilon_{2n}}(\beta_1,\dots,\beta_{2n})\}$ is not orthogonal. However, a base dual to $\{w_{\varepsilon_1,\dots,\varepsilon_{2n}}(\beta_1,\dots,\beta_{2n})\}$ can be easily constructed. Consider the covectors

$$\tilde{w}_{\varepsilon_1,\dots,\varepsilon_{2n}}(\beta_1,\dots,\beta_{2n}) = \langle 0| \prod_{p:\varepsilon_p=\frac{1}{2}} B'(\beta_p+\pi i) \ ,$$

where

$$B'(\beta_p+\pi i) = \lim_{\sigma\to\beta_p} B(\sigma+\pi i)(\sigma-\beta_p)\frac{\pi}{\xi}\left(\text{sh } \frac{\pi^2 i}{\xi}\right)^{-1} \ .$$

The base $\{\tilde{w}_{\varepsilon_1,\dots,\varepsilon_{2n}}(\beta_1,\dots,\beta_{2n})\}$ possesses properties similar to those of the base $\{w_{\varepsilon_1,\dots,\varepsilon_{2n}}(\beta_1,\dots,\beta_{2n})\}$, the most important property being the triangularity: in the decomposition

$$\tilde{w}_{\varepsilon_1,\dots,\varepsilon_{2n}}(\beta_1,\dots,\beta_{2n}) = \sum_{\varepsilon'} C^{\varepsilon'_1\dots\varepsilon'_{2n}}_{\varepsilon_1,\dots,\varepsilon_{2n}}(\beta_1,\dots,\beta_{2n}) e_{\varepsilon'_1,\dots,\varepsilon'_{2n}} \ ,$$

only those $C^{\varepsilon'_1,\dots,\varepsilon'_{2n}}_{\varepsilon_1,\dots,\varepsilon_{2n}}$ differ from zero for which $(\varepsilon'_1,\dots,\varepsilon'_{2n}) \geq (\varepsilon_1,\dots,\varepsilon_{2n})$. This triangularity is opposite to the triangularity of $\{w_{\varepsilon_1,\dots,\varepsilon_{2n}}(\beta_1,\dots,\beta_{2n})\}$. The following statement holds.

Lemma 5. The bases $\{w_{\varepsilon_1,\dots,\varepsilon_{2n}}(\beta_1,\dots,\beta_{2n})\}$ and $\{\tilde{w}_{\varepsilon_1,\dots,\varepsilon_{2n}}(\beta_1,\dots,\beta_{2n})\}$ are dual:

$$w_{\varepsilon_1,\dots,\varepsilon_{2n}}(\beta_1,\dots,\varepsilon_{2n})\tilde{w}^*_{\varepsilon'_1,\dots,\varepsilon'_{2n}}(\beta_1,\dots,\beta_{2n}) = \prod_{j=1}^{2n}\delta_{\varepsilon_j,\varepsilon'_j} \ . \tag{46}$$

Proof. From Eq. (41) it follows that

$$w^*_{\varepsilon_1,\dots,\varepsilon_{2n}}(\beta_1,\dots,\varepsilon_{2n}) = \prod_{p:\varepsilon_p=\frac{1}{2}}\prod_{\substack{j=1\\ j\neq p}}^{2n} \frac{\text{sh } \frac{\pi}{\xi}(\beta_p-\beta_j-\pi i)}{\text{sh } \frac{\pi}{\xi}(\beta_p-\beta_j)} \prod_{p:\varepsilon_p=\frac{1}{2}} C(\beta_p)|0\rangle \ .$$

Consider the scalar product (46). Let $(\varepsilon_1,\dots,\varepsilon_{2n})$ differ from $(\varepsilon'_1,\dots,\varepsilon'_{2n})$, then necessarily a number j exists for which $\varepsilon_j = -\frac{1}{2}, \varepsilon'_j = \frac{1}{2}$. Then the scalar product (46) can be presented as follows,

$$\langle 0| \prod_{p:\varepsilon_p=\frac{1}{2}} B(\beta_p)C(\beta_j) \prod_{q:\varepsilon_q=\frac{1}{2},q\neq j} C(\beta_q)|0\rangle \ .$$

Let us move $C(\beta_j)$ to the left using the commutation relation which follows from Eq. (43) (the second term in RHS can be omitted because Lemma 2 implies $\langle 0| \prod_{s\neq m} B(\beta_s)A(\beta_m) = 0$):

$$B(\beta_p)C(p_j) = C(\beta_j)B(\beta_p) - \frac{\mathrm{sh}\,\frac{\pi^2 i}{\xi}}{\mathrm{sh}\,\frac{\pi}{\xi}(\beta_p-\beta_j)}(A(\beta_p)D(\beta_j) - A(\beta_j)D(\beta_p))\ .$$

Hence

$$\langle 0| \prod_{p:\varepsilon_p=\frac{1}{2}} B(\beta_p)C(\beta_j) = \langle 0|C(\beta_j) \prod_{p:\varepsilon_p=\frac{1}{2}} B(\beta_p) = 0\ .$$

Thus the scalar product differs from zero only for $(\varepsilon_1,\dots,\varepsilon_{2n})$ coinciding with $(\varepsilon'_1,\dots,\varepsilon'_{2n})$. The calculation of the scalar product in this case is a technical problem we will not go into.
Q.E.D.

Now we are in a position to define the operation $\langle\ \rangle_n$.

Definition 1. Consider a function $F(\lambda_1,\dots,\lambda_n|\mu_1,\dots,\mu_n)$ which is invariant under independent permutations of elements in the sets $\lambda_1,\dots,\lambda_n$ and $\mu_1,\dots,\mu_n$. With every function of this kind is associated the following vectors from $H^*_{n,0}$:

$$\langle F\rangle_n(\beta_1,\dots,\beta_{2n}) = \sum_{\substack{\{1,\dots,2n\}\\=\{i_1,\dots,i_n\}\cup\{j_1,\dots,j_n\}}} F(\beta_{i_1},\dots,\beta_{in}|\beta_{j_1},\dots,\beta_{jn})$$
$$\times \prod_{p=1}^{n}\prod_{q=1}^{n} \frac{1}{\mathrm{sh}\,\frac{\pi}{\xi}(\beta_{i_p}-\beta_{j_q})} w_{\varepsilon_1,\dots,\varepsilon_{2n}}(\beta_1,\dots,\beta_{2n})\ ,$$

where $\varepsilon_{j_p} = \frac{1}{2}, \varepsilon_{i_p} = -\frac{1}{2}, p = 1,\dots,n$.

Lemmas 1, 2, 4 imply the following properties of the operation $\langle\ \rangle_n$.

1. $$\langle F\rangle_n(\beta_1,\dots,\beta_i,\beta_{i+1},\dots,\beta_{2n})\hat{S}_{i,i+1}(\beta_i-\beta_{i+1}) = \langle F\rangle_n(\beta_1,\dots,\beta_{i+1},\beta_i,\dots,\beta_{2n}) \tag{47}$$

2. Denoting the components of $\langle F\rangle_n(\beta_1,\dots,\beta_{2n})$ in the base $e_{\varepsilon_1,\dots,\varepsilon_{2n}}$ by $\langle F\rangle_n(\beta_1,\dots,\beta_{2n})_{\varepsilon_1,\dots,\varepsilon_{2n}}$ one has

$$\langle F\rangle_n(\beta_1,\dots,\beta_{2n})_{-\frac{1}{2},\dots,-\frac{1}{2},\frac{1}{2},\dots,\frac{1}{2}}$$
$$= \prod_{p=1}^{n}\prod_{q=1}^{n} \frac{1}{\mathrm{sh}\,\frac{\pi}{\xi}(\beta_{n+p}-\beta_q-\pi i)} F(\beta_1,\dots,\beta_n|\beta_{n+1},\dots,\beta_{2n})\ .$$

3. If $F(\lambda_1,\dots,\lambda_n|\mu_1,\dots,\mu_n)$ is an entire function if all its arguments then the only singularities of $\langle F\rangle_n(\beta_1,\dots,\beta_{2n})$ are the simple poles of the points $\beta_j=\beta_i+\pi i+i\xi k$ with residues

$$\begin{aligned}\operatorname{res}\langle F\rangle_n(\beta_1,\dots,\beta_{2n}) &= \frac{\xi}{\pi}(e_{i,\frac12}\otimes e_{j,-\frac12}+(-1)^k e_{i,-\frac12}\otimes e_{j,\frac12})\\ &\otimes\langle Q_{\beta_i}\rangle_{n-1}(\beta_1,\dots,\hat\beta_i,\dots,\hat\beta_j,\dots,\beta_{2n})\prod_{l\le i-1}\operatorname{sh}^{-1}\frac{\pi}{\xi}(\beta_l-\beta_i)\\ &\times\prod_{l>j}\operatorname{sh}^{-1}\frac{\pi}{\xi}(\beta_l-\beta_i-\pi i)\hat S_{j-1,i}(\beta_{j-1}-\beta_i)\dots\hat S_{i+1,i}(\beta_{i+1}-\beta_i)\ ,\\ &\qquad\beta_j=\beta_i+\pi i+i\xi k\ ,\end{aligned}\tag{48}$$

where

$$\begin{aligned}&Q_\tau(\lambda_1,\dots,\lambda_{n-1}|\mu_1,\dots,\mu_{n-1})\\ &\quad=F(\lambda_1,\dots,\lambda_{n-1},\tau|\mu_1,\dots,\mu_{n-1},\tau+\pi i+i\xi k)\ .\end{aligned}$$

At this point we finish the discussion of the properties of $\langle\ \rangle_n$ and pass on the second important object which is the integral transformation $\Phi_{n,\sigma}$.

II. The integral transformation $\Phi_{n,\sigma}(p)$. Consider the function

$$\varphi(\beta)=\left(\frac{\xi}{\pi}\right)^{\frac12}\times\exp\left(-\int_0^\infty\frac{2\left(\sin^2\left(\frac{\kappa\beta}{2}\right)\operatorname{sh}\left(\frac{\pi+\xi}{2}\kappa\right)+\operatorname{sh}^2\left(\frac{\pi\kappa}{4}\right)\operatorname{sh}\left(\frac{\pi-\xi}{2}\kappa\right)\right)}{\kappa\operatorname{sh}\left(\frac{\xi\kappa}{2}\right)\operatorname{sh}\pi\kappa}d\kappa\right).$$

Having in mind the importance of this function for what follows let us list its main properties.

1. $\varphi(\beta)\sim 2\exp\left(\mp\frac12\left(\frac{\pi}{\xi}+1\right)\beta\right)$ for $\beta\to\pm\infty$. (49)

2. $\varphi(\beta)$ is an even function with the following poles and zeros in the strip $0<\operatorname{Im}\beta<2\pi$: simple poles at the points $\beta=\frac{\pi}{2}i+i\xi k, k\ge 0$, simple zeros at the point $\beta=\frac{3\pi}{2}i+i\xi k, k\ge 1$.

3. $\varphi(\beta)$ satisfies the identities

$$
\begin{aligned}
&\varphi(\beta - i\xi) = \varphi(\beta)\frac{\operatorname{ch}\frac{1}{2}\left(\beta + \frac{\pi i}{2}\right)}{\operatorname{ch}\frac{1}{2}\left(\beta - \frac{\pi i}{2} - i\xi\right)}\,, \\
&\varphi(\beta - 2\pi i) = \varphi(\beta)\frac{\operatorname{sh}\frac{\pi}{\xi}\left(\beta - \frac{\pi i}{2}\right)}{\operatorname{sh}\frac{\pi}{\xi}\left(\beta - \frac{3\pi i}{2}\right)}\,, \\
&\varphi(\beta - \pi i + i\xi k)\varphi(\beta) \\
&\quad = i^{k-1}\frac{\prod_{j=0}^{k-1}\operatorname{sh}\frac{1}{2}\left(\beta + i\xi j - \frac{\pi i}{2}\right)}{2\operatorname{sh}\frac{\pi}{\xi}\left(\beta - \frac{\pi i}{2}\right)\operatorname{sh}\frac{1}{2}\left(\beta - \frac{\pi i}{2}\right)\prod_{j=0}^{k}\operatorname{ch}\frac{1}{2}\left(\beta + i\xi j - \frac{\pi i}{2}\right)}\,, \\
&\qquad\qquad\qquad\qquad\qquad\qquad\qquad\qquad k \geq 0\,.
\end{aligned}
$$

Take some polynomial $P(x)$ and consider the integral

$$
\int_{-\infty}^{\infty}\prod_{j=1}^{2n}\varphi(\alpha - \beta_j)P\left(\exp\left(\frac{2\pi}{\xi}\alpha\right)\right)\exp\left(\alpha k - \frac{\pi}{\xi}(n-2)\alpha\right)d\alpha\,, \tag{50}
$$

where $|k| \leq n-1$. From the asymptotics (49) it follows that the integral (50) is defined correctly when deg $(P) \leq n-1$. We shall define a certain regularization which will provide us with the possibility of dealing with the integral for deg $(P) \leq 2n-2$. The regularized integral will be denoted by $⨏$.

Definition 2. Let P be a polynomial of the degree $\leq 2n-2$, then

$$
\begin{aligned}
&⨏_{-\infty}^{\infty}\prod_{j=1}^{2n}\varphi(\alpha - \beta_j)P\left(\exp\left(\frac{2\pi\alpha}{\xi}\right)\right)\exp\left(\alpha k - \frac{\pi}{\xi}(n-2)\alpha\right)d\alpha \\
&\quad \equiv \int_{\mathbb{R} + \frac{\pi i}{2} - i0}\prod_{j=1}^{2n}\varphi(\alpha - \beta_j)\frac{1}{\operatorname{sh}\frac{\pi}{\xi}\left(\alpha - \beta_j - \frac{3\pi i}{2}\right)}\exp\left(-\frac{\pi}{\xi}\Sigma\beta_j\right) \\
&\quad \times P_1\left(\exp\left(\frac{2\pi\alpha}{\xi}\right)\right)\exp\left(\alpha k - \frac{\pi}{\xi}(3n-2))\right)d\alpha \\
&\quad + \int_{\Gamma}\prod_{j=1}^{2n}\varphi(\alpha - \beta_j)P_2\left(\tau^4\exp\left(\frac{2\pi\alpha}{\xi}\right)\right)\exp\left(\alpha k - \frac{\pi}{\xi}(n-2)\alpha\right)d\alpha\,,
\end{aligned} \tag{51}
$$

where Γ envelops the points $\alpha = \beta_j - \frac{\pi i}{2} - i\xi k, k = 0, \ldots, [\pi/\xi], \tau = \exp\left(\frac{\pi^2 i}{\xi}\right)$, and the polynomials P_1 and P_2 are defined by the relation

$$\prod_{j=1}^{2n}\left(x\tau^{-3} - \exp\left(\frac{2\pi}{\xi}\beta_j\right)\right) P(x) = P_1(x)$$
$$+ \prod_{j=1}^{2n}\left(x\tau^{2} - \tau^{-1}\exp\left(\frac{2\pi}{\xi}\beta_j\right)\right) P_2(x)$$
$$- \tau^{-2(n-2)}\prod_{j=1}^{2n}\left(x\tau^{-3} - \exp\left(\frac{2\pi}{\xi}\beta_j\right)\right) P_2(x\tau^4) \qquad (52)$$

and the requirement $\deg P_1 \leq 3n-1$ (evidently in this case RHS of Eq. (51) is defined correctly).

Let us prove two statements which established the selfconsistency of Definition 1.

Lemma 6. It is always possible to find polymials P_1 and P_2 satisfying (52) and the requirement $\deg P_1 \leq 3n-1$.

Proof. If one takes in the capacity of $P_2(x)$ the monomial x^k, the maximal degree of the two latter terms in the RHS of (52) is equal to $2n+k$ and the coefficient before x^{2n+k} is equal to $(\tau^{-4} - \tau^{4-8n+4k})$. This coefficient does not vanish for $k \geq n$ (recall that we consider ξ in generic position). That is why one can successively reduce the degree of the polynomial $\Pi\left(x - \exp\left(\frac{2\pi}{\xi}\beta_j\right)\right) P(x)$ by subtracting

$$\Pi\left(x\tau^{-2} - \tau^{-1}\exp\left(\frac{2\pi}{\xi}\beta_j\right)\right) x^k$$
$$- \Pi\left(x\tau^{-3} - \exp\left(\frac{2\pi}{\xi}\beta_j\right)\right) x^k\tau^{4k}$$

with the proper coefficients. It is clear that the process terminates when the degree of the remainder is equal to $3n-1$. It should be noticed that the polynomials P_1 and P_2 are defined ambiguously.
Q.E.D.

Lemma 7. The ambiguity of the definition of P_1 and P_2 does not affect the value of the RHS of Eq. (51).

Proof. Consider two different pairs of polynomials P_1, P_2 and P_1', P_2' satisfying (52) for the same P. Polynomials $Q_1 = P_1 - P_1'$ and $Q_2 = P_2 - P_2'$

satisfy the equation

$$Q_1(x) = \prod_{j=1}^{2n} \left(x\tau^{-3} - \exp\left(\frac{2\pi}{\xi}\beta_j\right)\right) Q_2(x\tau^4)\tau^{-2(n-2)}$$
$$- \prod_{j=1}^{2n} \left(x\tau^{-2} - \tau^{-1}\exp\left(\frac{2\pi}{\xi}\beta_j\right)\right) Q_2(x) \ , \qquad (53)$$

with deg $Q \leq 3n-1$. Eq. (53) implies (see the proof of Lemma 6) that deg $Q_2 \leq n-1$. To prove the uniqueness of this definition one has to show that

$$\int_{\mathbb{R}+\frac{\pi i}{2}-i0} \prod_{j=1}^{2n} \frac{\varphi(\alpha-\beta_j)}{\operatorname{sh}\frac{\pi}{\xi}\left(\alpha-\beta_j-\frac{3\pi i}{2}\right)} \exp\left(-\frac{\pi}{\xi}\Sigma\beta_i\right) Q_1\left(\exp\left(\frac{2\pi}{\xi}\alpha\right)\right)$$
$$\times \exp\left(\alpha k - \frac{\pi}{\xi}(3n-2)\alpha\right) + \int_\Gamma \prod_{j=1}^{2n} \varphi(\alpha-\beta_j)$$
$$\times \exp\left(\alpha k - \frac{\pi}{\xi}(n-2)\alpha\right) Q_2\left(\tau^4 \exp\left(\frac{2\pi}{\xi}\alpha\right)\right) d\alpha = 0 \ . \qquad (54)$$

The inequality deg $Q_2 < n-1$ and the fact that $\varphi(\alpha-\beta_j)$ has no singularities in the strip $-\frac{3\pi}{2} < \operatorname{Im}\beta < \frac{\pi}{2}$ except those enveloped by Γ imply the possibility of deformation of the contour Γ in the second integral in Eq. (54) to

$$\int_\Gamma \prod_{j=1}^{2n} \varphi(\alpha-\beta_j) Q_2\left(\tau^4 \exp\left(\frac{2\pi\alpha}{\xi}\right)\right) \exp\left(\alpha k - \frac{\pi}{\xi}(3n-2)\alpha\right) d\alpha$$
$$= -\int_{\mathbb{R}+\frac{\pi i}{2}i0} \prod_{j=1}^{2n} \varphi(\alpha-\beta_j) Q_2\left(\tau^4 \exp\left(\frac{2\pi\alpha}{\xi}\right)\right)$$
$$\times \exp\left(\alpha k - \frac{\pi}{\xi}(3n-2)\alpha\right) d\alpha$$
$$+ \int_{\mathbb{R}-\frac{3\pi i}{2}-i0} \prod_{j=1}^{2n} \varphi(\alpha-\beta_j) Q_2\left(\tau^4 \exp\left(\frac{2\pi\alpha}{\xi}\right)\right)$$
$$\times \exp\left(\alpha k - \frac{\pi}{\xi}(3n-2)\alpha\right) \ . \qquad (55)$$

Taking into account

$$\varphi(\alpha - 2\pi i) = \varphi(\alpha) \frac{\operatorname{sh}\frac{\pi}{\xi}\left(\alpha - \frac{\pi i}{2}\right)}{\operatorname{sh}\frac{\pi}{\xi}\left(\alpha - \frac{3\pi i}{2}\right)} \ , \qquad (56)$$

we transform the second integral in the RHS of (55) to

$$\int_{\mathbb{R}+\frac{\pi i}{2}-i0}\prod_{j=1}^{2n}\varphi(\alpha-\beta_i)\frac{\operatorname{sh}\frac{\pi}{\xi}\left(\alpha-\beta_j-\frac{\pi i}{2}\right)}{\operatorname{sh}\frac{\pi}{\xi}\left(\alpha-\beta_j-\frac{3\pi i}{2}\right)}\exp\left(\alpha k-\frac{\pi}{\xi}(n-2)\alpha\right)$$
$$\times\,\tau^{2(n-2)}Q_2\left(\exp\left(\frac{2\pi\alpha}{\xi}\right)\right).$$

Now Eq. (54) obviously follows from Eq. (53).
Q.E.D.

Lemma 7 clarifies the meaning of the regularization (51). Informally, it can be explained as follows: we move the contour of integration in RHS of (51) to the line $\operatorname{Im}\alpha=\frac{\pi}{2}-0$ (the singularities of $\varphi(\alpha-\beta_j)$ do not hinder making it). Then using the property (56) of $\varphi(\alpha)$ we rewrite the divergent part of the integral in the form $-\int_{\mathbb{R}+\frac{\pi i}{2}-i0}+\int_{\mathbb{R}-\frac{3\pi i}{2}-i0}$ (some function). Now ignoring the fact that both the integral over $\mathbb{R}+\frac{\pi i}{2}-i0$ and that over $\mathbb{R}-\frac{3\pi i}{2}-i0$ are divergent we deform the contour $(\mathbb{R}-\frac{3\pi i}{2}-i0)\cup(-(\mathbb{R}+\frac{\pi i}{2}-i0))$ to the contour Γ enveloping the singularities of the integrand. Definition 2 and Lemmas 6, 7 give the formal treatment of this procedure. A nice property of the regularization $⨎$ is that in many cases one can handle $⨎$ as the usual $\int$ deforming contours, etc. These manipulations can be justified post factum using the formal Definition 2.

Now we are in a position to define the integral transformations $\Phi_{n,\sigma}(P)$.

Definition 3. Consider an antisymmetric polynomial $P(x_1,\dots,x_{n-1})$ whose degree with respect to every x_j does not exceed $2n-2$. The transformation $\Phi_{n,\sigma}(\sigma=0,\pm1)$ associates with this polynomial a function of $\beta_1,\dots,\beta_{2n}$:

$$\Phi_{n,\sigma}(P)=\left(\frac{1}{2\pi i}\right)^{n-1}\frac{1}{(n-1)!}⨎_{-\infty}^{\infty}d\alpha_1\dots⨎_{-\infty}^{\infty}d\alpha_{n-1}\prod_{i=1}^{n-1}\prod_{j=1}^{2n}\varphi(\alpha_i-\beta_j)$$
$$\times\prod_{i<j}\operatorname{sh}(\alpha_i-\alpha_j)\exp\left(-\frac{\pi}{\xi}(n-2)\Sigma\alpha_i+\sigma\Sigma\alpha_i\right)$$
$$\times\,P\left(\exp\left(\frac{2\pi}{\xi}\alpha_1\right),\dots\exp\left(\frac{2\pi}{\xi}\alpha_{n-1}\right)\right).$$

III. Polynomial Δ_n. The third necessary object is the polynomial $\Delta_n(x_1,\dots,x_{n-1}|a_1,\dots,a_n|b_1,\dots,b_n)$ which we are going to describe.

This polynomial is antisymmetric with respect to $x_1, \dots, x_{n-1}$ and symmetric with respect to $a_1, \dots, a_n$ and $b_1, \dots, b_n$ separately.

Definition 4. Consider an $(n-1) \times (n-1)$ matrix with the following matrix elements:

$$A_{ij}^{(n)} = A_i^{(n)}(x_j|a_1 \dots a_n|b_1 \dots b_n) ,$$

where

$$A_i^{(n)}(x|a_1, \dots, a_n|b_1, \dots, b_n) = 2^{-n-i} \\ \times \left[\prod_{l=1}^{n} (x - a_l \tau) \left\{ \sum_{k=0}^{i-1} (1 - \tau^{2(i-k)})(-1)^k x^{i-k-1} \tau^k \sigma_k(b_1, \dots, b_n) \right\} \right. \\ \left. + \tau^{2i} \prod_{l=1}^{n} (x - b_l \tau^{-1}) \left\{ \sum_{k=0}^{i-1} (1 - \tau^{2(i-k)})(-1)^k x^{i-k-1} \tau^k \sigma_k(a_1, \dots, a_n) \right\} \right] . \tag{57}$$

σ_k is the elementary symmetric polynomial of degree k ($\sigma_k(a_1, \dots, a_n) = 0$ for $k > n$ or $k < 0$). Polynomial $\Delta_n(x_1, \dots, x_{n-1}|a_1, \dots, a_n |b_1, \dots, b_n)$ is the determinant of this matrix.

Evidontly Δ_n is symmetric with respect to $a_1, \dots, a_n$ and $b_1, \dots, b_n$ separately and antisymmetric with respect to $x_1, \dots, x_{n-1}$ polynomial. Its degree with respect to x_i is $2n-2$, its degree with respect to $a_i(b_i)$ is $n-1$. Polynomial Δ_n is a homogeneous polynomial of the total degree $n(n-1) + \frac{1}{2}(n-1)(n-2)$. The main property of Δ_n is described by the following statement.

Lemma 8.

$$\Delta_n(x_1, \dots, x_{n-1}|a_1, \dots, a_{n-1}, a|b_1, \dots, b_{n-1}, a\tau^2) \\ = 2^{-3(n-1)} \prod_{j=1}^{n-1} (x_j - a\tau) \left\{ \sum_{k=1}^{n-1} (-1)^k \Delta_{n-1}(x_1, \dots, \hat{x}_k, \dots, \right. \\ x_{n-1}|a_1, \dots, a_{n-1}|b_1, \dots, b_{n-1}) x_k^{-1} \\ \left. \times \left[\prod_{j=1}^{n-1} (x_k - b_j\tau)(x_k - a_j\tau) - \tau^{4(n-1)} \prod_{j=1}^{n-1} (x_k - b_j\tau^{-1})(x_k - a_j\tau^{-1}) \right] \right\} \tag{58}$$

Proof. Using the identity $\sigma_k(a_1, \dots, a_n) = \sigma_k(a_1, \dots, a_{n-1}) + a_n\sigma_{k-1}(a_1, \dots, a_{n-1})$ one can show that

$$A_i^{(n)}(x|a, \dots, a_{n-1}, a|b_1, \dots, b_{n-1}, a\tau^2)$$
$$= \left(\frac{x - a\tau}{2}\right)\left\{A_i^{(n-1)}(x|a_1, \dots, a_{n-1}|b_1, \dots, b_{n-1})\right.$$
$$\left. - \frac{1}{2}a\tau^2 A_{i-1}^{(n-1)}(x|a_1, \dots, a_{n-1}|b_1, \dots, b_{n-1})\right\}$$

while

$$A_{n-1}^{(n-1)}(x|a_1, \dots, a_{n-1}|b_1, \dots, b_{n-1}) = 2^{-2(n-1)}x^{-1}$$
$$\times \left\{\prod_{j=1}^{n-1}(x - a_j\tau)(x - b_j\tau) - \tau^{4(n-1)}\prod_{j=1}^{n-1}(x - a_j\tau^{-1})(x - b_j\tau^{-1})\right\} .$$

Now the proof reduces to the successive adding of the first row with the coefficient $a\tau^2/2$ to the second one, of the second row with the coefficient $a\tau^2/2$ to the third one, etc., and following expansion of determinant along the last row.
Q.E.D.

IV. The formulas for the soliton form factors. Consider the functions

$$F_n^{(\gamma)}(x_1, \dots, x_{n-1}|\lambda_1, \dots, \lambda_n|\mu_1, \dots, \mu_n) = \exp\left(\frac{\pi\gamma}{2\xi}\sum_{i=1}^{n}(\mu_i - \lambda_i - \pi i)\right)$$
$$\times \Delta_n\left(x_1, \dots, x_{n-1}|\exp\left(\frac{2\pi}{\xi}\lambda_1\right), \dots, \exp\left(\frac{2\pi}{\xi}\lambda_n\right)\Big|\exp\left(\frac{2\pi}{\xi}\mu_1\right)\right.$$
$$\left.\times \exp\left(\frac{2\pi}{\xi}\mu_n\right)\right)\exp\left(-\frac{\pi}{\xi}(n-1)\sum_{j=1}^{2n}\beta_j\right),$$
$$\gamma = \pm 1 \tag{59}$$

which are antisymmetric polynomials of $x_1, \dots, x_{n-1}$ and entire symmetric functions of $\lambda_1, \dots, \lambda_n$ and $\mu_1, \dots, \mu_n$. The form factors of topological current $j_\mu = \varepsilon_{\mu\nu}\frac{\beta}{2\pi}\partial_\nu u\,(f_\mu)$, of energy-momentum tensor $T_{\mu\nu}\,(f_{\mu\nu})$, of

$\exp(\pm i\beta\mu/2)\,(f^{\pm})$ are given by the formulas

$$f_\mu(\beta_1,\ldots\beta_{2n}) = \frac{\pi}{\xi} M d^n \prod_{i<j}\zeta(\beta_i-\beta_j)\Big\{\Phi_{n,1}(\langle F_n^{(1)} - F_n^{(-1)}\rangle_n)e^{-\frac{1}{2}\Sigma\beta_j}$$
$$+(-1)^{\mu-1}\Phi_{n,-1}(\langle F_n^{(1)} - F_n^{(-1)}\rangle_n)e^{\frac{1}{2}\Sigma\beta_j}\Big\},$$
$$f_{\mu\nu}(\beta_1,\ldots,\beta_{2n}) = M^2 d^n \prod_{i<j}\zeta(\beta_i-\beta_j)\sum_{j=1}^{2n}(e^{\beta_j}+(-1)^{\mu-1}e^{-\beta_j}j)$$
$$\times\Big\{\Phi_{n,1}(\langle F_n^{(1)} + F_n^{(-1)}\rangle_n)e^{-\frac{1}{2}\Sigma\beta_j} + (-1)^{\nu-1}\Phi_{n,-1}$$
$$\times(\langle F_n^{(1)} + F_n^{(-1)}\rangle_n)e^{\frac{1}{2}\Sigma\beta_j}\Big\},$$
$$f^{\pm}(\beta_1,\ldots,\beta_{2n}) = d^n\prod_{i<j}\zeta(\beta_i-\beta_j)\Phi_{n,0}(\langle F_n^{(\pm1)}\rangle_n) \tag{60}$$

where

$$\zeta(\beta) = c\,\mathrm{sh}\,\frac{\beta}{2}\exp\left(\int_0^\infty \frac{\sin^2\left(\frac{1}{2}\kappa(\beta+\pi i)\right)\mathrm{sh}\left(\frac{\pi-\xi}{2}\kappa\right)}{\kappa\mathrm{sh}\left(\frac{\xi\kappa}{2}\right)\mathrm{sh}(\pi\kappa)\mathrm{ch}\left(\frac{\pi\kappa}{2}\right)}d\kappa\right),$$
$$c = \left(\frac{4\pi}{\xi}\right)^{\frac{1}{4}}\exp\left(\frac{1}{4}\int_0^\infty \frac{\mathrm{sh}\left(\frac{\pi\kappa}{2}\right)\mathrm{sh}\left(\frac{\pi-\xi}{2}\kappa\right)}{\kappa\mathrm{sh}\left(\frac{\xi\kappa}{2}\right)\mathrm{ch}^2\left(\frac{\pi\kappa}{2}\right)}d\kappa\right),$$
$$d - c^{-1}\frac{\pi}{\xi}\frac{1}{2\pi}.$$

The components of $f_\mu(\beta_1,\ldots,\beta_{2n}), f_{\mu\nu}(\beta_1,\ldots,\beta_{2n}), f^{\pm}(\beta_1,\ldots,\beta_{2n})$ in the basis $e_{\varepsilon_1,\ldots,\varepsilon_{2n}}$ are denoted by $f_\mu(\beta_1,\ldots,\beta_{2n})_{\varepsilon_1,\ldots,\varepsilon_{2n}}, f_{\mu\nu}(\beta_1,\ldots,\beta_{2n})_{\varepsilon_1,\ldots,\varepsilon_{2n}}, f^{\pm}(\beta_1,\ldots,\beta_{2n})_{\varepsilon_1,\ldots,\varepsilon_{2n}}$.

The function ζ is regular in the strip $0 > \mathrm{Im}\beta > -2\pi$ and satisfies the identities

$$\zeta(\beta)S_0(\beta) = \zeta(-\beta), \quad \zeta(\beta-2\pi i) = \zeta(-\beta),$$
$$\zeta(\beta)\zeta(\beta-\pi i) = \varphi^{-1}\left(\beta+\frac{\pi i}{2}\right). \tag{61}$$

It behaves as $\frac{1}{\sqrt{2}}\exp\left(\pm\frac{1}{4}\left(\frac{\pi}{\xi}+1\right)\beta\right)$ for $\beta\to\pm\infty$.

It is useful to mention that the operation $\langle\ \rangle_n$ and the integral transformation $\Phi_{n,\sigma}$ can be rearranged in Eq. (60).

In the following sections it will be shown that the form factors satisfy all the requirements formulated in Sec. 1. It will be also explained why we associate them with the operators $j_\mu, T_{\mu\nu}, \exp\left(\pm i\frac{\beta u}{2}\right)$.

To finish this section let us write down explicitly the formulas for two-soliton form factors ($n = 1$) because they will be important later and the formulas (60) cannot be applied to ($n = 1$) directly:

$$f_{\mu\nu}(\beta_1,\beta_2) = \frac{M^2}{4}(e^{\frac{1}{2}(\beta_1+\beta_2)} + (-1)^\mu e^{-\frac{1}{2}(\beta_1+\beta_2)})(e^{\frac{1}{2}(\beta_1+\beta_2)}$$
$$+ (-1)^\nu e^{\frac{1}{2}(\beta_1+\beta_2)})d\zeta(\beta_1-\beta_2)\frac{\text{ch}\,\frac{1}{2}(\beta_1-\beta_2)}{\text{sh}\,\frac{\pi}{2\xi}(\beta_1-\beta_2+\pi i)}(e_{\frac{1}{2},-\frac{1}{2}} + e_{-\frac{1}{2},\frac{1}{2}})\,,$$

$$f_\mu(\beta_1,\beta_2) = M(e^{\frac{1}{2}(\beta_1+\beta_2)} + (-1)^\mu e^{-\frac{1}{2}(\beta_1+\beta_2)})\frac{\xi}{\pi}d\zeta(\beta_1-\beta_2)$$
$$\times \frac{1}{\text{ch}\,\frac{\pi}{2\xi}(\beta_1-\beta_2+\pi i)}(e_{\frac{1}{2},-\frac{1}{2}} - e_{-\frac{1}{2},\frac{1}{2}})\,,$$

$$f^\pm(\beta_1,\beta_2) = \frac{\xi}{\pi}d\zeta(\beta_1-\beta_2)\frac{1}{\text{sh}\,\frac{\pi}{\xi}(\beta_1-\beta_2+\pi i)}$$
$$\times\left(\exp\left(\pm\frac{\pi}{2\xi}(\beta_1-\beta_2+\pi i)\right)e_{\frac{1}{2},-\frac{1}{2}}\right.$$
$$\left.+\exp\left(\mp\frac{\pi}{2\xi}(\beta_1-\beta_2+\pi i)\right)e_{-\frac{1}{2},\frac{1}{2}}\right)\,.$$

4
THE MAIN PROPERTIES OF THE SOLITON FORM FACTORS

In the present section it is shown that the form factor $f_\mu, f_{\mu\nu}, f^\pm$ do satisfy Axioms 1,2,3. Corresponding statements are formulated as Theorems.

Theorem 1. Form factors $f_\mu, f_{\mu\nu}, f^\pm$ satisfy Axiom 1 which means that

$$\begin{aligned} &f_\mu(\beta_1, \ldots, \beta_{2n})_{\varepsilon_1,\ldots,\varepsilon_{2n}} S^{\varepsilon_i,\varepsilon_{i+1}}_{\varepsilon'_i,\varepsilon'_{i+1}}(\beta_i - \beta_{i+1}) \\ &\quad = f_\mu(\beta_1, \ldots, \beta_{i+1}, \beta_i, \ldots, \beta_{2n})_{\varepsilon_1,\ldots,\varepsilon'_{i+1},\varepsilon'_i,\ldots,\varepsilon_{2n}} \end{aligned} \tag{62}$$

the same equations are valid for $f_{\mu\nu}, f^\pm$.

Proof. Form factors are constructed through the functions $F_n^{(\gamma)}$ via the operation $\langle\ \rangle_n$ and the integral transformation Φ_n. The property of $\langle\ \rangle_n$ presented by Eq. (47) implies that

$$\begin{aligned} &\langle F_n^{(\gamma)}\rangle_n(\beta_1, \ldots, \beta_{2n})_{\varepsilon_1,\ldots,\varepsilon_{2n}} \hat{S}^{\varepsilon_i\varepsilon_{i+1}}_{\varepsilon'_i,\varepsilon'_{i+1}}(\beta_i - \beta_{i+1}) \\ &\quad = \langle F_n^{(\gamma)}\rangle_n(\beta_1, \ldots, \beta_{i+1}, \beta_i \ldots \beta_{2n})_{\varepsilon_1,\ldots,\varepsilon'_{i+1},\varepsilon'_i,\ldots,\varepsilon_{2n}} . \end{aligned}$$

The application of the transformation Φ_n does not disturb this relation. The consequent multiplication by $\Pi\zeta(\beta_i - \beta_j)$ changes (61) $\hat{S}(\beta_i - \beta_{i+1})$ to $\hat{S}(\beta_i - \beta_{i+1})S_0(\beta_i - \beta_{i+1}) = S(\beta_i - \beta_{i+1})$.
Q.E.D.

Now we pass to the proof of a considerably more difficult statement that the form factors satisfy Axiom 2. To this end we establish first some properties of function $\langle F_n^{(\gamma)}\rangle_n$.

Combining Eqs. (48), (58) one comes to the conclusion that the only singularities of $\langle F_n^{(\gamma)}\rangle_n$ are the simple poles at the points $\beta_j = \beta_i + \pi i + i\xi k\,(j > i, k \in Z)$ with residues

$$
\begin{aligned}
&\operatorname{res}\langle F_n^{(\gamma)}\rangle_n(x_1,\dots,x_{n-1}|\beta_1,\dots,\beta_{2n})\\
&\quad = \frac{\xi}{\pi}(e_{i,\frac12}\otimes e_{j,-\frac12} + (-1)^k e_{i,-\frac12}\otimes e_{j,\frac12})\\
&\quad \otimes \Big(\sum_{l=1}^{n-1}(-1)^l\langle F_{n-1}^{(\gamma)}\rangle_{n-1}(x_1,\dots,\hat{x}_l,\dots,x_{n-1}|\\
&\quad \times \beta_1,\dots,\hat{\beta}_i,\dots,\hat{\beta}_j,\dots,\beta_{2n})\\
&\quad \times 2^{-3(n-1)}\Big\{\prod_{p\neq i,j}\Big(x_l - e^{\frac{2\pi}{\xi}\beta_p}\tau\Big) - \tau^{4(n-1)}\\
&\quad \times \prod_{p\neq i,j}\Big(x_l - e^{\frac{2\pi}{\xi}\beta_p}\tau^{-1}\Big)\Big\}x_l^{-1}\Big)\\
&\quad \times \prod_{p=1}^{n-1}(x_p - e^{\frac{2\pi}{\xi}\beta_i}\tau)\prod_{q<i-1}\operatorname{sh}^{-1}\frac{\pi}{\xi}(\beta_q-\beta_i)\prod_{g>j}\operatorname{sh}^{-1}\frac{\pi}{\xi}(\beta_q-\beta_i-\pi i)\\
&\quad \times\hat{S}_{j-1,i}(\beta_{j-1}-\beta_i)\dots\hat{S}_{i+1,i}(\beta_{i+1}-\beta_i)\,,\\
&\qquad\qquad\qquad \beta_j = \beta_i + \pi i + i\xi k\,.
\end{aligned}
\tag{63}
$$

Lemma 1. The equation

$$
\begin{aligned}
&\langle F_n^{(\gamma)}\rangle_n(\beta_1,\dots,\beta_{2n})_{\varepsilon_1,\dots,\varepsilon_{2n}}\hat{S}^{\varepsilon_i,\varepsilon_{i+1}}_{\varepsilon'_i,\varepsilon'_{i+1}}(\beta_i-\beta_{i+1})\\
&\quad = \langle F_n^{(\gamma)}\rangle_n(\beta_i,\dots,\beta_{i+1},\beta_i,\dots,\beta_{2n})_{\varepsilon_1,\dots,\varepsilon'_{i+1},\varepsilon'_i,\dots\varepsilon_{2n}}
\end{aligned}
\tag{64}
$$

allows one to express all the components of $\langle F_n^{(\gamma)}\rangle_n$ through one of them. The components of $\langle F_n^{(\gamma)}\rangle_n$ can be presented as follows:

$$\langle F_n^{(\gamma)}\rangle_n(\beta_1,\dots,\beta_{2n})_{\varepsilon_1,\dots,\varepsilon_{2n}} = \exp\left(\gamma\frac{\pi}{\xi}\left(\sum_{j=1}^{2n}\varepsilon_j\beta_j + \sum_{j=1}^{n}\pi i\varepsilon_j\right)\right)$$
$$\times \exp\left(\frac{\pi}{\xi}\sum_{j=1}^{2n}\beta_j\right) R^{(\gamma)}_{\varepsilon_1,\dots,\varepsilon_{2n}}\left(x_1,\dots,x_{n-1}\right|$$
$$\times \exp\left(\frac{2\pi}{\xi}\beta_1\right),\dots,\exp\left(\frac{2\pi}{\xi}\beta_{2n}\right)\bigg) \tag{65}$$

where $R^{(\gamma)}_{\varepsilon_1,\dots,\varepsilon_{2n}}(x_1,\dots,x_{n-1}|b_1,\dots,b_{2n})$ is a rational function of all its arguments. The only singularities of this function are the simple poles at the points $b_j = b_i\tau^2, j > i, 2|\sum_{l=i}^{i+1}\varepsilon_l| \le j-i$. Besides $R^{(\gamma)}_{\varepsilon_1,\dots,\varepsilon_{2n}}$ is an antisymmetrical function with respect to $x_1,\dots,x_{n-1}$. It is homogeneous:

$$R^{(\gamma)}_{\varepsilon_1,\dots,\varepsilon_{2n}}(ax_1,\dots,ax_{n-1}|ab_1,\dots,ab_{2n})$$
$$= a^{\frac{1}{2}(n-1)(n-2)-n} R^{(\gamma)}_{\varepsilon_1,\dots,\varepsilon_{2n}}(x_1,\dots,x_{n-1}|b_1,\dots,b_{2n})\,. \tag{66}$$

Proof. The proof of the Lemma is very straightforward; that is why we present only the main formula it needs:

$$\langle F_n^{(\gamma)}\rangle_n(\beta_1,\dots,\beta_i,\beta_{i+1},\dots,\beta_{2n})_{\varepsilon_1,\dots,\varepsilon_{i-1},-\frac{1}{2},\frac{1}{2},\varepsilon_{i+2},\dots,\varepsilon_{2n}}$$
$$= \mathrm{sh}^{-1}\frac{\pi}{\xi}(\beta_i-\beta_{i+1})$$
$$\times\Big\{\langle F_n^{(\gamma)}\rangle_n(\beta_1,\dots,\beta_{i+1},\beta_i,\dots,\beta_{2n})_{\varepsilon_1,\dots,\varepsilon_{i-1},\frac{1}{2},-\frac{1}{2},\dots,\varepsilon_{2n}}$$
$$\times\,\mathrm{sh}\frac{\pi}{\xi}(\beta_i-\beta_{i+1}-\pi i)$$
$$+\,\mathrm{sh}\frac{\pi^2 i}{\xi}\langle F_n^{(\gamma)}\rangle_n(\beta_1,\dots,\beta_i,\beta_{i+1},\dots,\beta_{2n})_{\varepsilon_1,\dots,\varepsilon_{i-1},\frac{1}{2},-\frac{1}{2},\dots,\varepsilon_{2n}}\Big\}\,. \tag{67}$$

Equation (67) being a consequence of Eq. (64) shows that one can express $\langle F_n^{(\gamma)}\rangle_{n,\varepsilon_1,\dots,\varepsilon_{i-1},\frac{1}{2},-\frac{1}{2},\varepsilon_{i+2},\dots,\varepsilon_{2n}}$ through $\langle F_n^{(\gamma)}\rangle_{n,\varepsilon_1,\dots,\varepsilon_{i-1},-\frac{1}{2},\frac{1}{2},\dots,\varepsilon_{2n}}$. This fact is sufficient to prove the first statement of the Lemma. The possibility to present $\langle F_n^{(\gamma)}\rangle_n(\beta_1,\dots,\beta_{2n})_{\varepsilon_1,\dots,\varepsilon_{2n}}$ in the form (65) can be also established using Eq. (67). Singularities of $\langle F_n^{(\gamma)}\rangle_n(\beta_1,\dots,\beta_{2n})_{\varepsilon_1,\dots,\varepsilon_{2n}}$ can be easily found using (63); the homogeneity is quite obvious.
Q.E.D.

The function $R^{(\gamma)}_{\varepsilon_1,\dots,\varepsilon_{2n}}(x_1,\dots,x_{n-1}|b_1,\dots,b_{2n})$ is antisymmetrical with respect to $x_1,\dots,x_{n-1}$ and, consequently, contains as a multiplier the polynomial $\prod_{i<j}(x_i-x_j)$ whose degree is equal to $\frac{1}{2}(n-1)(n-2)$. This fact and Eq. (66) allow to reconstruct this function through its residues which can be taken in number less than the total number of poles by $n-1$. The most convenient form can be obtained for $R^{(\gamma)}_{-\frac{1}{2},\dots,-\frac{1}{2},\frac{1}{2},\dots,\frac{1}{2}}$, $R^{(\gamma)}_{\frac{1}{2},\dots,\frac{1}{2},-\frac{1}{2},\dots,-\frac{1}{2}}$, $R^{(\gamma)}_{-\frac{1}{2},\frac{1}{2},\dots,\frac{1}{2},-\frac{1}{2},\dots,-\frac{1}{2}}$ and $R^{(\gamma)}_{\frac{1}{2},-\frac{1}{2},\dots,-\frac{1}{2},\frac{1}{2},\dots,\frac{1}{2}}$.

Functions $R^{(\gamma)}_{\mp\frac{1}{2},\dots,\mp\frac{1}{2},\pm\frac{1}{2},\dots,\pm\frac{1}{2}}(x_1,\dots,x_{n-1}|b_1,\dots,b_{2n})$ are symmetrical with respect to $b_1\dots b_n$ and $b_{n+1}\dots b_{2n}$ separately. They have n^2 simple poles at the points $b_j=b_i\tau^2, j>n, i\le n$ with residues (63):

$$
\begin{aligned}
&\operatorname*{res}_{b_j=b_i\tau^2} R^{(\gamma)}_{\underbrace{\mp\frac{1}{2},\dots,\mp\frac{1}{2}}_{n},\underbrace{\pm\frac{1}{2},\dots,\pm\frac{1}{2}}_{n}}(x_1,\dots,x_{n-1}|b_1,\dots,b_{2n})\\
&=2^{-(n-2)}\prod_{\substack{k>n\\ k\neq j}}\frac{1}{b_k-b_i\tau^2}\prod_{\substack{k\le n\\ k\neq i}}\frac{1}{b_k-b_i}\prod_{l=1}^{n-1}(x_l-b_i\tau)\\
&\times\sum_{k=1}^{n-1}(-1)^k x_k^{-1}\left(\prod_{l\neq i,j}(x_k-\tau b_l)-\tau^{4(n-1)}\prod_{l\neq i,j}(x_k-\tau^{-1}b_l)\right)\\
&\times R^{(\gamma)}_{\underbrace{\mp\frac{1}{2},\dots,\mp\frac{1}{2}}_{n-1},\underbrace{\pm\frac{1}{2},\dots,\pm\frac{1}{2}}_{n-1}}(x_1,\dots,\hat{x}_k,\dots,x_{n-1}|\\
&\times b_1,\dots,\hat{b}_i,\dots,\hat{b}_j,\dots,b_{2n})\,.
\end{aligned}\tag{68}
$$

Equations (68) provide us with the possibility to construct $R^{(\gamma)}_{\mp\frac{1}{2},\dots,\mp\frac{1}{2},\pm\frac{1}{2},\dots,\pm\frac{1}{2}}$ recurrently. It can be shown that

$$
\begin{aligned}
R^{(1)}_{\frac{1}{2},-\frac{1}{2}}(b_1,b_2)=R^{(1)}_{-\frac{1}{2},\frac{1}{2}}(b_1,b_2)&=R^{(-1)}_{\frac{1}{2},-\frac{1}{2}}(b_1,b_2)\\
&=R^{(-1)}_{-\frac{1}{2},\frac{1}{2}}(b_1,b_2)=2(b_2\tau^{-1}-b_1\tau)^{-1}\,;
\end{aligned}
$$

consequently

$$
\begin{aligned}
&R^{(1)}_{\underbrace{-\frac{1}{2},\dots,-\frac{1}{2}}_{n},\underbrace{\frac{1}{2},\dots,\frac{1}{2}}_{n}}(x_1,\dots,x_{n-1}|b_1,\dots,b_{2n})\\
&=R^{(1)}_{\underbrace{\frac{1}{2},\dots,\frac{1}{2}}_{n},\underbrace{-\frac{1}{2},\dots,\frac{1}{2}}_{n}}(x_1,\dots,x_{n-1}|b_1,\dots,b_{2n})
\end{aligned}
$$

$$= R^{(-1)}_{\underbrace{-\frac{1}{2},\dots,-\frac{1}{2}}_{n},\underbrace{\frac{1}{2},\dots,\frac{1}{2}}_{n}}(x_1,\dots,x_{n-1}|b_1,\dots,b_{2n})$$

$$= R^{(-1)}_{\underbrace{\frac{1}{2},\dots,\frac{1}{2}}_{n},\underbrace{-\frac{1}{2},\dots,-\frac{1}{2}}_{n}}(x_1,\dots,x_{n-1}|b_1,\dots,b_{2n})$$

$$= 2^{n^2}\frac{\Delta_n(x_1,\dots,x_{n-1}|b_1,\dots,b_n|b_{n+1},\dots,b_{2n})}{\prod\limits_{j=n+1}^{2n}\prod\limits_{i=1}^{n}(b_j\tau^{-1}-b_i\tau)}\,. \tag{69}$$

The latter equality follows from Eq. (59) and Lemma 3 of Sec. 3.

Functions $R^{(\gamma)}_{\pm\frac{1}{2},\mp\frac{1}{2},\dots,\mp\frac{1}{2},\pm\frac{1}{2},\dots,\pm\frac{1}{2}}(x_1,\dots,x_{n-1}|b_1,\dots,b_{2n})$ are symmetric with respect to $b_2,\dots,b_{n+1}$ and $b_{n+2},\dots,b_{2n}$ separately. They have n^2+n-1 simple poles at the points $b_j=b_i\tau^2, j\geq n+2, 1\leq i\leq n+1, b_j=b_1\tau^2, 2\leq j\leq n+1$. For the recurrent reconstruction of these functions one has to indicate the residues at n^2 poles, the thing we are doing now (see (63)):

$$\operatorname*{res}_{\substack{b_j=b_1\tau^2\\ 2\leq j\leq n+1}} R^{(\gamma)}_{\pm\frac{1}{2},\underbrace{\mp\frac{1}{2},\dots,\mp\frac{1}{2}}_{n},\underbrace{\pm\frac{1}{2},\dots,\pm\frac{1}{2}}_{n-1}}(x_1,\dots,x_{n-1}|b_1,\dots,b_{2n})$$

$$= 2\prod_{\substack{l=2\\ l\neq j}}^{2n}(b_l-b_1\tau^2)^{-1}\prod_{l=n+2}^{2n}(b_l-b_1)^{-1}\prod_{l=1}^{n-1}(x_l-b_1\tau)$$

$$\times\sum_{k=1}^{n-1}(-1)^k x_k^{-1}\left\{\prod_{l\neq 1,j}(x_k-b_l\tau)-\tau^{4(n-1)}\prod_{l\neq 1,j}(x_k-b_l\tau^{-1})\right\}$$

$$\times R^{(\gamma)}_{\underbrace{\mp\frac{1}{2},\dots,\mp\frac{1}{2}}_{n-1},\underbrace{\mp\frac{1}{2},\dots,\pm\frac{1}{2}}_{n-1}}(x_1,\dots,\hat{x}_k,\dots,x_{n-1}|$$

$$b_2,\dots,\hat{b}_j,\dots,b_{2n})\,,$$

$$\operatorname*{res}_{\substack{b_j=b_i\tau^2\\ n+2\leq j\leq 2n, 2\leq i\leq n+1}} R^{(\gamma)}_{\pm\frac{1}{2},\underbrace{\mp\frac{1}{2},\dots,\mp\frac{1}{2}}_{n},\underbrace{\pm\frac{1}{2},\dots,\pm\frac{1}{2}}_{n-1}}(x_1,\dots,x_{n-1}|b_1,\dots,b_{2n})$$

$$= 8\prod_{l=2}^{n+1}(b_l-b_1\tau^2)^{-1}\prod_{\substack{l=1\\ l\neq i}}^{n-1}(b_i-b_l)^{-1}\prod_{l=n+2}^{2n}(b_l-b_i\tau^2)^{-1}\prod_{l=1}^{n-1}(x_l-b_i\tau)$$

$$\times \sum_{k=1}^{n-1}(-1)^k x_k^{-1}\left\{\prod_{l\neq i,j}(x_k-b_l\tau)-\tau^{4(n-1)}\prod_{l\neq i,j}(x_k-b_l\tau^{-1})\right\}$$

$$\times R^{(\gamma)}_{\underbrace{\pm\frac{1}{2},\mp\frac{1}{2},\dots,\mp\frac{1}{2}}_{n-1},\underbrace{\pm\frac{1}{2},\dots,\pm\frac{1}{2}}_{n-2}}(x_1,\dots,\hat{x}_k,\dots,x_{n-1}|$$

$$b_1,\dots,\hat{b}_i,\dots,\hat{b}_j,\dots,b_{2n}) \,. \tag{70}$$

Equations (70) indicate the residues n^2 poles and so are sufficient for the recurrent procedure. The initial conditions are the same:

$$\begin{aligned} R^{(1)}_{\frac{1}{2},-\frac{1}{2}}(b_1,b_2) = R^{(1)}_{-\frac{1}{2},\frac{1}{2}}(b_1,b_2) &= R^{(-1)}_{\frac{1}{2},-\frac{1}{2}}(b_1,b_2) \\ &= R^{(-1)}_{-\frac{1}{2},\frac{1}{2}}(b_1,b_2) = 2(b_2\tau^{-1}-b_1\tau)^{-1} \,; \end{aligned}$$

consequently

$$\begin{aligned} &R^{(1)}_{\underbrace{\frac{1}{2},-\frac{1}{2},\dots,-\frac{1}{2}}_{n},\underbrace{\frac{1}{2},\dots,\frac{1}{2}}_{n-1}}(x_1,\dots,x_{n-1}|b_1,\dots,b_{2n}) \\ &= R^{(1)}_{\underbrace{-\frac{1}{2},\frac{1}{2},\dots,\frac{1}{2}}_{n},\underbrace{-\frac{1}{2},\dots,-\frac{1}{2}}_{n-1}}(x_1,\dots,x_{n-1}|b_1,\dots,b_{2n}) \\ &= R^{(-1)}_{\underbrace{\frac{1}{2},-\frac{1}{2},\dots,-\frac{1}{2}}_{n},\underbrace{\frac{1}{2},\dots,\frac{1}{2}}_{n-1}}(x_1,\dots,x_{n-1}|b_1,\dots,b_{2n}) \\ &= R^{(-1)}_{\underbrace{-\frac{1}{2},\frac{1}{2},\dots,\frac{1}{2}}_{n},\underbrace{-\frac{1}{2},\dots,-\frac{1}{2}}_{n-1}}(x_1,\dots,x_{n-1}|b_1,\dots,b_{2n}) \\ &= 2^{n^2+2n-2}\frac{\tilde{\Delta}_n(x_1,\dots,x_{n-1}|b_1|b_2,\dots,b_{n+1}|b_{n+2},\dots,b_{2n})}{\prod_{j=2}^{2n}(b_j\tau^{-1}-b_1\tau)\prod_{j=n+2}^{2n}\prod_{i=1}^{n+1}(b_j\tau^{-1}-b_i\tau)} \,. \end{aligned}$$

The last equality defines $\tilde{\Delta}_n$. The following Lemma provides us with an explicit expression for $\tilde{\Delta}_n$.

Lemma 2. The polynomial $\tilde{\Delta}_n(x_1,\dots,x_{n-1}|b_n|a_1,\dots,a_n|b_1,\dots,b_{n-1})$ is equal to $2^{-n+1}\prod_{l=1}^{n-1}(b_n-b_l\tau^2)\det\|B^{(n)}_{ij}\|$, where $B^{(n)}$ is a $(n-1)\times(n-1)$ matrix with the following matrix elements:

$$B^{(n)}_{ij} = B^{(n)}_i(x_j|b_n|a_1,\dots,a_n|b_1,\dots,b_{n-1}) \,,$$

where

$$B_i^{(n)}(x|b_n|a_1,\ldots,a_n|b_1,\ldots,b_{n-1}) = \frac{1}{x-b_n\tau^3}$$
$$\times \Big\{A_i^{(n)}(x|a_1,\ldots,a_n|b_1,\ldots,b_{n-1},b_n\tau^4)(x-b_n\tau)$$
$$- b_n x^{-1}\tau^{6(n-1)}\prod_{l-1}^{n-1}(b_n-b_l\tau^{-2})^{-1}\prod_{l=1}^{n}(b_n-a_l\tau^{-2})^{-1}$$
$$\times\Bigg[\prod_{l=1}^{n}(x-a_l\tau)(x-b_l\tau)-\tau^{4(n-1)}(x-b_n\tau^3)$$
$$\prod_{l=1}^{n}(x-a_l\tau^{-1})\prod_{l=1}^{n-1}(x-b_l\tau^{-1})\Bigg]$$
$$\times A_i^{(n)}(b_n\tau^3|a_1,\ldots a_n|b_1,\ldots,b_{n-1},b_n\tau^4)\Big\}\ . \tag{71}$$

Proof. It is evident that $\det\|B_{ij}^{(nj)}\|$ is a function antisymmetric with respect to $x_1,\ldots,x_{n-1}$ and symmetric with respect to $a_1,\ldots,a_n$ and $b_1,\ldots,b_{n-1}$. This function is a homogeneous function of the proper degree (66). It is also evident that the poles at the points $b_n = a_l\tau^{-2}$ which occur in (71) are fictitious because $A_i^{(n)}(b_n\tau^3|a_1,\ldots,a_k|b_1,\ldots,b_{n-1},b_n\tau^4)$ contains the multiplier $\Pi(b_n - a_l\tau^{-2})$. The pole at the point $x = b_n\tau^3$ is also fictitious. At first sight it is not quite clear whether the poles of $\det\|B_{ij}^{(n)}\|$ at the points $b_n = b_l\tau^{-2}(l \le n-1)$ which appear in all the matrix elements are actually simple ones. Let us clarify the fact. Let b_l for some $1 \le l \le n-1$ be equal to $b_l = b_n\tau^2 + \Delta, \Delta \ll 1$, then $B_{ij}^{(n)}$ can be presented as follows:

$$B_{ij}^{(n)} = \Delta^{-1} f_i g_j + \mathrm{O}(1)\ . \tag{72}$$

The explicit expressions for f_i and g_j can be obtained from Eq. (71). It is easy to show that the determinant of a matrix with matrix elements of this kind is $\mathrm{O}(\Delta^{-1})$, which means that the pole at the points $b_l = b_n\tau^2$ is simple. Consequently the function $2^{-n+1}\Pi(b_n - b_l\tau^{-2})\det\|B_{ij}^{(n)}\|$ is a polynomial.

Thus $2^{-n+1}\Pi(b_n - b_l\tau^{-2})\det\|B_{ij}^{(n)}\|$ satisfies all the structural requirements on $\tilde{\Delta}_n$. We have now to show that it satisfies the recurrent relations following Eq. (70).

The calculation of $\det\|B_{ij}^{(n)}\|$ at the point $a_n = b_n\tau^2 (a_i = b_n\tau^2 (i \le n-1)$ can be treated due to the symmetry) is based on the use of the identities

$$\begin{aligned}
&B_i^{(n)}(x|b_n|a_1,\dots,a_{n-1},b_n\tau^2|b_1,\dots,b_{n-1})\\
&\quad+\frac{1}{2}b_n\tau^4 B_{i-1}^{(n)}(x|b_n|a_1,\dots,a_{n-1},b_n\tau^2|b_1,\dots,b_{n-1})\\
&\quad=\frac{1}{2}(x-b_n\tau)\{A_i^{(n-1)}(x|a_1,\dots,a_{n-1}|b_1,\dots,b_{n-1})\\
&\quad-x^{-1}\left[\prod_{l=1}^{n-1}(x-a_l\tau)(x-b_l\tau)-\tau^{4(n-1)}\prod_{l=1}^{n-1}(x-a_l\tau^{-1})(x-b_l\tau^{-1})\right]\\
&\quad\times 2^{-2(n-1)}C_i(b_n|a_1,\dots,a_{n-1}|b_1,\dots,b_{n-1})\}\,,\quad i\le n-2\,;
\end{aligned}$$

$$\begin{aligned}
&B_{n-1}^{(n)}(x|b_n|a_1,\dots,a_{n-1},b_n\tau^2|b_1,\dots,b_{n-1})\\
&\quad+\frac{1}{2}b\tau^4 B_{n-2}^{(n)}(x|b_n|a_1,\dots,a_{n-1},b_n\tau^2|b_1,\dots b_{n-1})\\
&\quad=\frac{1}{2}(x-b_n\tau)\prod_{l=1}^{n-1}(b_n-b_l\tau^{-4})(b_n-b_l\tau^{-2})^{-1}2^{-2(n-1)}x^{-1}\\
&\quad\times\left[\prod_{l=1}^{n-1}(x-a_l\tau)(x-b_l\tau)-\tau^{4(n-1)}\prod_{l=1}^{n1}(x-a_l\tau^{-1})(x-b_l\tau^{-1})\right]\,,
\end{aligned}$$

where C_i is some function. Adding successively the i-th row of the determinant with coefficient $\frac{1}{2}b_n\tau^4$ to the $(i+1)$-th row, and then subtracting from the $1,2,\dots,n-2$ rows the $(n-1)$-th one with the coefficients $C_i\Pi(b_n-b_l\tau^{-2})(b_n-b_l\tau^{-4})^{-1}$ and expanding the determinant along the last row, one obtains the required relation.

The calculation of $\det\|B_{ij}^{(n)}\|$ at the point $b_1 = a_n\tau^2$ (at $b_j = a_i\tau^2, j\le n-1, i\le n-1$, due to the symmetry) can be performed in a similar manner, being based on use of identities

$$\begin{aligned}
&B_i^{(n)}(x|b_n|a_1,\dots,a_n|a_n\tau^2,b_2,\dots,b_{n-1})\\
&\quad+\frac{1}{2}a_n\tau^4 B_{i-1}^{(n)}(x|b_n|a,\dots,a_n|a_n\tau^2,b_2,\dots,b_{n-1})\\
&\quad=\frac{1}{2}(x-a_n\tau)\{B_i^{(n-1)}(x|b_n|a_1,\dots,a_{n-1}|b_1,\dots,b_{n-1})-2^{-2(n-1)}x^{-1}
\end{aligned}$$

$$\times \left[\prod_{l=1}^{n-1}(x-a_l\tau)(x-b_{l+1}\tau)-\tau^{4(n-1)}\prod_{l=1}^{n-1}(x-a_l\tau^{-1})(x-b_{l+1}\tau^{-1})\right]$$
$$\times \tilde{C}_i(b_n|a_1,\dots,a_{n-1}|b_2,\dots,b_{n-1})\} , \quad i\le n-2 ;$$

$$\begin{aligned}
&B^{(n)}_{n-1}(x|b_n|a_1,\dots,a_n|a_n\tau^2,b_2,\dots,b_{n-1})\\
&\quad + a\tau^4\frac{1}{2}B^{(n)}_{n-2}(x|b_n|a_1,\dots,a_n|a_n\tau^2,b_2,\dots,b_{n-1})\\
&\quad = \frac{1}{2}(x-a_n\tau)(b_n-a\tau^{-2})(b_n-a_n)^{-1}2^{-2(n-1)}x^{-1}\\
&\quad \times \left[\prod_{l=1}^{n-1}(x-a_l\tau)(x-b_{l+1}\tau)-\tau^{4(n-1)}\prod_{l=1}^{n-1}(x-a_l\tau^{-1})(x-b_{l+1}\tau^{-1})\right] .
\end{aligned}$$

Q.E.D.

Now we are in a position to show that the form factors satisfy Axiom 2.

Theorem 2. Form factors $f_\mu, f_{\mu\nu}, f^\pm$ satisfy Axiom 2, which means

$$\begin{aligned}
f_\mu(\beta_1,\dots,\beta_{2n}+2\pi i)_{\varepsilon_1,\dots,\varepsilon_{2n}} &= f_\mu(\beta_{2n},\beta_1,\dots,\beta_{2n-1})_{\varepsilon_{2n},\varepsilon_1,\dots,\varepsilon_{2n-1}} ,\\
f_{\mu\nu}(\beta_1,\dots,\beta_{2n}+2\pi i)_{\varepsilon_1,\dots,\varepsilon_{2n}} &= f_{\mu\nu}(\beta_{2n},\beta_1,\dots,\beta_{2n-1})_{\varepsilon_{2n},\varepsilon_1,\dots,\varepsilon_{2n-1}}\\
f^\pm(\beta_1,\dots,\beta_{2n}+2\pi i)_{\varepsilon_1,\dots,\varepsilon_{2n}} &= -f^\pm(\beta_{2n},\beta_1,\dots,\beta_{2n-1})_{\varepsilon_{2n},\varepsilon_1,\dots,\varepsilon_{2n-1}} .
\end{aligned} \tag{73}$$

These identities are to be understood in the sense of analytic continuation with respect to β_{2n}.

Proof. From the definition of $f_\mu, f_{\mu\nu}, f^\pm$ and the identity

$$\zeta(\beta-2\pi i)=\zeta(-\beta) ,$$

it follows that Eqs. (73) are equivalent to

$$\begin{aligned}
&\Phi_{n,\sigma}(\langle F^{(\gamma)}_n\rangle_n)(\beta_1,\dots,\beta_{2n-1},\beta_{2n}+2\pi i)_{\varepsilon_1,\dots,\varepsilon_{2n-1},\varepsilon_{2n}}\\
&\quad = \Phi_{n,\sigma}(\langle F^{(\gamma)}_n\rangle_n)(\beta_{2n},\beta_1,\dots,\beta_{2n-1})_{\varepsilon_{2n},\varepsilon_1,\dots,\varepsilon_{2n-1}}
\end{aligned} \tag{74}$$

for $\gamma=\pm1, \sigma=0, \pm1$.

As was explained in the proof of Lemma 1 using the relation

$$\begin{aligned}
&\Phi_{n,\sigma}(\langle F^{(\gamma)}_n\rangle_n)(\beta_1,\dots,\beta_i,\beta_{i+1},\dots,\beta_{2n})_{\varepsilon_1,\dots,\varepsilon_i,\varepsilon_{i+1},\dots,\varepsilon_{2n}}\\
&\quad \times \hat{S}^{\varepsilon_i,\varepsilon_{i+1}}_{\varepsilon'_i,\varepsilon'_{i+1}}(\beta_i-\beta_{i+1})\\
&\quad = \Phi_{n,\sigma}(\langle F^{(\gamma)}_n\rangle_n)(\beta_1,\dots,\beta_{i+1},\beta_i,\dots,\beta_{2n})_{\varepsilon_1,\dots,\varepsilon'_{i+1},\varepsilon'_i,\dots,\varepsilon_{2n}} ,
\end{aligned}$$

one can change the order of $\varepsilon_1 \dots \varepsilon_{2n-1}$ in Eq. (74) and thus reduce the proof of Eq. (73) for arbitrary $\varepsilon_1 \dots \varepsilon_{2n-1}$ to the proof of the two identities:

$$\begin{aligned}&\Phi_{n,\sigma}(\langle F_n^{(\gamma)}\rangle_n)(\beta_1,\dots,\beta_{2n}+2\pi i)_{-\frac{1}{2},\dots,-\frac{1}{2},\frac{1}{2},\dots,\frac{1}{2}}\\ &= \Phi_{n,\sigma}(\langle F_n^{(\gamma)}\rangle_n)(\beta_{2n},\beta_1,\dots,\beta_{2n-1})_{\frac{1}{2},-\frac{1}{2},\dots,-\frac{1}{2},\frac{1}{2},\dots,\frac{1}{2}}\,,\\ &\Phi_{n,\sigma}(\langle F_n^{(\gamma)}\rangle_n)(\beta_1,\dots,\beta_{2n}+2\pi i)_{\frac{1}{2},\dots,\frac{1}{2},-\frac{1}{2},\dots,-\frac{1}{2}}\\ &= \Phi_{n,\sigma}(\langle F_n^{(\gamma)}\rangle_n)(\beta_{2n},\beta_1,\dots,\beta_{2n-1})_{-\frac{1}{2},\frac{1}{2},\dots,\frac{1}{2},-\frac{1}{2},\dots,-\frac{1}{2}}\,.\end{aligned} \tag{75}$$

Equations (42), (69), (71) and the definition of $\Phi_{n,\sigma}$ imply that the proof of Eqs. (75) reduces to the proof of the following statement. Continuing analytically with respect to $\beta_{2n} : \beta_{2n} \to \beta_{2n} + 2\pi i$ the integral

$$\begin{aligned}I_1(\beta_1,\dots,\beta_{2n}) &= \frac{1}{(n-1)!}\fint_{-\infty}^{\infty} d\alpha_1 \dots \fint_{-\infty}^{\infty} d\alpha_{n-1} \prod_{i=1}^{n-1}\prod_{j=1}^{2n}\varphi(\alpha_i-\beta_j)\\ &\times \prod_{i<j}\operatorname{sh}(\alpha_i-\alpha_j)\exp\left(\sigma\Sigma\alpha_i - \frac{\pi}{\xi}(n-2)\Sigma\alpha_i\right)\\ &\times \Delta_n\left(\exp\frac{2\pi}{\xi}\alpha_1,\dots,\exp\frac{2\pi}{\xi}\alpha_{n-1}\Big|\exp\frac{2\pi}{\xi}\beta_1,\dots,\exp\frac{2\pi}{\xi}\beta_n\Big|\right.\\ &\left.\times\exp\frac{2\pi}{\xi}\beta_{n+1},\dots,\exp\frac{2\pi}{\xi}\beta_{2n}\right)\,,\end{aligned}$$

one obtains the integral

$$\begin{aligned}I_2(\beta_1,\dots,\beta_{2n}) &= \frac{1}{(n-1)!}\fint_{-\infty}^{\infty} d\alpha_1 \dots \fint_{-\infty}^{\infty} d\alpha_{n-1} \prod_{i=1}^{n-1}\prod_{j=1}^{2n}\varphi(\alpha_i-\beta_j)\\ &\times \prod_{i\le j}\operatorname{sh}(\alpha_i-\alpha_j)\exp\left(\sigma\Sigma\alpha_i - \frac{\pi}{\xi}(n-2)\Sigma\alpha_i\right)\\ &\times \tilde{\Delta}_n\left(\exp\frac{2\pi}{\xi}\alpha_1,\dots,\exp\frac{2\pi}{\xi}\alpha_{n-1}\Big|\exp\frac{2\pi}{\xi}\beta_{2n}\Big|\exp\frac{2\pi}{\xi}\beta_1,\dots,\exp\frac{2\pi}{\xi}\beta_n\Big|\right.\\ &\left.\times\exp\frac{2\pi}{\xi}\beta_{n+1},\dots,\exp\frac{2\pi}{\xi}\beta_{2n-1}\right)2^{n-1}\\ &\times\prod_{j=n+1}^{2n-1}\left(\exp\frac{2\pi}{\xi}\beta_j - \tau^2\exp\frac{2\pi}{\xi}\beta_{2n}\right)^{-1}\,.\end{aligned}$$

Let us prove it. During the proof we shall manipulate $\fint$ as with the usual $\int$. The necessary rigour can be achieved by using Definition 2 of Sec. 3, but in that case the proof becomes very bulky.

Notice that the antisymmetry of $\Delta_n(x_1, \ldots, x_{n-1}|\ldots)$ and $\tilde{\Delta}_n(x_1, \ldots, x_{n-1}|\ldots)$ with respect to $x_1, \ldots, x_{n-1}$ allows one to replace in I_1, I_2 the product $\frac{1}{(n-1)!}\Pi\mathrm{sh}\,(\alpha_i - \alpha_j)$ by $2^{-\frac{(n-1)(n-2)}{2}}\exp(\Sigma\alpha_i(n-2i))$. Now noting that every column of matrices $\det\|A^{(n)}\|$, $\det\|B^{(n)}\|$ depends only on one α_k we can integrate the columns separately and rewrite I_1, I_2 in the form

$$I_1 = 2^{-\frac{(n-1)(n-2)}{2}}\det\|\mathfrak{A}_{ij}\|$$
$$I_2 = 2^{-\frac{(n-1)(n-2)}{2}}\det\|\mathfrak{B}_{ij}\|$$

where

$$\mathfrak{A}_{ij} = \int_{-\infty}^{\infty} d\alpha \prod_{p=1}^{2n} \varphi(\alpha - \beta_p)\exp\left(\alpha(n-2j) + \sigma\alpha - \frac{\pi}{\xi}(n-2)\alpha\right)$$
$$\times A_i^{(n)}\left(\exp\frac{2\pi}{\xi}\alpha \middle| \exp\frac{2\pi}{\xi}\beta_1, \ldots, \exp\frac{2\pi}{\xi}\beta_n \middle| \right.$$
$$\left.\times \exp\frac{2\pi}{\xi}\beta_{n+1}, \ldots, \exp\frac{2\pi}{\xi}\beta_{2n}\right),$$

$$\mathfrak{B}_{ij} = \int_{-\infty}^{\infty} d\alpha \prod_{p=1}^{2n} \varphi(\alpha - \beta_p)\exp\left(\alpha(n-2j) + \sigma\alpha - \frac{\pi}{\xi}(n-2)\alpha\right)$$
$$\times B_i^{(n)}\left(\exp\frac{2\pi}{\xi}\alpha \middle| \exp\frac{2\pi}{\xi}\beta_{2n} \middle| \exp\frac{2\pi}{\xi}\beta_1, \ldots \exp\frac{2\pi}{\xi}\beta_n \middle| \exp\frac{2\pi}{\xi}\beta_{n+1}\ldots\right.$$
$$\left.\times \exp\frac{2\pi}{\xi}\beta_{2n-1}\right).$$

We shall show that $\mathfrak{A}_{ij}$ being continued analytically over $\beta_{2n} : \beta_{2n} \to \beta_{2n} + 2\pi i$ gives $\mathfrak{B}_{ij}$. Let us first deform the contours of integration for $\mathfrak{A}_{ij}$ and $\mathfrak{B}_{ij}$ to the line $\mathrm{Im}\,\alpha - \frac{\pi}{2} - 0$ ($\varphi(\alpha)$ has no singularities for $0 \leq \mathrm{Im}\,\alpha < \pi/2$). Now let us begin shifting β_{2n} in the function $\mathfrak{A}_{ij}(\beta_1, \ldots, \beta_{2n})$ upward along the imaginary axis. The picture of poles of the integrand is shown in Fig. 2a.

When the imaginary part of β_{2n} becomes close to the nearest pole of the integrand it meets the contour of integration. When shifting further we have to deform the contour. When the continuation is finished, we obtain

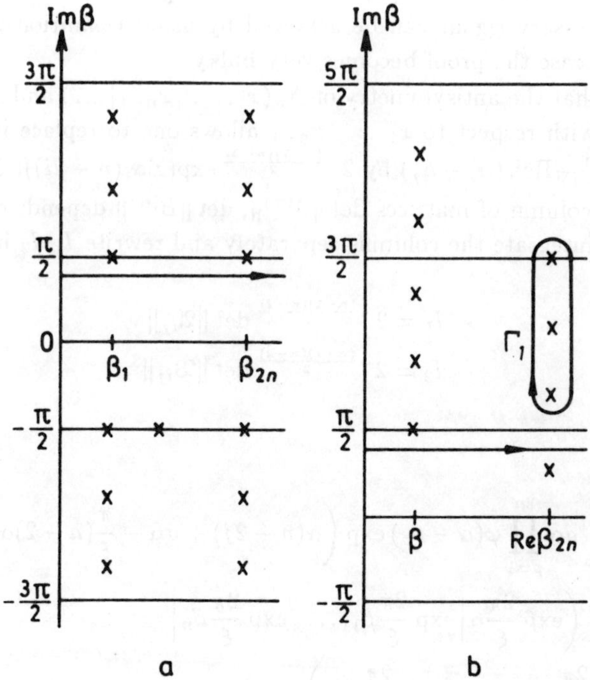

Fig. 2.

from (56) an integral of

$$\prod_{l=1}^{2n}\varphi(\alpha-\beta_l)\exp\left((n-2j)\alpha+\sigma\alpha-\frac{\pi}{\xi}(n-2)\alpha\right)$$
$$\times A_i^{(n)}\left(\exp\frac{2\pi}{\xi}\alpha|\exp\frac{2\pi}{\xi}\beta_1,\ldots,\exp\frac{2\pi}{\xi}\beta_n|\exp\frac{2\pi}{\xi}\beta_{n+1},\ldots,\right.$$
$$\left.\times\,\tau^4\exp\frac{2\pi}{\xi}\beta_{2n}\right)\mathrm{sh}\,\frac{\pi}{\xi}\left(\alpha-\beta_{2n}-\frac{\pi i}{2}\right)\mathrm{sh}^{-1}\frac{\pi}{\xi}\left(\alpha-\beta_{2n}-\frac{3\pi i}{2}\right),$$

taken over the joint of contours represented on Fig. 2b. Let us transform the integral over Γ_1. It can be evaluated through its residues. Notice that poles of the integrand are spaced by $i\xi$. This fact implies the possibility of replacing of the integrand by

$$\prod_{j=1}^{2n}\varphi(\alpha-\beta_j)\prod_{j=1}^{2n-1}\left(\frac{\exp\frac{2\pi}{\xi}\alpha-\tau\exp\frac{2\pi}{\xi}\beta_j}{\left(\exp\frac{2\pi}{\xi}\beta_{2n}\right)\tau^3-\left(\exp\frac{2\pi}{\xi}\beta_j\right)\tau}\right)$$
$$\times\frac{\operatorname{sh}\frac{\pi}{\xi}\left(\alpha-\beta_{2n}-\frac{\pi i}{2}\right)}{\operatorname{sh}\frac{\pi}{\xi}\left(\alpha-\beta_{2n}-\frac{3\pi i}{2}\right)}$$
$$\times\exp\left(\alpha(n-2j)+\sigma\alpha-\frac{\pi}{\xi}(n-2)\alpha+\frac{2\pi}{\xi}\left(\beta_{2n}-\alpha+\frac{3\pi i}{2}\right)\right)$$
$$\times A_i^{(n)}\left(\tau^3\exp\frac{2\pi}{\xi}\beta_n\middle|\exp\frac{2\pi}{\xi}\beta_1,\dots,\exp\frac{2\pi}{\xi}\beta_n\middle|\exp\frac{2\pi}{\xi}\beta_{n+1},\dots,\tau^4\right.$$
$$\left.\times\exp\frac{2\pi}{\xi}\beta_{2n}\right)\ . \tag{76}$$

It is clear that the function (76) has no other singularities in the strip $\frac{\pi}{2}-0\le\operatorname{Im}\alpha\le\frac{5\pi}{2}-0$ but those enveloped by Γ_1; hence Γ_1 can be deformed to $\left(\mathbb{R}+\frac{5\pi i}{2}-i0\right)\cup\left(-\left(\mathbb{R}+\frac{\pi i}{2}-i0\right)\right)$. Using the identity

$$\varphi(\alpha+2\pi i)=\varphi(\alpha)\frac{\operatorname{sh}\frac{\pi}{\xi}\left(\alpha+\frac{\pi i}{2}\right)}{\operatorname{sh}\frac{\pi}{\xi}\left(\alpha+\frac{3\pi i}{2}\right)}$$

we transform the integral over $\mathbb{R}+\frac{5\pi i}{2}-i0$ to the integral over $\mathbb{R}+\frac{\pi i}{2}-i0$. As a result the integral over Γ_1 transforms to

$$-\int_{\mathbb{R}+\frac{\pi i}{2}-i0}\prod_{j=1}^{2n}\varphi(\alpha-\beta_j)\exp\left(\alpha(n-2j)+\sigma\alpha-\frac{\pi}{\xi}(n-2)\alpha\right)$$
$$\times A_i^{(n)}\left(\tau^3\exp\frac{2pi}{\xi}\beta_{2n}\middle|\exp\frac{2\pi}{\xi}\beta_1,\dots,\exp\frac{2\pi}{\xi}\beta_n\middle|\exp\frac{2\pi}{\xi}\beta_{n+1},\dots,\tau^4\right.$$
$$\left.\times\exp\frac{2\pi}{\xi}\beta_{2n}\right)$$
$$\times\prod_{j=1}^{2n-1}\left(\exp\frac{2\pi}{\xi}\beta_{2n}-\tau^{-2}\exp\frac{2\pi}{\xi}\beta_j\right)^{-1}\left(\exp\frac{2\pi}{\xi}\alpha-\tau^3\exp\frac{2\pi}{\xi}\beta_{2n}\right)^{-1}$$
$$\left[\prod_{j=1}^{2n}\left(\exp\frac{2\pi}{\xi}\alpha-\tau\exp\frac{2\pi}{\xi}\beta_j\right)-\tau^{4(n-1)}\left(\exp\frac{2\pi}{\xi}\alpha-\tau^3\exp\frac{2\pi}{\xi}\beta_{2n}\right)\right.$$
$$\left.\times\prod_{j=1}^{2n-1}\left(\exp\frac{2\pi}{\xi}\alpha-\tau^{-1}\exp\frac{2\pi}{\xi}\beta_j\right)\right]\exp\frac{2\pi}{\xi}(\beta_{2n}-\alpha)\tau^{-6(n-1)}d\alpha\ . \tag{77}$$

Adding to (77)

$$
\begin{aligned}
&\oint_{\mathbb{R}+\frac{\pi i}{2}-i0} \prod_{j=1}^{2n} \varphi(\alpha-\beta_j) \exp\left(\alpha(n-2j)+\sigma\alpha-\frac{\pi}{\xi}(n-2)\alpha\right) \\
&\quad \times A_i^{(n)}\left(\exp\frac{2\pi}{\xi}\alpha \middle| \exp\frac{2\pi}{\xi}\beta_1,\dots,\exp\frac{2\pi}{\xi}\beta_n \middle| \right. \\
&\quad \left. \times \exp\frac{2\pi}{\xi}\beta_{n+1},\dots,\tau^4\exp\frac{2\pi}{\xi}\beta_{2n}\right) \\
&\quad \times \left(\frac{\exp\frac{2\pi}{\xi}\alpha-\tau\exp\frac{2\pi}{\xi}\beta_{2n}}{\exp\frac{2\pi}{\xi}\alpha-\tau^3\exp\frac{2\pi}{\xi}\beta_{2n}}\right) d\alpha\,,
\end{aligned}
$$

one obtains

$$
\begin{aligned}
&\oint_{\mathbb{R}+\frac{\pi i}{2}-i0} \prod \varphi(\alpha-\beta_j) \exp\left(\alpha(n-2j)+\sigma\alpha-\frac{\pi}{\xi}(n-2)\alpha\right) \\
&\times B_i^{(n)}\left(\exp\frac{2\pi}{\xi}\alpha \middle| \exp\frac{2\pi}{\xi}\beta_{2n} \middle| \exp\frac{2\pi}{\xi}\beta_1,\dots,\exp\frac{2\pi}{\xi}\beta_n \middle| \exp\frac{2\pi}{\xi}\beta_{n+1},\dots,\right. \\
&\left. \times \exp\frac{2\pi}{\xi}\beta_{2n-1}\right) d\alpha\,.
\end{aligned}
$$

Q.E.D.

Remark. The theorem remains valid if $\exp(\alpha(n-2j)+\sigma\alpha)$ in the integral over α_j is replaced by $\exp(k_j\alpha)$, where $k_1,\dots,k_{n-1}$ are arbitrary integers which do not exceed $n-1$ for $\forall i$.

Let us now pass on to the calculation of the residue of the form factors at the point $\beta_j=\beta_i+\pi i, j>i$. From the symmetry property (62) it follows that we can restrict ourselves to considering the case $j=2n, i=2n-1$.

Theorem 3. Form factors $f_\mu(\beta_1,\dots,\beta_{2n}), f_{\mu\nu}(\beta_1,\dots,\beta_{2n})$, $f^\pm(\beta_1,\dots,\beta_{2n})$ have no other singularities in the strip $0\le \operatorname{Im}\beta_{2n}\le\pi$ but the simple poles at the points $\beta_{2n}=\beta_j+\pi i-i\xi k$, $j\le 2n-1, 0\le k\le\left[\frac{\pi}{\xi}\right]$. The residues at the point $\beta_{2n}=\beta_{2n-1}+\pi i$ are given by the formulae

$$2\pi i \,\mathrm{res}\, f_\mu(\beta_1,\dots,\beta_{2n})_{\varepsilon_1,\dots,\varepsilon_{2n}} = f_\mu(\beta_1,\dots,\beta_{2n-2})_{\varepsilon'_1,\dots,\varepsilon'_{2n-2}}\delta_{\varepsilon_{2n}-\varepsilon'_{2n-1}}$$
$$\times\left\{\delta_{\varepsilon_1}^{\varepsilon'_1}\dots\delta_{\varepsilon_{2n-1}}^{\varepsilon'_{2n-1}} - S_{\tau_1,\varepsilon_1}^{\varepsilon'_{2n-1},\varepsilon'_1}(\beta_{2n-1}-\beta_1)\dots\right.$$
$$\left.S_{\varepsilon_{2n-1},\varepsilon_{2n-2}}^{\tau_{2n-3},\varepsilon'_{2n-2}}(\beta_{2n-1}-\beta_{2n-2})\right\}\,,\quad \beta_{2n}=\beta_{2n-1}+\pi i\,,$$
$$2\pi i \,\mathrm{res}\, f^+(\beta_1,\dots,\beta_{2n})_{\varepsilon_1,\dots,\varepsilon_{2n}} = f^+(\beta_1,\dots,\beta_{2n-2})_{\varepsilon'_1,\dots,\varepsilon'_{2n-2}}\delta_{\varepsilon_{2n},-\varepsilon'_{2n-1}}$$
$$\times\left\{\delta_{\varepsilon_1}^{\varepsilon'_1}\dots\delta_{2n-1}^{\varepsilon'_{2n-1}} + S_{\tau_1,\varepsilon_1}^{\varepsilon'_{2n-1},\varepsilon'_1}(\beta_{2n-1}-\beta_1)\dots\right.$$
$$\left.S_{\varepsilon_{2n-1},\varepsilon_{2n-2}}^{\tau_{2n-3},\varepsilon'_{2n-2}}(\beta_{2n-1}-\beta_{2n-2})\right\}\,,\quad \beta_{2n}=\beta_{2n-1}+\pi i\,. \tag{78}$$

The formula for the residues of $f_{\mu\nu}$ is quite similar to the formula for the residue of f_μ.

Proof. Function $\zeta(\beta)$ is regular in the strip $0 < \mathrm{Im}\,\beta < 2\pi$. This is why singularities of form factors are caused by singularities of the function $\Phi_{n,\sigma}(\langle F_n^{(\gamma)}\rangle_n)$. Singularities of the latter are of double origin. First, the poles at the points $\beta_{2n}=\beta_j+\pi i+i\xi k$ are implied in the definition of the operation $\langle\ \rangle_n$. The residue of $\langle F_n^{(\gamma)}\rangle_n$ at the point $\beta_{2n}=\beta_{2n-1}+\pi i+i\xi k$ is given by Eq. (63):

$$\mathrm{res}\langle F_n^{(\gamma)}\rangle_n\left(\exp\frac{2\pi}{\xi}\alpha_1,\dots,\exp\frac{2\pi}{\xi}\alpha_{n-1}\middle|\beta_1,\dots,\beta_{2n}\right)$$
$$= i^k\frac{\xi}{\pi}(e_{2n-1,-\frac12}\otimes e_{2n,\frac12}+(-1)^k e_{2n-1,\frac12}\otimes e_{2n,-\frac12})$$
$$\otimes\left\{\sum_{l=1}^{n-1}(-1)^l\langle F_{n-1}^{(\gamma)}\rangle_{n-1}\left(\exp\frac{2\pi}{\xi}\alpha_1,\dots,\widehat{\exp\frac{2\pi}{\xi}\alpha_l},\dots,\right.\right.$$
$$\left.\times\exp\frac{2\pi}{\xi}\alpha_{n-1}\middle|\beta_1,\dots,\beta_{2n-2}\right)$$
$$\times\left[\prod_{p=1}^{2n-2}\mathrm{sh}\,\frac{\pi}{\xi}\left(\alpha_l-\beta_p-\frac{\pi i}{2}\right)-\tau^{2(n-1)}\prod_{p=1}^{2n-2}\mathrm{sh}\,\frac{\pi}{\xi}\left(\alpha_l-\beta_p+\frac{\pi i}{2}\right)\right]$$
$$\left.\times\exp\left(\frac{\pi}{\xi}(2n-3)\alpha_l+\frac{\pi}{\xi}\sum_{i\neq l}\alpha_i-\frac{\pi}{\xi}(n-1)\beta_{2n-1}\right)\right\}$$
$$\times\prod_{l=1}^{n-1}\mathrm{sh}\frac{\pi}{\xi}\left(\alpha_l-\beta_{2n-1}+\frac{\pi i}{2}\right)\prod_{p=1}^{2n-2}\mathrm{sh}^{-1}\frac{\pi}{\xi}(\beta_{2n-1}-\beta_p)\,,$$
$$\beta_{2n}=\beta_{2n-1}+\pi i+i\xi k\,.$$

Second, the singularities can appear during the analytic continuation with respect to β_{2n} of the integrals over $\alpha_1, \ldots, \alpha_{n-1}$. The fact is that for $\operatorname{Re}\beta_{2n} \to \beta_j$ and certain imaginary part of β_{2n} the contour of integration appears to be clutched between two singularities of the integrand. The points nearest to the real axis where this situation is realized are $\beta_{2n} = \beta_j + \pi i$ (see Fig. 3). Now it is clear that the poles at the point $\beta_{2n} = \beta_j + \pi i + i\xi k$, $k = -1, \ldots, -\left[\frac{\pi}{\xi}\right]$, are caused only by the poles of $\langle F_n^{(\gamma)} \rangle_n$. Residues at these poles are connected with breather form factors and will be considered in the next section. The poles at the points $\beta_{2n} = \beta_j + \pi i$ have to be considered individually.

Let us consider the pole at the point $\beta_{2n} = \beta_{2n-1} + \pi i$. The situation of the poles of the integrand on the plane of the variable α_j, $\operatorname{Im}\beta_{2n} = \pi$ for $j = 1 \ldots n-1$, is indicated in Fig. 3a.

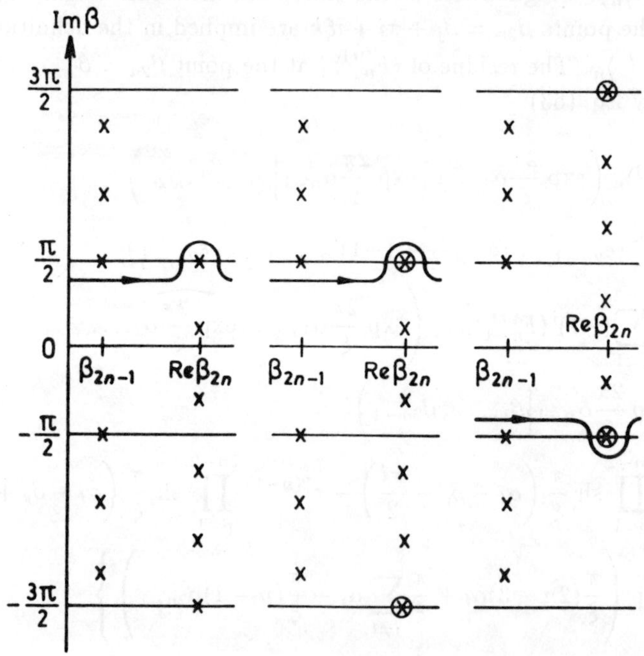

Fig. 3.

It is clear that when β_{2n} tends to $\beta_{2n-1}+\pi i$ the contour appears to be clutched between two simple poles of the integrand. Hence the integral over α_j, considered as a function of β_{2n}, has a simple pole at the point $\beta_{2n}=\beta_{2n-1}+\pi i$. At first glance the pole of $\Phi_{n,\sigma}(\langle F_n^{(\gamma)}\rangle_n)$ at the point $\beta_{2n-1}+\pi i$ is of the n-th degree ($(n-1)$ integrations plus the pole of $\langle F_n^{(\gamma)}\rangle_n$). However this is not the case. The pole is simple. Let us explain this fact. First, it is clear that $\frac{1}{(n-1)!}\prod_{i<j}^{n-1}\mathrm{sh}\,(\alpha_i-\alpha_j)$ in the definition of $\Phi_{n,\sigma}$ can be replaced by $\frac{1}{(n-2)!}\prod_{i<j}^{n-2}\mathrm{sh}\,(\alpha_i-\alpha_j)\prod_{i=j}^{n-2}\mathrm{ch}\,(\alpha_i+\beta_{2n})$ due to the antisymmetry of Δ_n with respect to $x_1,\ldots,x_{n-1}$. After the replacement, the poles at the point $\beta_{2n}=\beta_{2n-1}+\pi i$ of the integrals over $\alpha_1,\ldots,\alpha_{n-2}$ disappear. The poles remain which are caused by the integration over α_{n-1} and by the pole of $\langle\ \rangle_n$. These poles do not interfere because the RHS of Eq. (63) contains a factor $\mathrm{sh}\,\frac{\pi}{\xi}\left(\alpha-\beta_{2n-1}-\frac{\pi i}{2}\right)$ (see Fig. 3a). Thus $\Phi_{n,\sigma}(\langle F_n^{(\gamma)}\rangle)$ has a simple pole at the point $\beta_{2n}=\beta_{2n-1}+\pi i$ and we have only to calculate the residue. To this end the following trick is proposed. Consider the function

$$
\begin{aligned}
G_\sigma^{(\gamma)}(\lambda|\beta_1,\ldots,\beta_{2n}) &= \left(\frac{1}{2\pi i}\right)^{n-1}\fint_{-\infty}^{\infty}d\alpha_1\ldots\\
&\times\fint_{-\infty}^{\infty}d\alpha_{n-1}\prod_{i=1}^{n-1}\prod_{j=1}^{2n}\varphi(\alpha_i-\beta_j)\\
&\times\frac{1}{2}(1-\exp(\alpha_{n-1}\ \ \sigma))\exp\left(-\frac{\pi}{\xi}(n-2)\sum_{i=1}^{n-1}\alpha_i+\sigma\sum_{i=1}^{n-1}\alpha_i\right)\\
&\times\frac{1}{(n-2)!}\prod_{i<j}^{n-2}\mathrm{sh}\,(\alpha_i-\alpha_j)\prod_{i=1}^{n-2}\mathrm{ch}\,(\alpha_i-\beta_{2n})\\
&\times\langle F_n^{(\gamma)}\rangle_n\left(\exp\frac{2\pi}{\xi}\alpha_1,\ldots,\exp\frac{2\pi}{\xi}\alpha_{n-1}\Big|\beta_1,\ldots,\beta_{2n}\right).
\end{aligned}
$$

Evidently,

$$
\begin{aligned}
\Phi_{n,\sigma}(\langle F_n^{(\gamma)}\rangle_n) &= G_\sigma^{(\gamma)}\left(\beta_{2n}-\frac{\pi i}{2}\Big|\beta_1,\ldots,\beta_{2n}\right)\\
&\quad+G_\sigma^{(\gamma)}\left(\beta_{2n}+\frac{\pi i}{2}\Big|\beta_1,\ldots,\beta_{2n}\right).
\end{aligned}
$$

Notice now that the pole of $G_\sigma^{(\gamma)}\left(\beta_{2n}-\frac{\pi i}{2}|\beta_1,\ldots,\beta_{2n}\right)$ at the point $\beta_{2n}=\beta_{2n-1}+\pi i$ is caused only by the pole of $\langle\ \rangle_n$, because $1-\exp\big(\alpha_{n-1}-\beta_{2n}$

$+\frac{\pi i}{2})$ cancels the poles at the points $\alpha_{n-1} = \beta_{2n} - \frac{\pi i}{2}, \alpha_{n-1} = \beta_{2n} - \frac{5\pi i}{2}$ (on Fig. 3 these poles are encircled). Let us calculate the residue of $G_\sigma^{(\gamma)}(\beta_{2n} - \frac{\pi i}{2}|\beta_1, \dots, \beta_{2n})$ at the point $\beta_{2n} = \beta_{2n-1} + \pi i$:

$$\operatorname{res} G_\sigma^{(\gamma)}\left(\beta_{2n} - \frac{\pi i}{2}\middle|\beta_1, \dots, \beta_{2n}\right)$$

$$= \frac{\xi}{\pi}\sum_{k=1}^{n-1}(-1)^k \left(\frac{1}{2\pi i}\right)^{n-1} \fint_{-\infty}^{\infty} d\alpha_1 \dots \fint_{-\infty}^{\infty} d\alpha_{n-1} \prod_{i=1}^{n-1}\prod_{j=1}^{2n-2} \varphi(\alpha_i - \beta_j)$$

$$\frac{1}{2}\mathrm{sh}^{-1}\frac{1}{2}\left(\alpha_{n-1} - \beta_{2n-1} + \frac{\pi i}{2}\right)\frac{1}{(n-2)!}\prod_{i<j}^{n-2}\mathrm{sh}\,(\alpha_i - \alpha_j)$$

$$\times \exp\left(-\frac{\pi}{\xi}(n-2)\sum_{i=1}^{n-1}\alpha_i + \sigma\sum_{i=1}^{n-1}\alpha_i + \frac{\pi}{\xi}(n-1)\alpha_k\right.$$

$$\left.-\frac{\pi}{\xi}(n-1)\beta_{2n-1} + \frac{1}{2}\left(\alpha_{n-1} - \beta_{2n-1} - \frac{\pi i}{2}\right)\right)\langle F_{n-1}^{(\gamma)}\rangle_{n-1}$$

$$\left(\exp\frac{2\pi}{\xi}\alpha_1, \dots, \widehat{\exp\frac{2\pi}{\xi}\alpha_k}, \dots, \exp\frac{2\pi}{\xi}\alpha_{n-1}\middle|\beta_1, \dots, \beta_{2n-2}\right)$$

$$\otimes\left(e_{2n,\frac{1}{2}} \otimes e_{2n-1,-\frac{1}{2}} + e_{2n,-\frac{1}{2}} \otimes e_{2n-1,\frac{1}{2}}\right)\prod_{i=1}^{2n-2}\mathrm{sh}^{-1}\frac{\pi}{\xi}(\beta_{2n-1} - \beta_i)$$

$$\times\left[\prod_{l=1}^{2n-2}\mathrm{sh}\,\frac{\pi}{\xi}\left(\alpha_k - \beta_l - \frac{\pi i}{2}\right) - \tau^{2(n-1)}\prod_{l=1}^{2n-2}\mathrm{sh}\,\frac{\pi}{\xi}\left(\alpha_k - \beta_l + \frac{\pi i}{2}\right)\right],$$

$$\beta_{2n} = \beta_{2n-1} + \pi i\,.$$

In deriving this relation we have used the identity

$$\varphi(\alpha)\varphi(\alpha - \pi i) = \frac{1}{\mathrm{sh}\,\frac{\pi}{\xi}\left(\alpha - \frac{\pi i}{2}\right)\mathrm{ch}\,\alpha}\,.$$

Transform the integral over α_k in k-th term using (56):

$$\frac{1}{2\pi i}\left(-\int_{-\infty}^{\infty} + \int_{-\infty-2\pi i}^{\infty-2\pi i}\right)\left(\prod_{j=1}^{2n-2}\varphi(\alpha_k - \beta_j)\mathrm{sh}\,\frac{\pi}{\xi}\left(\alpha_k - \beta_j + \frac{\pi i}{2}\right)\right.$$

$$\left.\times\prod_{j\neq k}^{n-2}\mathrm{sh}\,(\alpha_k - \alpha_j)\exp\left(\frac{\pi}{\xi}(n-1)\alpha_k + \sigma\alpha_k\right)d\alpha_k\right), \quad k \leq n-2\,;$$

$$\frac{1}{2\pi i}\left(-\int_{-\infty}^{\infty}+\int_{-\infty-2\pi i}^{\infty-2\pi i}\right)\left(\prod_{j=1}^{2n-2}\varphi(\alpha_{n-1}-\beta_j)\operatorname{sh}\frac{\pi}{\xi}\left(\alpha_{n-1}-\beta_j+\frac{\pi i}{2}\right)\right.$$

$$\times\exp\left(\frac{\pi}{\xi}(n-1)\alpha_{n-1}+\frac{1}{2}\alpha_{n-1}+\sigma\alpha_{n-1}\right)$$

$$\left.\times\frac{1}{2\operatorname{sh}\frac{1}{2}\left(\alpha_{n-1}-\beta_{2n-1}+\frac{\pi i}{2}\right)}d\alpha_{n-1}\right)\ ,\qquad k=n-1\ .$$

The integrals from the first group are equal to zero because the integrands have no singularities in the strip $-2\pi<\operatorname{Im}\alpha_k<0$. The last integral is equal to

$$\prod_{j=1}^{2n-2}\varphi\left(\beta_{2n-1}-\beta_j-\frac{\pi i}{2}\right)\exp\left(\beta_{2n-1}\left(\frac{\pi}{\xi}n-\frac{\pi}{\xi}+\frac{1}{2}+\sigma\right)\right)$$

$$\times\operatorname{sh}\frac{\pi}{\xi}(\beta_{2n-1}-\beta_j)\ .$$

Hence

$$\operatorname{res}G_\sigma^{(\gamma)}\left(\beta_{2n}-\frac{\pi i}{2}\middle|\beta_1,\dots,\beta_{2n}\right)=\frac{\xi}{\pi}e^{\sigma\beta_{2n-1}}\prod_{j=1}^{2n-2}\varphi\left(\beta_{2n-1}-\beta_j-\frac{\pi i}{2}\right)$$

$$\times\Phi_{n-1,\sigma}\left(\langle F_{n-1}^{(\gamma)}\rangle_{n-1}\right)\otimes\left(e_{2n,\frac{1}{2}}\otimes e_{2n-1,-\frac{1}{2}}+e_{2n,-\frac{1}{2}}\otimes e_{2n-1,\frac{1}{2}}\right)\ .$$

Let us consider now $G_\sigma^{(\gamma)}\left(\beta_{2n}+\frac{\pi i}{2}\middle|\beta_1,\dots,\beta_{2n}\right)$. The remark to Theorem 2 implies

$$G_\sigma^{(\gamma)}\left(\beta_{2n}+\frac{\pi i}{2}\middle|\beta_1,\dots,\beta_{2n}\right)$$

$$=G_\sigma^{(\gamma)}\left(\beta_{2n}+\frac{\pi i}{2}\middle|\beta_{2n},-2\pi i,\beta_1,\dots,\beta_{2n-1}\right)\ .$$

Continuing RHS with respect to $\beta_{2n}:\beta_{2n}\to\beta_{2n}+\pi i$ one obtains the picture of poles presented on Fig. 3b. The poles which are cancelled by $1-\exp\left(\alpha_{n-1}-\beta_{2n}-\frac{\pi i}{2}\right)$ are encircled. It is clear that the pole of $G_\sigma^{(\gamma)}(\beta_{2n}+\frac{\pi i}{2}|\beta_{2n}-2\pi i,\beta_1,\dots,\beta_{2n-1})$ is caused by the pole of $\langle F_n^{(\gamma)}\rangle_n(\beta_{2n}-2\pi i,\beta_1,\dots,\beta_{2n-1})$. The calculation of the residue is similar to the calculation for

$\mathrm{res}\, G(\beta_{2n} - \frac{\pi i}{2}|\dots)$ presented above and leads to the following result:

$$\mathrm{res}\ G_\sigma^{(\gamma)}\left(\beta_{2n} - \frac{\pi i}{2}\Big|\beta_1, \dots, \beta_{2n}\right)$$
$$= \frac{\xi}{\pi}\prod_{j=1}^{2n-2}\varphi\left(\beta_{2n-1} - \beta_j + \frac{\pi i}{2}\right)\exp(\sigma(\beta_{2n-1} + \pi i))$$
$$\Phi_{n-1,\sigma}(\langle F_{n-1}^{(\gamma)}\rangle_{n-1})(\beta_1, \dots, \beta_{2n-2})(e_{2n,\frac{1}{2}} \otimes e_{2n-1,-\frac{1}{2}} + e_{2n,-\frac{1}{2}} \otimes e_{2n-1,\frac{1}{2}})$$
$$\times \hat{S}_{2n-1,1}(\beta_{2n-1} - \beta_1)\dots\hat{S}_{2n-1,2n-2}(\beta_{2n-1} - \beta_{2n-2}),$$
$$\beta_{2n} = \beta_{2n-1} + \pi i .$$

Hence

$$\mathrm{res}\,\Phi_{n,\sigma}(\langle F_n^{(\gamma)}\rangle_n)(\beta_1, \dots, \beta_{2n}) = \prod_{j=1}^{2n-2}\varphi\left(\beta_{2n-1} - \beta_j - \frac{\pi i}{2}\right)$$
$$\times \Phi_{n-1,\sigma}(\langle F_{n-1}^{(\gamma)}\rangle_{n-1})(\beta_1, \dots, \beta_{2n-2})$$
$$\otimes (e_{2n,\frac{1}{2}} \otimes e_{2n-1,-\frac{1}{2}} + e_{2n-1,\frac{1}{2}} \otimes e_{2n,-\frac{1}{2}})$$
$$\times (I - (-1)^\sigma S_0(\beta_{2n-1} - \beta_1)S_{2n-1,1}(\beta_{2n-1} - \beta_1)\dots$$
$$\times S_0(\beta_{2n-1} - \beta_{2n-2})\hat{S}_{2n-1,2n-2}(\beta_{2n-1} - \beta_{2n-2})),$$
$$\beta_{2n} = \beta_{2n-1} + \pi i . \tag{79}$$

In deriving Eq. (79) we have used the identity

$$S_0(\alpha) = \frac{\varphi\left(\alpha + \frac{\pi i}{2}\right)}{\varphi\left(\alpha - \frac{\pi i}{2}\right)} .$$

Now, using the identitites

$$\zeta(\beta)\zeta(\beta - \pi i) = \varphi^{-1}\left(\beta + \frac{\pi i}{2}\right), \quad d = \zeta^{-1}(\pi i)\frac{1}{2\pi i}\frac{\pi}{\xi}$$

one obtains formulae (78).

Q.E.D.

We have shown that soliton form factors as defined in the previous section do satisfy Axioms 1, 2, 3. For $\xi < \pi$ it should be shown that the breather form factors defined by the soliton ones satisfy Axiom 4. We shall make it in Sec. 5.

Remark. The reasonings presented after the proof of Lemma 1 provides us with the possibility of the calculation of form factor norms which will be important in Secs. 9, 10. In the previous section the base $\tilde{w}_{\varepsilon_1,\dots,\varepsilon_{2n}}(\beta_1,\dots,\beta_{2n})$ has been defined which is dual to $w_{\varepsilon_1,\dots,\varepsilon_{2n}}(\beta_1,\dots,\beta_{2n})$. This base possesses the triangularity opposite to that of the base $w_{\varepsilon_1,\dots,\varepsilon_{2n}}(\beta_1,\dots,\beta_{2n})$. This fact and the formula (69) imply the existence of another representation for $\langle F_n^{(\gamma)}\rangle_n$:

$$\langle F_n^{(\gamma)}\rangle_n = \langle 0| \sum_{\substack{\{1\dots 2n\}\\=\{i_1\dots i_n\}\cup\{j_1,\dots j_n\}}} F_n^{(-\gamma)}(\beta_{i_1},\dots,\beta_{i_n}|\beta_{j_1}\dots\beta_{j_n})$$
$$\times \prod_{p=1}^{n}\prod_{q=1}^{n} \mathrm{sh}^{-1}\frac{\pi}{\xi}(\beta_{i_p}-\beta_{j_q}+\pi i)\tilde{w}_{\varepsilon_1,\dots,\varepsilon_{2n}}(\beta_1,\dots,\beta_{2n}) ,$$
$$\varepsilon_{i_p}=\frac{1}{2} , \quad \varepsilon_{j_p}=-\frac{1}{2} .$$

Evidently,

$$\overline{F_n^{(\gamma)}(\beta_1,\dots,\beta_n|\beta_{n+1},\dots,\beta_{2n})} = F_n^{(\gamma)}(\beta_{n+1},\dots,\beta_{2n}|\beta_1,\dots,\beta_n) .$$

Using these facts and Eq. (46), one obtains

$$\begin{aligned}(\langle F_n^{(\gamma)}\rangle_n, \langle F_n^{(\gamma')}\rangle_n) &= \sum_{\{1,\dots,2n\}=\{i_1\dots i_n\}\cup\{j_1,\dots,j_n\}} F_n^{(\gamma)}(\beta_{i_1},\dots,\beta_{in}|\beta_{j_1}\dots\beta_{j_n})\\ &\times F_n^{(-\gamma')}(\beta_{i_1},\dots,\beta_{in}|\beta_{j_1},\dots,\beta_{jn})\\ &\times \prod_{p=1}^{n}\prod_{q=1}^{n}\frac{1}{\mathrm{sh}\,\frac{\pi}{\xi}(\beta_{ip}-\beta_{jq})\mathrm{sh}\,\frac{\pi}{\xi}(\beta_{i_p}-\beta_{j_q}+\pi i)} .\end{aligned} \tag{80}$$

This formula allows us to calculate the norms of form factors.

5

BREATHERS FORM FACTORS IN SG MODEL

Formula (17) implies that the form factors corresponding to both solitons and breathers can be obtained as the residues of soliton form factors. In the present section we consider for the sake of brevity only the form factors f^+. Other form factors can be treated in a similar way.

According to Eq. (17) for the calculation of form factors corresponding to $2n-2$ solitons and a breather, one has to calculate the residues of $f^+(\beta_1, \ldots, \beta_{2n})$ at the point $\beta_{2n} = \beta_{2n-1} + \pi i - i\xi m$, the rapidity of breather being equal to $\beta_{2n-1} + i(\frac{\pi}{2} - m\frac{\xi}{2})$. Let us calculate the residue. As it has been established in the previous section the pole at the point $\beta_{2n} = \beta_{n-1} + \pi i - i\xi m, m \geq 1$ is caused only by the pole of $\langle\ \rangle_n$. Applying Eq. (63) one obtains

$$\operatorname{res} f^+(\beta_1, \ldots, \beta_{2n}) = a_m d^{n-1} d_m \frac{1}{(n-1)!} \prod_{i<j}^{2n-2} \zeta(\beta_i - \beta_j)$$

$$\times \prod_{i=1}^{2n-2} \zeta_{s,m}\left(\beta_i - \beta_{2n-1} - \frac{\pi i - i\xi m}{2}\right)$$

$$\times (e_{2n,\frac{1}{2}} \otimes e_{2n-1,-\frac{1}{2}} + (-1)^m e_{2n,-\frac{1}{2}} \otimes e_{2n-1,\frac{1}{2}})$$

$$\otimes (n-1) \fint_{-\infty}^{\infty} d\alpha_1 \ldots \fint_{-\infty}^{\infty} d\alpha_{n-1} \prod_{i=1}^{n-1} \prod_{j=1}^{2n-2} \varphi(\alpha_i - \beta_j)$$

$$\times \prod_{i=1}^{n-1} \varphi_m\left(\alpha_i - \beta_{2n-1} - \frac{\pi i}{2} + \frac{\xi}{2} m\right) \prod_{i<j} \operatorname{sh}(\alpha_i - \alpha_j)$$

$$\times \langle F^{(1)}_{n-1}\rangle_{n-1}\left(\exp\frac{2\pi}{\xi}\alpha_1,\ldots,\exp\frac{2}{\xi}\alpha_{n-2}|\beta_1,\ldots,\beta_{2n-2}\right)$$
$$\times\left[\prod_{l=1}^{2n-2}\left(\exp\frac{2\pi}{\xi}\alpha_{n-1}-\tau\exp\frac{2\pi}{\xi}\beta_l\right)-\tau^{4(n-1)}\right.$$
$$\left.\times\prod_{l=1}^{2n-2}\left(\exp\frac{2\pi}{\xi}\alpha_{n-1}-\tau^{-1}\exp\frac{2\pi}{\xi}\beta_l\right)\right]$$
$$\times\exp\left(-\frac{\pi}{\xi}(n-3)\sum_{i=1}^{n-1}\alpha_i\right)\prod_{l=1}^{2n-2}\mathrm{sh}^{-1}\frac{\pi}{\xi}(\beta_{2n-1}-\beta_l)\,,$$
$$\beta_{2n}=\beta_{2n-1}+\pi i-i\xi m$$

where

$$\varphi_m(\alpha)=\frac{\prod_{l=1}^{2n-2}\mathrm{sh}\frac{1}{2}\left(\alpha-i\frac{\xi m}{2}+i\xi k\right)}{\prod_{k=0}^{m}\mathrm{ch}\frac{1}{2}\left(\alpha-i\frac{\xi m}{2}+i\xi k\right)}$$
$$\zeta_{s,m}(\beta)=\zeta\left(\beta-\frac{\pi i}{2}+\frac{i\xi m}{2}\right)\zeta\left(\beta+\frac{\pi i}{2}-\frac{i\xi m}{2}\right)$$
$$=C\exp\left(2\int_0^\infty\frac{\sin^2\frac{1}{2}(\beta+\pi i)\mathrm{ch}\xi\frac{k}{2}\mathrm{sh}\frac{\xi mk}{2}}{k\,\mathrm{sh}\,\pi k\mathrm{ch}\,\frac{\pi k}{2}\mathrm{sh}\frac{\xi k}{2}}dk\right)$$
$$\times\frac{\varphi^{-1}\left(\beta+\pi i-i\frac{\xi m}{2}\right)}{\prod_{j=0}^{m-1}\mathrm{ch}\frac{1}{2}\left(\beta+\frac{\pi i}{2}-\frac{i\xi(m-2j)}{2}\right)}\,,$$
$$C=\exp\left(\int_0^\infty\frac{m\mathrm{sh}\frac{\xi k}{2}\mathrm{ch}\,\frac{\pi k}{2}-\mathrm{ch}\,\frac{\xi k}{2}\mathrm{sh}\,\frac{\xi mk}{2}}{k\,\mathrm{sh}\,\frac{\xi k}{2}\mathrm{sh}\,\pi k\,\mathrm{ch}\,\frac{\pi k}{2}}dk\right)\,,$$
$$d_m=\zeta(\pi i-i\xi m)da_m^{-1}\,.$$

A more explicit formula for d_m will be given later. Now let us apply the familiar trick with the transformation of the contour of integration over α_{n-1} and obtain the result

$$\mathrm{res}\,f^+(\beta,\ldots\beta_{2n})=a_m d^{n-1}d_m\frac{1}{(n-2)!}\prod_{i<j}^{2n-2}\zeta(\beta-\beta_j)$$
$$\times\zeta_{s,m}\left(\beta_i-\beta_{2n-1}-\frac{\pi i-i\xi m}{2}\right)$$
$$\times\left(e_{2n,\frac{1}{2}}\otimes e_{2n-1,-\frac{1}{2}}+(-1)^m e_{2n,-\frac{1}{2}}\otimes e_{2n-1,\frac{1}{2}}\right)$$

$$\times \fint_{-\infty}^{\infty} d\alpha_1 \ldots \fint_{-\infty}^{\infty} d\alpha_{n-2} \int_{\Gamma'_m} d\alpha_{n-1} \prod_{i=1}^{n-1} \prod_{j=1}^{2n-2} \varphi(\alpha_i - \beta_j)$$

$$\times \varphi_m \left(\alpha_i - \beta_{2n-1} - \frac{\pi i}{2} + \frac{i\xi m}{2}\right)$$

$$\times \prod_{i<j}^{n-1} \mathrm{sh}(\alpha_i - \alpha_j) \exp\left(-\frac{\pi}{\xi}(n-3)\sum_{i=1}^{n-2} \alpha_i\right)$$

$$\times \langle F_{n-1}^{(1)} \rangle_{n-1} \left(\exp\frac{2\pi}{\xi}\alpha_1, \ldots, \exp\frac{2\pi}{\xi}\alpha_{n-2} | \beta_1, \ldots, \beta_{2n-2}\right),$$

$$\beta_{2n} = \beta_{2n-1} + \pi i - i\xi m,$$

where Γ'_m is the contour enveloping the point $\beta_{2n-1} - \frac{\pi i}{2} - i\xi m + i\xi j, j = 0, \ldots, m$.

Continuing along the same line one can obtain a formula for the factors corresponding to arbitrary numbers of solitons and breathers. We present only the final result denoting, to avoid a confusion of terms, the rapidities of breathers by θ. To this end it is useful to introduce an integral transformation $\phi_{n,m_1,\ldots,m_k}$.

Definition 1. Transformation $\phi_{n,m_1,\ldots,m_k;\sigma}$ prescribes for every antisymmetric polynomial $P(x_1, \ldots x_{n-1})$ a function of $\beta_1, \ldots, \beta_{2n}, \theta_1, \ldots, \theta_k$ by the following rule:

$$\phi_{n,m_1,\ldots m_k;\sigma}(P) = \frac{1}{(n-1)!}\left(\frac{1}{2\pi i}\right)^{n+k-1} \fint_{-\infty}^{\infty} d\alpha_1 \ldots \fint_{-\infty}^{\infty} d\alpha_{n-1} \int_{\Gamma_1} d\alpha_n$$

$$\to \int_{\Gamma_k} d\alpha_{n+k-1} \prod_{i=1}^{n+k-1} \prod_{j=1}^{2n} \varphi(\alpha_i - \beta_j) \prod_{i=1}^{n+k-1} \prod_{j=1}^{k} \varphi_{m_k}(\alpha_i - \theta_j)$$

$$\times \exp\left(\sigma \sum_{i=1}^{n+k-1} \alpha_i - \frac{\pi}{\xi}(n-2)\sum_{i=1}^{n-1} \alpha_i\right) \prod_{i<j}^{n+k-1} \mathrm{sh}(\alpha_i - \alpha_j)$$

$$\times P\left(\exp\frac{2\pi}{\xi}\alpha_1, \ldots, \exp\frac{2\pi}{\xi}\alpha_{n-1}\right),$$

where Γ_k are the contours enveloping the points $\theta_k - \pi i - i\frac{\xi}{2}m_k + i\xi j, j = 0, \ldots, m_k$.

Let us introduce the functions

$$\zeta_{m,s}(\beta) = \zeta_{s,m}(\beta) \frac{\mathrm{sh}\frac{\pi}{\xi}(\beta - \frac{\pi i}{2} + \frac{i\xi m}{2})}{\mathrm{sh}\frac{\pi}{\xi}(\beta + \frac{\pi i}{2} - \frac{i\xi m}{2})},$$

and for $m_1 > m_2$

$$\zeta_{m_1,m_2}(\beta) = \zeta\left(\beta - \frac{i\xi}{2}(m_1 - m_2)\right)\zeta\left(\beta + \frac{i\xi}{2}(m_1 - m_2)\right)$$

$$\times \zeta\left(\beta + \pi i - i\frac{\xi}{2}(m_1 + m_2)\right)\zeta\left(\beta - \pi i + \frac{i\xi}{2}(m_1, + m_2)\right)$$

$$\times \operatorname{sh}^{-1}\frac{\pi}{\xi}\left(\beta + \frac{i\xi}{2}(m_1 + m_2) + \pi i\right)$$

$$= c_{m_1,m_2} \exp\left(4\int_0^\infty \frac{\sin^2 \frac{1}{2}(\beta + \pi i)k \ \operatorname{ch}(\frac{\pi}{2} - \frac{\xi}{2}m_1)k \ \operatorname{sh} \frac{\xi m_2 k}{2} \ \operatorname{ch} \frac{\xi k}{2}}{k \ \operatorname{sh} \frac{\xi k}{2} \operatorname{ch} \frac{\pi k}{2} \ \operatorname{sh} \pi k} d_k\right)$$

$$\times \frac{2 \prod\limits_{j=0}^{m_1 - m_2} \operatorname{sh} \frac{1}{2}(\beta + \frac{i\xi}{2}(m_1 - m_2 - 2j))}{\prod\limits_{l=1}^{m_1 + m_2 - 1} \operatorname{ch} \frac{1}{2}(\beta + \frac{i\xi}{2}(m_1 + m_2 - 2l))},$$

$$c_{m_1,m_2}$$

$$= \exp\left(2\int_0^\infty \frac{m_2 \operatorname{ch} \frac{\pi k}{2} \operatorname{sh} \frac{\xi k}{2} - \operatorname{ch}(\frac{\pi}{2} - \frac{\xi m_1}{2})k \ \operatorname{sh} \frac{\xi m_2 k}{2} \operatorname{ch} \frac{\xi k}{2}}{k \ \operatorname{sh} \pi k \ \operatorname{sh} \frac{\xi k}{2} \ \operatorname{ch} \frac{\pi k}{2}} dk\right),$$

$$\zeta_{m_2,m_1}(\beta) = \zeta_{m_1,m_2}(\beta).$$

Form factors corresponding to arbitrary numbers of solitons and breathers are given by

$$f^+(\beta_1, \ldots, \beta_{j_1}\theta_1\beta_{j_1+1} \ldots \beta_{j_2}\theta_2 \ldots \theta_k\beta_{j_k+1} \ldots \beta_{2n})_{m_1,\ldots m_k}$$

$$= d^n \prod_{j=1}^{k} d_{m_j} \prod_{i<j} \zeta(\beta_i - \beta_j) \prod_{l<k} \zeta_{m_l,m_k}(\theta_l - \theta_k) \prod_{l=1}^{k}\prod_{i=1}^{j_l} \zeta_{m_l,s}(\theta_l - \beta_i)$$

$$\times \prod_{l=1}^{k} \prod_{i=j_l+1}^{2n} \zeta_{sm_l}(\beta_i - \theta_l)\phi_{n,m_1\ldots m_k;0}(\langle F_n^{(1)}\rangle_n),$$

$$d_m = \prod_{l=1}^{m-1} \sin\frac{\xi l}{2}\left(\sin \xi m c_{m,m}^{-1}\frac{2}{\pi}\right)^{\frac{1}{2}}, \tag{81}$$

where the operation $\langle\ \rangle_n$ is applied of course only to solitons. The component of (81) in the decomposition with respect to the base $e_{\varepsilon_1,\ldots\varepsilon_{2n}}$ is denoted by

$$f^+(\beta_1, \ldots, \beta_{j_1}\theta_1\beta_{j_1+1} \ldots \beta_{j_2}\theta_2 \ldots \theta_k\beta_{j_k+1} \ldots \beta_{2n})_{\varepsilon_1,\ldots\varepsilon_{j_1} m_1, \varepsilon_{j_1+1},\ldots,\varepsilon_{2n}}.$$

Our next goal is to show that these form factors possess the necessary properties.

Theorem 1. Breather form factors satisfy Axiom 1, which means

$$\begin{aligned}
&f^{+}(\ldots,\beta_j,\theta_l,\ldots,)_{\ldots,\varepsilon_j,m_l,\ldots}S_{s,m_l}(\beta_j-\theta_l)\\
&\quad=f^{+}(\ldots,\theta_l,\beta_j,\ldots)_{\ldots,m_l,\varepsilon_j,\ldots},\\
&f^{+}(\ldots,\theta_l,\theta_{l+1},\ldots)_{\ldots,m_l,m_{l+1},\ldots}S_{m_l,m_{l+1}}(\theta_l-\theta_{l+1})\\
&\quad=f^{+}(\ldots,\theta_{l+1},\theta_l,\ldots)_{\ldots,m_{l+1},m_l,\ldots}.
\end{aligned}$$

Proof. The function $\phi_{n,m_1,\ldots,m_k;0}\ (\langle F_n^{(1)}\rangle_n)$ is invariant with respect to permutations of breathers with breathers and breathers with solitons. That is why the statement of the theorem is equivalent to

$$\begin{aligned}
\zeta_{m,s}(\beta)S_{m,s}(\beta)&=\zeta_{s,m}(-\beta)\ ,\\
\zeta_{m_1,m_2}(\beta)S_{m_1,m_2}(\beta)&=\zeta_{m_2,m_1}(-\beta)\ .
\end{aligned}$$

These identities can be checked directly.
Q.E.D.

Theorem 2. Breather form factors satisfy Axiom 2, which means

$$f^{+}(\ldots\theta_k+2\pi i)_{\ldots,m_k}=f^{+}(\theta_k,\ldots)_{m_k,\ldots}\ .$$

Proof. The simplest way to prove the theorem is to cite the corresponding formula for soliton form factors and to use the identity

$$\begin{aligned}
&\operatorname{res} f^{+}(\ldots,\beta_{2n-1},\beta_{2n})_{\ldots,\varepsilon_{2n-1},\varepsilon_{2n}}\\
&\quad=a_m\delta_{\varepsilon_{2n-1},-\varepsilon_{2n}}f\left(\ldots,\beta_{2n-1}+\frac{\pi i-i\xi m}{2}\right)_{\ldots,m},\\
&\qquad\qquad\beta_{2n}=\beta_{2n-1}+\pi i-i\xi m\ .
\end{aligned}$$

Actually, everything is much simpler. It can be shown that $\phi_{n,m_1,\ldots,m_k;0}$ $(\langle F_n^{(1)}\rangle_n)\prod\varphi^{-1}(\beta_j-\sigma_k+\pi i-i\frac{\xi m_k}{2})$ is a periodic function of breather rapidities with the period equal to $2\pi i$, and that functions $\zeta_{s,m},\zeta_{m_1,m_2}$ satisfy the identities

$$\begin{aligned}
\zeta_{s,m}(\beta-2\pi i)\varphi\left(\beta-\pi i-i\frac{\xi m}{2}\right)&=\zeta_{m,s}(-\beta)\varphi\left(-\beta+\pi i-\frac{i\xi m}{2}\right)\ ,\\
\zeta_{m_1,m_2}(\beta-2\pi i)&=\zeta_{m_2,m_1}(-\beta)\ .
\end{aligned}$$

Q.E.D.

Let us pass onto the analytic properties of breather form factors. Consider $f(\dots,\theta_k)\dots_{,m_k}$ as a function of θ_k. The function $\zeta_{s,m}(\beta)$ is regular in the strip $-\pi < \mathrm{Im}\beta < 0$ and has simple zeros at the points $\beta = -\frac{\pi i}{2} + i\frac{\xi}{2}(m+2j); j = 0,\dots$. The function $\zeta_{m_1,m_2}(\beta)$ has simple poles at the points $\beta = -\pi i + i\frac{\xi}{2}(m_1+m_2) - i\xi j, j = 1,\dots$, and simple zeros at the points $\beta = -\frac{i\xi}{2}|m_1 - m_2| + i\xi j, j = 0,\dots$. Now we have to understand what singularities appear during the analytical continuation of $\phi_{n,m_1,\dots,m_k;0}$. The singularities of the integrand with respect to $\alpha_1,\dots,\alpha_{n-1}$ are presented on Fig. 4a.

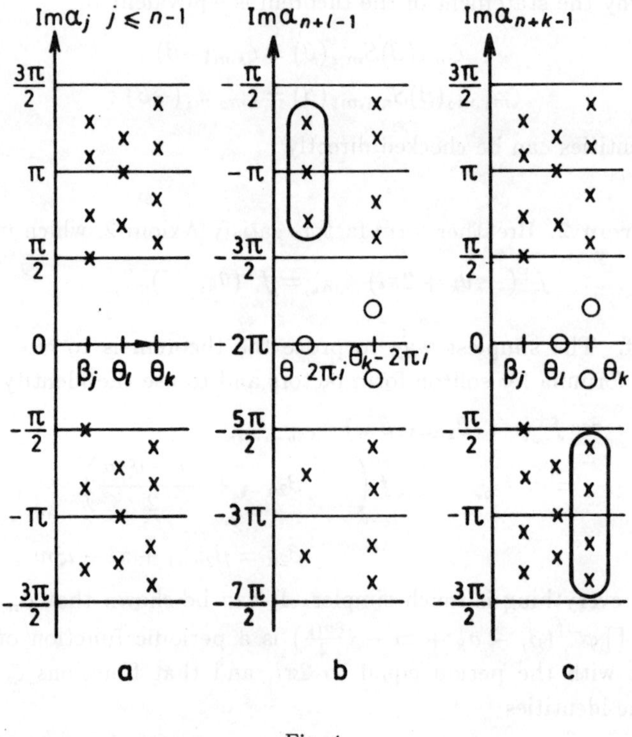

Fig. 4.

It is clear that during the analytical continuation into the strip $0 < \mathrm{Im}\,\theta_k < \pi$, the contour of integration over $\alpha_1,\dots\alpha_{n-1}$ cannot be clutched between singularities of the integrand. Hence the integrals over $\alpha_1, \ldots,$ α_{n-1} give no singularities in the strip $0 < \mathrm{Im}\,\theta_k < \pi$. Consider now the integral over α_{n+l-1}, $l < k$. It is clear that the singularities of this integral considered as a function of θ_k can be situated on the line $\mathrm{Re}\,\theta_k = \mathrm{Re}\,\theta_l$. The contour of integration, poles and zeros of the integral are presented on Fig. 4b. Due to the antisymmetry of the integrand with respect to permutation of α_{n+l-1} and α_{n+k-1} the contour of integration can be replaced by $\Gamma_l \cup \Gamma_k$. After this replacement the integral has no singularities for $\mathrm{Re}\,\theta_k = \mathrm{Re}\,\theta_l$ because both the poles on the line $\mathrm{Re}\,\alpha_{n+l-1} = \mathrm{Re}\,\theta_l$ and those on the line $\mathrm{Re}\,\alpha_{n+l-1} = \mathrm{Re}\,\theta_k$ are situated inside the contour. And, what is more, the integral over α_{n+l-1} vanishes for $m_k > m_l + 1$ and $\theta_k = \theta_l + \pi i - \frac{i\xi}{2}(m_l + m_k - 2j), j = m_l + 1 \ldots,$ because all the singularities inside $\Gamma_l \cup \Gamma_k$ for these values of θ_k are cancelled by zeros. These zeros of the integral over α_{n+l-1} cancel a part of the poles of $\zeta_{m_l,m_k}(\theta_l - \theta_k)$. Finally, let us consider the integral over α_{n+k-1}. The locations of the poles and zeros of the integrand are shown on Fig. 4c. It is clear that for $\theta_k = \beta_j + \frac{\pi i}{2} + \frac{i\xi}{2}(m_k - 2p)$, $j = 1, \ldots, 2n, p = 0, \ldots, m_k - 1$, and $\theta_k = \theta_l + \frac{i\xi}{2}(m_k + m_l - 2p)$ $l = 1, \ldots, k-1, p = 0, \ldots, m_k + m_l$, the contour of integration appears to be clutched between two poles of the integrand, hence there are simple poles of the integral at these points. A part of these poles is cancelled after the multiplication by $\zeta_{s,m_k}(\beta_j - \theta_k), \zeta_{m_l,m_k}(\theta_l - \theta_k)$. For $m_l > m_k + 1$, the integrand vanishes when $\theta_k = \theta_l + \pi i - \frac{i\xi}{2}(m_k + m_l - 2j), j = m_k + 1, \ldots$. Summarizing all the above reasonings we obtain the following picture of poles of the form factor: form factor $f(\beta_1, \ldots, \theta_1, \ldots, \theta_k)$ has simple poles at the points $\theta_k = \beta_j + \frac{\pi i}{2} - i\frac{\xi}{2}(m_k - 2p), p = 0, \ldots, m_k - 1; \theta_k = \theta_l + \frac{i\xi}{2}(m_k + m_l) - i\xi p, p = 0, \ldots, \min(m_l, m_k) - 1; \theta_k = \theta_l + \pi i - \frac{i\xi}{2}(m_k + m_l - 2p), p = 1, \ldots, \min(m_l, m_k)$. This picture is in agreement with that required in Sec. 1. We have now to calculate the residues at these poles.

Theorem 3. Form factor $f^+(\ldots, \theta_k)_{\ldots, m_k}$ satisfies Axioms 3, 4, which means than

$$2\pi i \operatorname{res} f^+(\ldots, \theta_{k-1}, \theta_k)_{\ldots, m_{k-1}, m_k}$$
$$= \delta_{m_k, m_{k-1}} f^+(\ldots) \left[1 - \prod_{i=1}^{2n} S_{m_k, s}(\theta_{k-1} - \beta_j) \prod_{p=1}^{k-2} S_{m_k, m_p}(\theta_{k-1} - \theta_p)\right],$$
$$\theta_k = \theta_{k-1} + \pi i,$$

$$\operatorname{res} f^+(\dots,\beta_{2n},\theta_k)_{\dots,\varepsilon_{2n},m_k}$$
$$= a_m^p f^+\left(\dots,\beta_{2n}+i\xi(m_k-p),\beta_{2n}+\frac{\pi i-i\xi p}{2}\right)_{\dots,\varepsilon_{2n},p},$$
$$\theta_k=\beta_{2n}+\frac{\pi i}{2}+\frac{i\xi}{2}(m_k-2p),p=0\dots m_k-1\,,$$
$$\operatorname{res} f^+(\dots,\theta_{k-1},\theta_k)_{\dots,m_{k-1},m_k}$$
$$= a_{m_1,m_2}^p f^+\left(\dots\theta_{k-1}+\frac{i\xi}{2}(m_k-p),\theta_{k-1}\right.$$
$$\left.+\frac{i\xi}{2}(m_{k-1}-p)\right)_{\dots,m_k+m_{k-1}-p,p},$$
$$\theta_k=\theta_{k-1}+\frac{i\xi}{2}(m_k+m_{k-1}-2p),p=0,\dots,\min(m_{k-1},m_k)-1\,,$$
$$\operatorname{res} f^+(\dots,\theta_{k-1},\theta_k)\dots m_{k-1},m_k$$
$$= \tilde{a}_{m_1,m_2}^p f^+\left(\dots,\theta_{k-1}+\frac{i\xi}{2}p,\theta_{k-1}+\pi i\right.$$
$$\left.-\frac{i\xi}{2}(m_k+m_{k-1}-2p)\right)_{\dots m_{k-1}-p,m_k-p},$$
$$\theta_k=\theta_{k-1}+\pi i-\frac{i\xi}{2}(m_k+m_{k-1}-2p),p=1,\dots,\min(m_k,m_{k-1})\,. \tag{82}$$

Proof. The proofs of the above identities are quite similar; that is why we shall prove only the first one which seems to be the most difficult. We leave the proofs of other identities to the reader.

The pole of $f^+(\dots,\theta_{k-1},\theta_k)_{\dots m_{k-1},m_k}$ at the point $\theta_k=\theta_{k-1}+\pi i$ exist only for $m_k=m_{k-1}$, being caused by the pole of the function $\zeta_{m_{k-1},m_k}(\theta_{k-1}-\theta_k)$. Let us evaluate the residues at $\theta_k=\theta_{k-1}+\pi i$:

$$2\pi i\operatorname{res} f^+(\beta_1,\dots,\beta_{j_1}\theta_1,\dots,\theta_{k-1},\theta_k)_{\varepsilon_1,\dots,m_{k-1},m_k}$$
$$=\sin\xi m_k\delta_{m_{k-1},m_k}\prod_{i<j}^{2n}\zeta(\beta_i-\beta_j)\prod_{p<q}^{k-2}\zeta_{m_p,m_q}(\theta_p-\theta_q)$$
$$\times\prod_{l=1}^{k-2}\prod_{i=1}^{j_l}\zeta_{s,m_l}(\beta_i-\theta_0)\prod_{l=1}^{k-2}\prod_{i=j_l+1}^{2n}\zeta_{m_l,s}(\theta_l-\beta_i)$$
$$\times\prod_{i=1}^{2n}\varphi^{-1}\left(\theta_{k-1}-\beta_i-\frac{i\xi m_k}{2}\right)\varphi^{-1}\left(\theta_{k-1}-\beta_i+\frac{i\xi m_k}{2}-\pi i\right)$$

$$\times \prod_{p=1}^{k-2} \varphi_{m_p}^{-1}\left(\theta_{k-1} - \theta_p - \frac{i\xi m_k}{2}\right) \varphi_{m_p}^{-1}\left(\theta_{k-1} - \theta_p + \frac{i\xi m_k}{2} - \pi i\right)$$

$$\times \fint_{-\infty}^{\infty} d\alpha_1 \ldots \fint_{-\infty}^{\infty} d\alpha_{n-1} \int d\alpha_n \ldots \int d\alpha_{n+k-1} \prod \varphi(\alpha_i - \beta_j)$$

$$\times \prod_{p=1}^{k-2} \varphi_{m_p}(\alpha_i - \theta_p) \exp\left(-\frac{\pi}{\xi}(n-2)\sum_{i=1}^{n-1} \alpha_i\right) \prod_{i<j} \mathrm{sh}\,(\alpha_i - \alpha_j)$$

$$\times \prod_{i=1}^{n+k-1} \mathrm{sh}^{-1}\left(\alpha_i - \theta_{k-1} - \frac{i\xi m_k}{2}\right) \mathrm{sh}^{-1}\left(\alpha_i - \theta_{k-1} + \frac{i\xi m_k}{2}\right)$$

$$\times \langle F_n^{(1)}\rangle_n \left(\exp\frac{2\pi}{\xi}\alpha_1, \ldots, \exp\frac{2\pi}{\xi}\alpha_{n-1} | \beta_1, \ldots \beta_{2n}\right) , \tag{83}$$

where we have used the identitites

$$\zeta_{s,m}(\beta)\zeta_{s,m}(\beta - \pi i) = \varphi^{-1}\left(-\beta - \frac{i\xi}{2}m\right)\varphi^{-1}\left(-\beta - \pi i + \frac{i\xi m}{2}\right) ,$$

$$\zeta_{m_1,m_2}(\beta)\zeta_{m_1,m_2}(\beta - \pi i)$$
$$= \varphi_{m_1}^{-1}\left(-\beta - \frac{i\xi m_2}{2}\right)\varphi_{m_1}^{-1}\left(-\beta - \pi i + \frac{i\xi m_2}{2}\right) ,$$

$$\varphi_m(\alpha)\varphi_m(\alpha - \pi i) = \mathrm{sh}^{-1}\left(\alpha - \frac{i\xi m}{2}\right)\mathrm{sh}^{-1}\left(\alpha + \frac{i\xi m}{2}\right) ,$$

$$2\pi i d_m^2 \zeta_{m,m}(-\pi i) = \sin m\xi .$$

Consider the integrals over α_{n+k-1} and α_{n+k-2},

$$\frac{1}{2\pi i}\int_{\Gamma_{k-1}} d\alpha_{n+k-2} \prod_{j=1}^{2n} \varphi(\alpha_{n+k-2} - \beta_j) \prod_{p=1}^{k-2} \varphi_{m_p}(\alpha_{n+k-2} - \theta_p)$$
$$\times \prod_{i<n+k-2} \mathrm{sh}(\alpha_{n+k-2} - \alpha_i)$$
$$\times \frac{1}{\mathrm{sh}(\alpha_{n+k-2} - \theta_{k-1} - \frac{i\xi m_k}{2})\mathrm{sh}\,(\alpha_{n+k-2} - \theta_{k-1} + \frac{i\xi m_1}{2})}$$

$$\frac{1}{2\pi}\int_{\Gamma_k} d\alpha_{n+k-1} \prod_{j=1}^{2n} \varphi(\alpha_{n+k-1} - \beta_j) \prod_{p=1}^{k-2} \varphi_{m_p}(\alpha_{n+k-1} - \theta_p)$$
$$\times \prod_{i<n+k-1} \mathrm{sh}\,(\alpha_{n+k-1} - \alpha_i)$$
$$\times \frac{1}{\mathrm{sh}(\alpha_{n+k-1} - \theta_{k-1} - \frac{i\xi m_k}{2})\ \mathrm{sh}(\alpha_{n+k-1} - \theta_{k-1} + \frac{i\xi m_k}{2})} ,$$

where the contours Γ_{k-1}, Γ_k envelop the points $\alpha_{n+k-2} = \theta_{k-1} \pm \frac{i\xi m_k}{2} - \pi i$ and $\alpha_{n+k-1} = \theta_{k-1} \pm \frac{i\xi m_k}{2}$ respectively (other poles being cancelled). Calculate the integral over α_{n+k-1}:

$$
\begin{aligned}
&\frac{1}{sin\xi m}\left\{\prod_{j=1}^{2n}\varphi\left(\theta_{k-1}-\beta_j+\frac{i\xi m_k}{2}\right)\prod_{p=1}^{k-2}\varphi_{m_p}\left(\theta_{k-1}-\theta_p+\frac{i\xi m_k}{2}\right)\right.\\
&\times\prod_{i<n+k-2}\mathrm{sh}\left(\theta_{k-1}-\alpha_i+\frac{i\xi m_k}{2}\right)\frac{1}{2\pi i}\int_{\Gamma_{k-1}}d\alpha_{n+k-2}\prod_{j=1}^{2n}\varphi(\alpha_{n+k-2}-\beta_j)\\
&\times\prod_{p=1}^{k-2}\varphi_{m_p}(\alpha_{n+k-2}-\theta_p)\prod_{i<n+k-2}\mathrm{sh}(\alpha_{n+k-2}-\alpha_i)\\
&\times\frac{1}{\mathrm{sh}\left(\alpha_{n+k-2}-\theta_{k-1}+\frac{i\xi m_k}{2}\right)}\\
&-\prod_{j=1}^{2n}\varphi\left(\theta_{k-1}-\beta_j-\frac{i\xi m_k}{2}\right)\prod_{p=1}^{k-2}\varphi_{m_p}\left(\theta_{k-1}-\theta_p-\frac{i\xi m_k}{2}\right)\\
&\times\prod_{i<n+k-2}\mathrm{sh}\left(\theta_{k-1}-\alpha_i-\frac{i\xi m_k}{2}\right)\frac{1}{2\pi i}\int_{\Gamma_{k-1}}d\alpha_{n-k-2}\prod_{j=1}^{2n}\varphi(\alpha_{n+k-2}-\beta_j)\\
&\times\prod_{p=1}^{k-2}\varphi_{m_p}(\alpha_{n+k-2}-\theta_p)\prod_{i<n+k-2}\mathrm{sh}(\alpha_{n+k-2}-\alpha_i)\\
&\left.\times\frac{1}{\mathrm{sh}(\alpha_{n+k-2}-\theta_{k-1}-\frac{i\xi m_k}{2})}\right\}.
\end{aligned}\tag{84}
$$

It is notable that one singularity under the integral over α_{n+k-2} has disappeared in each term in Eq. (84). Take the integral over α_{n+k-2}:

$$
\begin{aligned}
&\frac{1}{\sin\xi m_k}\prod_{j=1}^{2n}\varphi\left(\theta_{k-1}-\beta_j-\frac{i\xi m_k}{2}\right)\varphi\left(\theta_{k-1}-\beta_j-\pi i+\frac{i\xi m_k}{2}\right)\\
&\times\prod_{p=1}^{k-2}\varphi_{m_p}\left(\theta_{k-1}-\theta_p-\pi i+\frac{i\xi m_k}{2}\right)\varphi_{m_p}\left(\theta_{k-1}-\theta_p-\frac{i\xi m_k}{2}\right)
\end{aligned}
$$

$$\times \prod_{i<n+k-2} \left| \mathrm{sh}\left(\alpha_i - \theta_{k-1} - \frac{i\xi}{2} m_k\right) \right|^2$$
$$\times \left\{ 1 - \prod_{i=1}^{2n} S_{m_k s}(\theta_{k-1} - \beta_j) \prod_{p=1}^{k-2} S_{m_k,p}(\theta_{k-1} - \theta_p) \right\} \qquad (85)$$

where we used the identities

$$S_{m,S}(\beta) = \frac{\varphi(\beta - \pi i - \frac{i\xi m}{2})\varphi(\beta + \frac{i\xi m}{2})}{\varphi(\beta - \pi i + \frac{i\xi m}{2})\varphi(\beta - \frac{i\xi m}{2})} \, ,$$
$$S_{m_1,m_2}(\beta) = \frac{\varphi_{m_1}(\beta - \pi i - \frac{i\xi m}{2})\varphi_{m_1}(\beta + \frac{i\xi m}{2})}{\varphi_{m_1}(\beta - \pi i + \frac{i\xi m}{2})\varphi_{m_1}(\beta - \frac{i\xi m}{2})} \, .$$

Substituting (85) to (86) we obtain the required identity.
Q.E.D.

Finally, in order to check the identity of Axiom 4 completely one has to show that form factor $f(\beta_1, \ldots, \beta_{2n})_{\ldots, \varepsilon_n}$ possesses proper analytical properties in the presence of breathers. We formulate the corresponding theorem but do not prove it as the proof is quite similar to that of Theorem 3.

Theorem 4. Besides the poles at the points $\beta_{2n} = \beta_j + \pi i - i\xi k, k = 0, \ldots, [\frac{\pi}{\xi}]$, the form factor $f^+(\beta_1, \ldots, \theta_l, \ldots, \beta_{2n})_{\varepsilon_1, \ldots, m_l, \ldots, \varepsilon_{2n}}$ has simple poles at the points $\beta_{2n} = \theta_l + \frac{\pi i}{2} + \frac{i\xi}{2}(m_l - 2p), p = 0, 1, \ldots, m_l - 1$, the residues being

$$\begin{aligned} &\mathrm{res}\, f^+(\ldots, \theta_k, \beta_{2n})_{\ldots, m_k, \varepsilon_{2n}} \\ &= a^p_m f^+\left(\ldots, \theta_k + \frac{i\xi}{2} j(m_k - p), \theta_k + \frac{\pi i}{2} - \frac{i\xi m_k}{2}\right)_{\ldots, p, \varepsilon_{2n}} \end{aligned}$$
$$\beta_{2n} = \theta_k + \frac{\pi i}{2} + \frac{i\xi}{2}(m_k - 2p), \quad p = 1, \ldots, m_k - 1 \, .$$

6

PROPERTIES OF THE OPERATORS $j_\mu, T_{\mu\nu}, \exp(\pm i\frac{\beta\mu}{2})$ IN SG MODEL

In Sec. 3 formulae have been presented for the soliton form factors $f_\mu, f_{\mu\nu}, f^\pm$. Their properties, which entail local commutativity, have been established in Secs. 4, 5. In the present section we established some additional properties of the form factors which demonstrates that it is natural to identify the operators defined by these form factors with $j_\mu, T_{\mu\nu}, \exp(\pm i\frac{\beta u}{2})$. We consider only soliton form factors; a generalization to form factors corresponding to both solitons and breathers is straightforward.

Theorem 1. Soliton form factors $f_\mu(\beta_1, \dots, \beta_{2n}), f_{\mu\nu}(\beta_1, \dots, \beta_{2n})$, $f^\pm(\beta_1, \dots, \beta_{2n})$ possess the properties

$$\begin{aligned} f^\mu(\beta_1, +\Delta, \dots, \beta_{2n} + \Delta) &= L_{\mu\nu}(\Delta) f_\nu(\beta_1, \dots, \beta_{2n}) \,, \\ f_{\mu\nu}(\beta_1 + \Delta, \dots, \beta_{2n} + \Delta) &= L_{\mu\mu'}(\Delta) L_{\nu\nu'}(\Delta) f_{\mu'\nu'}(\beta_1, \dots \beta_{2n}) \,, \\ f^\pm(\beta_1 + \Delta, \dots, \beta_{2n} + \Delta) &= f^+(\beta_1, \dots, \beta_{2n}) \,. \end{aligned} \tag{86}$$

Equations (86) means that with respect to Lorentz transformations j_μ is a vector, $T_{\mu\nu}$ is an (1, 1)-tensor, and $\exp(\pm i\frac{\beta u}{2})$ are scalars.

Proof. Eqs. (86) follow evidently from Eqs. (60) and the identities

$$\begin{aligned} &\phi_{n,\sigma}(\langle F_n^{(\gamma)} \rangle_n)(\beta_1 + \Delta, \dots, \beta_{2n} + \Delta) \\ &\quad = \exp((n-1)\Delta\sigma) \phi_{n,\sigma}(\langle F_n^{(\gamma)} \rangle_n)(\beta_1, \dots, \beta_{2n}) \,. \end{aligned}$$

A Lorentz transformation corresponds to a simultaneous shift of rapidities; that is why Eqs. (86) imply that $j_\mu, T_{\mu\nu}$ and $\exp(\pm i\frac{\beta u}{2})$ are vector, (1, 1)-tensor and scalar respectively.
Q.E.D.

It is possible to introduce an operator of charge conjugation postulating its action in the space of states:

$$C|ph\rangle = |ph\rangle, \quad CZ^*_\varepsilon(\beta)C^{-1} = Z^*_{-\varepsilon}(\beta), \quad \varepsilon = \pm\frac{1}{2},$$
$$CZ^*_m(\beta)C^{-1} = (-1)^m Z^*_m(\beta) .$$

The following theorem establishes the properties of operators defined through $f_\mu, f_{\mu\nu}, f^\pm$ with respect to charge conjugation.

Theorem 2. Soliton form factors satisfy the identities

$$f_\mu(\beta_1, \ldots, \beta_{2n})_{\varepsilon_1,\ldots,\varepsilon_{2n}} = -f_\mu(\beta_1, \ldots, \beta_{2n})_{-\varepsilon_1,\ldots,-\varepsilon_{2n}},$$
$$f_{\mu\nu}(\beta_1, \ldots, \beta_{2n})_{\varepsilon_1,\ldots,\varepsilon_{2n}} = f_{\mu\nu}(\beta_1, \ldots, \varepsilon_{2n})_{-\varepsilon_1,\ldots,-\varepsilon_{2n}},$$
$$f^+(\beta_1, \ldots, \beta_{2n})_{\varepsilon_1,\ldots,\varepsilon_{2n}} = f^-(\beta_1, \ldots, \beta_{2n})_{-\varepsilon_1,\ldots,-\varepsilon_{2n}},$$

which mean that

$$Cj_\mu C^{-1} = -j_\mu, \quad CT_{\mu\nu}C^{-1} = T_{\mu\nu},$$
$$C\exp\left(\pm i\frac{\beta u}{2}\right)C^{-1} = \exp\left(\mp i\frac{\beta u}{2}\right) .$$

Proof. From the reasonings presented in Sec. 4 it follows that

$$\begin{aligned}\langle F_n^{(\gamma)}\rangle_n(x_1, \ldots, x_{n-1}|\beta_1, \ldots, \beta_{2n})_{-\frac{1}{2},\ldots,-\frac{1}{2},\frac{1}{2},\ldots\frac{1}{2}} \\ = \langle F_n^{(-\gamma)}\rangle_n(x_1, \ldots, x_{n-1}|\beta_1, \ldots, \beta_{2n})_{\frac{1}{2},\ldots,\frac{1}{2},-\frac{1}{2},\ldots,-\frac{1}{2}} .\end{aligned} \tag{87}$$

Successively used (64) one can obtain from (87) a relation

$$\begin{aligned}\langle F_n^{(\gamma)}\rangle_n(x_1, \ldots, x_{n-1}|\beta_1, \ldots, \beta_{2n})_{\varepsilon,\ldots,\varepsilon_{2n}} \\ = \langle F_n^{(-\gamma)}\rangle_n(x_1, \ldots, x_{n-1}|\beta_1, \ldots, \beta_{2n})_{-\varepsilon_1,\ldots,-\varepsilon_{2n}},\end{aligned} \tag{88}$$

where $\varepsilon_1, \ldots, \varepsilon_{2n}$ is an arbitrary set of indices.

The statement of the theorem follows immediately from this relation.
Q.E.D.

Theorem 3. Soliton form factors satisfy the identities

$$\left(\sum_{j=1}^{2n} \operatorname{ch} \beta_j\right) f_0(\beta_1, \dots, \beta_{2n}) = \left(\sum_{j=1}^{2n} \operatorname{sh} \beta_j\right) f_1(\beta_1, \dots, \beta_{2n}) ,$$

$$f_{01}(\beta_1, \dots, \beta_{2n}) = f_{10}(\beta_1, \dots, \beta_{2n}) , \tag{89}$$

which mean that

$$\partial_\mu j_\mu = 0 , \quad T_{\mu\nu} = T_{\nu\mu} .$$

Proof. From the definition of $f_\mu, f_{\mu\nu}$ it follows that Eqs. (89) are equivalent to the following relation:

$$\exp\left(-\frac{1}{2}\sum \beta_j\right)\left(\sum e^{\beta_j}\right)\phi_{n,1}(\langle F_n^{(\gamma)}\rangle_n)$$
$$= \exp\left(\frac{1}{2}\sum \beta_j\right)\left(\sum e^{-\beta_j}\right)\phi_{n,-1}(\langle F_n^{(\gamma)}\rangle_n) ,$$

which, in turn, is equivalent to

$$\exp\left(-\frac{1}{2}\sum \beta_j\right)\left(\sum e^{\beta_j}\right)\phi_{n,1}(\Delta_n)$$
$$= \exp\left(\frac{1}{2}\sum \beta_j\right)\left(\sum e^{-\beta_j}\right)\phi_{n,-1}(\Delta_n) . \tag{90}$$

Let us prove Eq. (90). Present, as in the proof of Theorem 4 of Sec. 4, the function $\phi_{n,\sigma}(\Delta_n)$ in the form

$$\phi_{n,\sigma}(\Delta_n) = \det \|\mathfrak{A}_{ij}^{(\sigma)}\| ,$$

where $\mathfrak{A}_{ij}^{(\sigma)}$ is an $(n-1)\times(n-1)$ matrix with the following matrix elements.

$$\mathfrak{A}_{ij}^{(\sigma)} = \fint_{-\infty}^{\infty} d\alpha \prod \varphi(\alpha - \beta_j) \exp\left(-(n-2)\frac{\pi}{\xi}\alpha + (n - 2j + \sigma)\alpha\right)$$
$$\times A_i\left(\exp\frac{2\pi}{\xi}\alpha \middle| \exp\frac{2\pi}{\xi}\beta_1, \dots, \exp\frac{2\pi}{\xi}\beta_n \middle| \exp\frac{2\pi}{\xi}\beta_{n+1}, \dots, \exp\frac{2\pi}{\xi}\beta_{2n}\right) .$$

Obviously the first $n-2$ columns of $\mathfrak{A}^{(-)}$ coincide with the last $n-2$ columns of $\mathfrak{A}^{(+)}$; hence the difference (90) can be presented as a determinant of the matrix whose first $(n-2)$ columns coincide with those of $\mathfrak{A}^{(-)}$, the last column being

$$\tilde{\mathfrak{A}}_{i,n-1} = \fint_{-\infty}^{\infty} d\alpha \prod \varphi(\alpha-\beta_j) \exp\left(-(n-2)\frac{\pi}{\xi}\alpha\right)$$
$$\times A_i\left(\exp\frac{2\pi}{\xi}\alpha \Big| \exp\frac{2\pi}{\xi}\beta_1, \ldots, \exp\frac{2\pi}{\xi}\beta_n \Big| \frac{2\pi}{\xi}\beta_{n+1}, \ldots, \exp\frac{2\pi}{\xi}\beta_{2n}\right)$$
$$\times \left[\exp\left(-\frac{1}{2}\sum\beta_j + (n-1)\alpha\right)\left(\sum e^{\beta_j}\right)\right.$$
$$\left. -(-1)^n \exp\left(\frac{1}{2}\sum\beta_j - (n-1)\alpha\right)\left(\sum e^{-\beta_j}\right)\right],$$

where $(-1)^n$ appears because of the permutation of the columns in $\mathfrak{A}^{(+)}$. Notice now that the column $\tilde{\mathfrak{A}}_{i,n-1}$ can be changed to

$$\tilde{\tilde{\mathfrak{A}}}_{i,n-1} = \fint_{-\infty}^{\infty} d\alpha \prod \varphi(\alpha-\beta_j) \exp\left(-(n-2)\frac{\pi}{\xi}\alpha\right)$$
$$\times \left[\prod \operatorname{ch}\frac{1}{2}\left(\alpha-\beta_j+\frac{\pi i}{2}\right) - (-1)^n \prod \operatorname{ch}\frac{1}{2}\left(\alpha-\beta_j-\frac{\pi i}{2}\right)\right]$$
$$\times A_i\left(\exp\frac{2\pi}{\xi}\alpha \Big| \frac{2\pi}{\xi}\beta_1, \ldots, \exp\frac{2\pi}{\xi}\beta_n \Big| \exp\frac{2\pi}{\xi}\beta_{n+1}, \ldots, \exp\frac{2\pi}{\xi}\beta_{2n}\right), \tag{91}$$

because the difference $\tilde{\tilde{\mathfrak{A}}}_{i,n-1} - \tilde{\mathfrak{A}}_{i,n-1}$ is a linear combination of the first $n-2$ columns. We want to show that $\tilde{\tilde{\mathfrak{A}}}_{i,n-1} = 0$. At first we shall present a "wrong" proof to clarify the idea. Then we shall give the rigorous proof.

The idea of the proof is to use the identity

$$\varphi(\alpha - i\xi) = -\varphi(\alpha)\frac{\operatorname{ch}\frac{1}{2}(\alpha+\frac{\pi i}{2})}{\operatorname{ch}\frac{1}{2}(\alpha-\frac{\pi i}{2}-i\xi)} \tag{92}$$

and to transform the integral (91) in the following way:

$$\tilde{\tilde{\mathfrak{A}}}_{i,n-1} = \left(\int_{\mathbb{R}} - \int_{\mathbb{R}-i\xi}\right) \prod \varphi(\alpha-\beta_j) \exp\left(-(n-2)\frac{\pi}{\xi}\alpha\right)$$
$$\times \prod \operatorname{ch}\frac{1}{2}\left(\alpha-\beta_j+\frac{\pi i}{2}\right) A_i\left(\exp\frac{2\pi}{\xi}\alpha \Big| \exp\frac{2\pi}{\xi}\beta_1, \ldots\right). \tag{93}$$

Notice now that the integrand in (93) has no singularities in the strip $0 > \operatorname{Im}\alpha > -\xi$; hence if we deal with a usual integral the identity $\approx\!\!\!\!\mathfrak{A}_{i,n-1} = 0$ would be evident. But the fact is that we are dealing with a regularized integral $⨍$ and it is essential here; as it will be clear later, only specific properties of the function A_i allow us to transform this "wrong" proof to a rigorous one.

Recall the definition of $⨍$:

$$⨍_{-\infty}^{\infty} \prod \varphi(\alpha-\beta_j) \exp\left(-(n-2)\frac{\pi}{\xi}\alpha\right) P\left(\exp\frac{2\pi}{\xi}\alpha\right) e^{k\alpha} d\alpha$$
$$\equiv \int_{\mathbb{R}+\frac{\pi i}{2}-i0} \prod \varphi(\alpha-\beta_j) \frac{1}{\operatorname{sh}\frac{\pi}{\xi}(\alpha-\beta_j-\frac{3\pi i}{2})} \exp\left(-\frac{\pi}{\xi}\sum\beta_j\right)$$
$$\times \exp\left(\alpha k - \frac{\pi}{\xi}(3n-2)\alpha\right) P_1\left(\exp\frac{2\pi}{\xi}\alpha\right) d\alpha$$
$$+ \int_{\Gamma} \prod \varphi(\alpha-\beta_j) \exp\left(\alpha k - \frac{\pi}{\xi}(n-2)\alpha\right) P_2\left(\tau^4 \exp\frac{2\pi}{\xi}\alpha\right) d\alpha \tag{94}$$

where $|k| \le n-1$, $\deg P \le 2n-2$, P_1 and P_2 are polynomials defined by

$$\prod(x\tau^{-3} - e^{\frac{2\pi}{\xi}\beta_j})P(x) = P_1(x) + \prod(x\tau^{-2} - \tau^{-1}e^{\frac{2\pi}{\xi}\beta_j})P_2(x)$$
$$- \tau^{-2(n-2)}\prod(x\tau^{-3} - e^{\frac{2\pi}{\xi}\beta_j})P_2(x\tau^4), \deg P_1 \le 3n-1\,. \tag{95}$$

The condition $|k| \le n-1$ is essential because for $\deg P_1 = 3n-1$ the integrand of the integral over $\mathbb{R} + \frac{\pi i}{2} - i0$ behaves as $\exp(k-n)\alpha$ when $\alpha \to \infty$. That is why, generally speaking it is impossible to divide the integral (91) into the sum of two integrals containing $\Pi \operatorname{ch}\frac{1}{2}(\alpha - \beta_j + \frac{\pi i}{2})$ and $\Pi \operatorname{ch}\frac{1}{2}(\alpha-\beta_j-\frac{\pi i}{2})$. The only case that it is possible is when it appears possible to find such a decomposition as (95) for which $\deg P_1 = 3n-2$. This requirement is a restriction on the polynomial P. We are going to show that the polynomial $A_i(x|\ldots)$ does satisfy this requirement.

Recall the definition

$$A_i(x|a_1,\ldots a_n|b_1,\ldots b_n) = 2^{-n-i}\prod_{l=1}^{n}(x-a_l\tau)\Big\{\sum_{k=0}^{i-1}(1-\tau^{2(i-k)})$$
$$\times(-1)^k x^{i-k-1}\tau^k\sigma_k(b_1,\ldots,b_n)\Big\} + \tau^{2i}\prod_{l=1}^{n}(x-b_l\tau^{-1})$$
$$\times\left\{\sum_{k=0}^{i-1}(1-\tau^{2(i-k)})(-1)^k x^{i-k-1}\tau^k\sigma_k(a_1,\ldots,a_n)\right\}.$$

Notice that for $x \to \infty$,

$$\sum_{k=0}^{i-1}(\tau^k - \tau^{2i-k})(-1)^k x^{i-k-1}\sigma_k(b_1,\dots,b_n)$$
$$= x^{-n+i-1}\left[\prod(x-\tau b_j) - \tau^{2i}\prod(x-\tau^{-1}b_j)\right] + \mathrm{O}(x^{-2}) .$$

Hence

$$A_i(x|a_1,\dots,a_n|b_1,\dots,b_n) = 2^{-n-i}\left[\prod_{l=1}^{n}(x-\tau a_l)(x-\tau b_l)\right.$$
$$\left. - \prod_{l=1}^{n}(x-\tau^{-1}a_l)(x-\tau^{-1}b_l)\right] x^{-n+i-1} + \mathrm{O}(x^{n-2}) . \qquad (96)$$

Take as $P_2(x)$ the following polynomial:

$$P_2(x) = 2^{-n-i}\sum_{k=0}^{n-i+1}\tau^{3k}\sigma_k\left(\exp\frac{2\pi}{\xi}\beta_1,\dots,\exp\frac{2\pi}{\xi}\beta_{2n}\right)x^{n-k+i-1} . \quad (97)$$

Evidently

$$P_2(x) = 2^{-n-i}\prod_{j=1}^{2n}\left(x - \tau^3\exp\left(\frac{2\pi}{\xi}\beta_j\right)\right)x^{-n+i-1} + \mathrm{O}(x^{-1})$$

for $x \to \infty$. Substituting Eqs. (97) and (96) in Eq. (95) one obtains

$$\prod_{j=1}^{2n}(x\tau^{-3} - e^{\frac{2\pi}{\xi}\beta_j})\mathrm{O}(x^{n-2}) = P_1(x)$$
$$+ \prod_{j=1}^{2n}(x\tau^{-2} - \tau^{-1}e^{\frac{2\pi}{\xi}\beta_j})\mathrm{O}(x^{-1}) - \tau^{2(n-2)}\prod_{j=1}^{2n}(x\tau^{-3} - e^{\frac{2\pi}{\xi}\beta_j})\mathrm{O}(x^{-1}) ,$$

and consequently

$$P_1(x) = \mathrm{O}(x^{3n-2}) ,$$

which means that it is possible to give meaning to $⨏$ for A_i taking for P_2 in Eq. (94) the polynomial (97).

Let us return to Eq. (91):

$$
\begin{aligned}
\tilde{\tilde{\mathfrak{A}}}_{i,n=1} &= \fint_{-\infty}^{\infty} d\alpha \prod \varphi(\alpha-\beta_j) \exp\left(-(n-2)\frac{\pi}{\xi}\alpha\right) \\
&\times A_i\left(\exp\frac{2\pi}{\xi}\alpha \middle| \exp\frac{2\pi}{\xi}\beta_1, \dots, \middle| \exp\frac{2\pi}{\xi}\beta_{n+1}, \dots\right) \\
&\times \left[\prod \mathrm{ch}\frac{1}{2}\left(\alpha-\beta_j+\frac{\pi i}{2}\right) - (-1)^n \prod \mathrm{ch}\frac{1}{2}\left(\alpha-\beta_j-\frac{\pi i}{2}\right)\right] \\
&= \int_{\mathbb{R}+\frac{\pi i}{2}-i0} \prod \varphi(\alpha-\beta_j) \frac{1}{\mathrm{sh}\frac{\pi}{\xi}(\alpha-\beta_j-\frac{3\pi i}{2})} \exp\left(-\frac{\pi}{\xi}\sum\beta_j\right) \\
&\times \exp\left(-\frac{\pi}{\xi}(3n-2)\alpha\right) P_1\left(\exp\frac{2\pi}{\xi}\alpha\right)\left[\prod \mathrm{ch}\frac{1}{2}\left(\alpha-\beta_j+\frac{\pi i}{2}\right)\right. \\
&\left. -(-1)^n \,\mathrm{ch}\frac{1}{2}\left(\alpha-\beta_j-\frac{\pi i}{2}\right)\right] d\alpha \\
&+ \int_{\Gamma} \prod \varphi(\alpha-\beta_j) \exp\left(\frac{\pi}{\xi}(n-2)\alpha\right) P_2\left(\exp\left(\frac{2\pi}{\xi}\alpha\right)\tau^4\right) \\
&\times \left[\prod \mathrm{ch}\frac{1}{2}\left(\alpha-\beta_j+\frac{\pi i}{2}\right) - (-1)^n \prod \mathrm{ch}\left(\alpha-\beta_j-\frac{\pi i}{2}\right)\right],
\end{aligned} \tag{98}
$$

where P_2 is given by (97), P_1 is defined by (95). Now all the problems concerning convergence have been overcome. Hence it is possible to rewrite Eq. (98), using Eq. (56), as follows:

$$
\begin{aligned}
\tilde{\tilde{\mathfrak{A}}}_{i,n-1} &= \left(\int_{\mathbb{R}+\frac{\pi i}{2}-i0} - \int_{\mathbb{R}+\frac{\pi i}{2}-i\xi-i0}\right) \prod \frac{\varphi(\alpha-\beta_j)\,\mathrm{ch}\frac{1}{2}(\alpha-\beta_j+\frac{\pi i}{2})}{\mathrm{sh}\frac{\pi}{\xi}(\alpha-\beta_j-\frac{3\pi i}{2})} \\
&\times \exp\left(-\frac{\pi}{\xi}\sum\beta_j\right) \exp\left(-\frac{\pi}{\xi}(3n-2)\alpha\right) P_1\left(\exp\frac{2\pi}{\xi}\alpha\right) \\
&+ \left(\int_{\Gamma} - \int_{\Gamma-i\xi}\right) \prod \varphi(\alpha-\beta_j)\,\mathrm{ch}\frac{1}{2}\left(\alpha-\beta_j+\frac{\pi i}{2}\right) \exp\left(-\frac{\pi}{\xi}(n-2)\right) \\
&\times P_2\left(\tau^4 \exp\frac{2\pi}{\xi}\alpha\right).
\end{aligned}
$$

The integrand of the first integral has $2n$ singularities in the strip $\frac{\pi}{2} - 0 >$ $\mathrm{Im}\alpha > \frac{\pi}{2} - 0 - \xi$, which are simple poles at the points $\alpha = \beta_j + \frac{3\pi i}{2}$

$-i([\frac{\pi}{\xi}]+1)\xi$, the corresponding residues being

$$\prod_{i\neq j}\frac{\varphi(\beta_j-\beta_i+\frac{3\pi i}{2}-i([\frac{\pi}{\xi}]+1)\xi)}{\operatorname{sh}\frac{\pi}{\xi}(\beta_j-\beta_i)}\exp\left(-\frac{\pi}{\xi}(3n-2)\beta_j-\frac{\pi}{\xi}\sum\beta_i\right)$$
$$\times P_1\left(\tau^3\exp\frac{2\pi}{\xi}\beta_j\right).$$

The integrand of the second integral has $2n$ singularities between the contours $\Gamma, \Gamma-i\xi$, which are simple poles at the points $\alpha=\beta_j-\frac{\pi i}{2}-i\xi([\frac{\pi}{\xi}]+1)$, the corresponding residues being

$$-\prod_{i\neq j}\varphi\left(\beta_j-\beta_i-\frac{\pi i}{2}-i\left(\left[\frac{\pi}{\xi}\right]+1\right)\xi\right)$$
$$\times\exp\left(-\frac{\pi}{\xi}(n-2)\beta_j\right)P_2\left(\tau^3\exp\frac{2\pi}{\xi}\beta_j\right).$$

As it follows from these formulae for residues and Eq. (95), the first and second integrals cancel each other; hence $\approx\!\!\!\!\mathfrak{A}_{i,n-1}=0$.
Q.E.D.

Theorem 4. From factors $f_\mu, f_{\mu\nu}$ can be rewritten as follows:

$$f_\mu(\beta_1,\dots,\beta_{2n})=\sum_i(e^{\beta_j}-(-1)^\mu e^{-\beta_j})g_-(\beta_1,\dots,\beta_{2n}), \tag{99}$$

$$f_{\mu\nu}(\beta_1,\dots,\beta_{2n})=\sum_{i,j}(e^{\beta_j}-(-1)^\mu e^{-\beta_j})(e^{\beta_j}-(-1)^\nu e^{-\beta_j})$$
$$\times g_+(\beta_1,\dots\beta_{2n}), \tag{100}$$

where

$$g_\pm(\beta_1,\dots,\beta_{2n})=(\langle G_n^{(1)}\rangle_n\pm\langle G_n^{(-1)}\rangle_n)\prod_{i<j}\zeta(\beta_i-\beta_j),$$

$$G_n^{(\pm1)}(\beta_1,\dots,\beta_n|\beta_{n+1},\dots,\beta_{2n})$$
$$=2^{\frac{-n(n-1)}{2}}\exp\left(\pm\frac{\pi}{2\xi}\sum_{j=1}^n(\beta_{n+j}-\beta_j-\pi i)\right)$$
$$\times G_n(\beta_1,\dots,\beta_n|\beta_{n+1},\dots,\beta_{2n})\exp\left(-\frac{\pi}{\xi}(n-1)\sum\beta_j\right).$$

$G_n(\beta_1,\dots,\beta_n|\beta_{n+1},\dots\beta_{2n})$ is the determinant of a $n\times n$ matrix $\mathfrak{G}_{ij}(i=0,\dots,n-1,j=0,\dots,n-1)$ with the following matrix elements:

$$\mathfrak{G}_{ij}=\frac{1}{2\pi i}\not\int_{-\infty}^{\infty}\prod\varphi(\alpha-\beta_j)\exp\left(-(n-2)\frac{\pi}{\xi}\alpha+(-n+1+2j)\alpha\right)$$
$$\times A_i\left(\exp\frac{2\pi}{\xi}\alpha\middle|\exp\frac{2\pi}{\xi}\beta_1,\dots,\exp\frac{2\pi}{\xi}\beta_n\middle|\exp\frac{2\pi}{\xi}\beta_{n+1},\dots,\exp\frac{2\pi}{\xi}\beta_{2n}\right),$$

where $A_i, i\geq 1$ are given by the usual formulae (57), $A_0(x|a_1,\dots,a_n|b_1,\dots,b_n)$ is an arbitrary polynomial of x whose degree is equal to $n-1$ and the higher coefficient is equal to $\frac{2\pi}{\xi}$.

Proof. Multiply $G(\beta_1,\dots,\beta_n|\beta_{n+1},\dots,\beta_{2n})$ by $\Sigma\exp(-\beta_j)$. The result can be presented as a determinant whose first column is multiplied by $\Sigma\exp(-\beta_j)$, while others remain unchanged. Now add to the first column the second one with coefficient $\sigma_3(\exp(-\beta_1),\dots,\exp(-\beta_{2n}))$, the third one with coefficient $\sigma_5(\exp(-\beta_1),\dots,\exp(-\beta_{2n}))$ etc., (recall that we denote by σ_k the elementary symmetrical polynomials). Consequently the first column changes to

$$\tilde{\mathfrak{G}}_{i0}=\frac{1}{2\pi i}\not\int_{-\infty}^{\infty}d\alpha\prod\varphi(\alpha-\beta_j)\exp\left(-(n-2)\frac{\pi}{\xi}\alpha\right)$$
$$\times A_i\left(\exp\frac{2\pi}{\xi}\alpha\middle|\exp\frac{2\pi}{\xi}\beta_1,\dots,\exp\frac{2\pi}{\xi}\beta_n\middle|\exp\frac{2\pi}{\xi}\beta_{n+1},\dots,\exp\frac{2\pi}{\xi}\beta_{2n}\right)$$
$$\times 2^{2n-1}\left[\prod\mathrm{ch}\frac{1}{2}\left(\alpha-\beta_j+\frac{\pi i}{2}\right)-(-1)^n\prod\mathrm{ch}\frac{1}{2}\left(\alpha-\beta_j-\frac{\beta i}{2}\right)\right]$$
$$\times\exp\left(-\frac{1}{2}\sum\beta_i\right).$$

As it follows from the proof of Theorem 3, $\tilde{\mathfrak{G}}_{i0}=0, i\geq 1$. Consider $\tilde{\mathfrak{G}}_{00}$. The integral in the formula for $\tilde{\mathfrak{G}}_{00}$ is to be understood as a usual one because the degree of A_0 is equal to $n-1$. Using Eq. (92) one obtains

$$\tilde{\mathfrak{G}}_{00}=\lim_{\Lambda\to\infty}\frac{1}{2\pi i}\int_{-\infty}^{\Lambda}d\alpha\prod\varphi(\alpha-\beta_j)\exp\left(-(n-2)\frac{\pi}{\xi}\alpha\right)$$
$$\times A_0\left(\exp\frac{2\pi}{\xi}\alpha\middle|\exp\frac{2\pi}{\xi}\beta_1,\dots\middle|\exp\frac{2\pi}{\xi}\beta_{n+1},\dots\right)\exp\left(-\frac{1}{2}\sum\beta_j\right)$$
$$\times 2^{2n-1}\left[\prod\mathrm{ch}\frac{1}{2}\left(\alpha-\beta_j+\frac{\pi i}{2}\right)-(-1)^n\prod\mathrm{ch}\frac{1}{2}\left(\alpha-\beta_j-\frac{\pi i}{2}\right)\right]$$
$$=\lim_{\Lambda\to\infty}\frac{1}{2\pi i}\left(\int_{-\infty}^{\Lambda}-\int_{-\infty-i\xi}^{\Lambda-i\xi}\right)\prod\varphi(\alpha-\beta_j)\exp\left(-(n-2)\frac{\pi}{\xi}\alpha\right)$$

$$\times A_0\left(\exp\frac{2\pi}{\xi}\alpha\Big|\dots\right)2^{2n-1}\prod \mathrm{ch}\frac{1}{2}\left(\alpha-\beta_j+\frac{\pi i}{2}\right)\exp\left(-\frac{1}{2}\sum\beta_j\right)$$

$$= \lim_{\Lambda\to\infty}\frac{1}{2\pi i}\int_{\Lambda}^{\Lambda-i\xi}\prod\varphi(\alpha-\beta_j)$$

$$\times\exp\left(-\frac{\pi}{\xi}\alpha(n-2)\right)A_0\left(\exp\frac{2\pi}{\xi}\alpha\Big|\dots\right)2^{2n-1}\prod\mathrm{ch}\frac{1}{2}\left(\alpha-\beta_j+\frac{\pi i}{2}\right)$$

$$\times\exp\left(-\frac{1}{2}\sum\beta_j\right)=2^{n-1}\exp\left(-\frac{1}{2}\sum\beta_j\right),$$

where we also used the fact that the higher coefficient is equal to $\frac{2\pi}{\xi}$. Now we have only to expand the determinant along the first column, obtaining

$$\left(\sum e^{-\beta_j}\right)\langle G_n^{(\gamma)}\rangle_n=\exp\left(-\frac{1}{2}\sum\beta_j\right)\phi_{n,1}(\langle F_n^{(\gamma)}\rangle_n)\,. \tag{101}$$

The equation

$$\left(\sum e^{\beta_j}\right)\langle G_n^{(\gamma)}\rangle_n=\exp\left(\frac{1}{2}\sum\beta_j\right)\phi_{n,-1}(\langle F_n^{(\gamma)}\rangle_n) \tag{102}$$

can be proven in a similar way. Eqs. (101), (102) are equivalent to Eqs. (99), (100).
Q.E.D.

Remark. As functions $G_n^{(\gamma)}$ do not depend on the particular form of A but only on its higher coefficient, it is not necessary to choose A_0 as symmetric with respect to $a_1,\dots,a_n,b_1,\dots,b_n$.

What are the analytical properties of $g_\pm(\beta_1,\dots,\beta_{2n})$, considered as functions of β_{2n}? For $n\geq 2$, one can choose the polynomial $A_0(\exp\frac{2\pi}{\xi}\alpha|\dots)$ in the form $(\exp\frac{2\pi}{\xi}\alpha-\exp\frac{2\pi}{\xi}(\beta_{2n}+\frac{\pi i}{2}))P(\exp\frac{2\pi}{\xi}\alpha)$ where P is a polynomial of degree $n-2$. The multiplier $(\exp\frac{2\pi}{\xi}\alpha-\exp\frac{2\pi}{\xi}(\beta_{2n}+\frac{\pi i}{2}))$ cancels the pole of $\varphi(\alpha-\beta_{2n})$ at the point $\alpha=\beta_{2n}+\frac{\pi i}{2}$ (see Fig. 3). Moreover, for this choice of A_0 the following relation is satisfied for the matrix $A_{ij}=A_i(x_j|a_1,\dots,a_n|b_1,\dots,b_n), i,j=0,\dots$:

$$\det\|A_{ij}\|_{n\times n}|_{b_n=a_n\tau^2}=2^{-3(n-1)}\prod_{j=0}^{n-1}(x_j-\tau^{-1}a_n)$$

$$\times\sum_{k=0}^{n-1}(-1)^k\det\|\tilde{A}_{ij}\|_{(n-1)\times(n-1)}(x_0,\dots,\hat{x}_k,\dots,$$

$$\times x_n|a_1,\dots,a_{n-1}|b_1,\dots,b_{n-1})$$

$$\times\left[\prod(x_k-a_j\tau)(x_k-b_j\tau)-\tau^{4(n-1)}\prod(x_k-a_j\tau^{-1})(x_k-b_j\tau^{-1})\right].$$

These facts taken together allow us to conclude that the function $g_\pm(\beta_1, \ldots, \beta_{2n})$ has for $n \geq 2$ the same singularities as $f_\mu(\beta_1, \ldots, \beta_{2n}), f_{\mu\nu}(\beta_1, \ldots, \beta_{2n})$, i.e., simple poles at the points $\beta_{2n} = \beta_j + \pi i - i\xi k$, where the residues are given by the formulae similar to (78) and (17). The case $n = 1$ needs a special consideration. For $n = 1$, one has

$$g_-(\beta_1, \beta_2)_{\varepsilon_1, \varepsilon_2} = d\zeta(\beta_1 - \beta_2) \frac{(-1)^{\frac{2\varepsilon_j+1}{2}}}{2\,\mathrm{ch}\frac{1}{2}(\beta_1 - \beta_2)\mathrm{ch}\frac{\pi}{2\xi}(\beta_1 - \beta_2 + \pi i)} \delta_{\varepsilon_1 - \varepsilon_2} ,$$

$$g_+(\beta_1, \beta_2)_{\varepsilon_1, \varepsilon_2} = d\zeta(\beta_1 - \beta_2) \frac{\pi}{4\xi} \frac{1}{\mathrm{ch}\,\frac{1}{2}(\beta_1 - \beta_2)\mathrm{sh}\frac{\pi}{2\xi}(\beta_1 - \beta_2 + \pi i)} \delta_{\varepsilon_1 - \varepsilon_2} ,$$

which means that g_- has a simple pole at the point $\beta_2 = \beta_1 + \pi i$ and g_+ has a double pole at the same point.

The following theorem establishes the most important properties of the operator $j_\mu, T_{\mu\nu}, \exp(\pm i\frac{\beta u}{2})$.

Theorem 5. Operators $Q = \int_{-\infty}^{\infty} j_0(x)dx, P_\mu = \int_{-\infty}^{\infty} T_{\mu 0}(x)dx$, and $w\!-\!\lim\limits_{x\to\infty} \exp(\pm i\dfrac{\beta u}{2})$ represent the charge, energy momentum and $\exp(\pi i Q)$, which means that

$$QZ^*_{\varepsilon_1}(\beta_1)\ldots Z^*_{\varepsilon_n}(\beta_n)|ph\rangle = 2\left(\sum_{j=1}^{n} \varepsilon_j\right) Z^*_{\varepsilon_1}(\beta_1)\ldots Z^*_n(\beta_n)|ph\rangle$$

$$P_\mu Z^*_{\varepsilon_1}(\beta_1)\ldots Z^*_{\varepsilon_n}(\beta_n)|ph\rangle = M\left(\sum(e^{\beta_j} + (-1)^\mu e^{-\beta_j})\right) \times Z^*_{\varepsilon_1}(\beta_1)\ldots Z^*_{\varepsilon_n}(\beta_n)|ph\rangle$$

$$\exp(\pi i Q)Z^*_{\varepsilon_1}(\beta_1)\ldots Z^*_{\varepsilon_n}(\beta_n)|ph\rangle = (-1)^n Z^*_{\varepsilon_1}(\beta_1)\ldots Z^*_{\varepsilon_n}(\beta_n)|ph\rangle \; .$$

Proof. Construct all the matrix elements of $j_\mu, T_{\mu\nu}, \exp(\pm i\frac{\beta u}{2})$ through $f_\mu, f_{\mu\nu}, f_x^\pm$ using Eq. (28). It is clear that the matrix elements of $\int_{-\infty}^{x} j_0(x)dx, \int_{-\infty}^{x} T_{\mu 0}(x)dx$ are constructed using Eq. (28) through g_-, $(\Sigma\exp(\beta_j) - (-1)^\mu \exp(-\beta_j))g_+$ respectively. As was to be expected,

$$\int_{-\infty}^{x} j_0(x)dx \xrightarrow{w} 0, \; \int_{-\infty}^{x} T_{\mu 0}(x)dx \xrightarrow{w} 0, \; \exp\left(\pm i\frac{\beta u}{2}\right) \xrightarrow{w} 1 \; ,$$
$$x \to -\infty$$

because the way of understanding the poles in Eq. (28) is in agreement with the sign of the exponent $\exp(i(p^1(A) - p^1(B))x)$. We want to consider the limit of the above operators for $x \to \infty$. To this end it is natural to use Eq. (29) in which the poles are understood in the opposite way. The equivalence Lemma (Lemma of Sec. 1) gives Eq. (29) for $\exp(\pm i\frac{\beta u}{2})$ with $(-1)^{n(B_2)}$ changed to $(-1)^{n(B)}$ (see the Remark to this Lemma). Taking the limit $x \to \infty$ we obtain

$$\lim_{x\to\infty} \langle \overrightarrow{A} | \exp\left(\pm\frac{i\beta\mu(x)}{2}\right) \Big| \overleftarrow{B} \rangle = (-1)^{n(B)} \Delta(A, B) \ .$$

The Equivalence Lemma is almost applicable to the operators $\int_{-\infty}^{x} j_0(x)dx$ and $\int_{-\infty}^{x} T_{\mu 0}(x)dx$, the only discrepancy between Eq. (28) and Eq. (29) being caused by the existence of the annihilation poles of the functions $g_-(\alpha|\beta), (e^\alpha - e^\beta - (-1)^\mu(e^{-\alpha} - e^{-\beta}))g_+(\alpha|\beta)$. For the matrix element of the operators $\int_{-\infty}^{x} j_0(x)dx$ and $\int_{-\infty}^{x} T_{\mu 0}(x)dx$ this discrepancy is $\sigma^3(A)\Delta(A|B)$ and $P_\mu(A)\Delta(A|B)$ respectively. In Eq. (29) one can take the limit $x \to \infty$ and obtain zero. That is why

$$\langle \overrightarrow{A} | \int_{-\infty}^{\infty} T_{\mu 0}(x) dx | \overleftarrow{B} \rangle = \Delta(A|B) P_\mu(A) \ ,$$

$$\langle \overrightarrow{A} | \int_{-\infty}^{\infty} j_0(x) dx | \overleftarrow{B} \rangle = \Delta(A|B) \sigma^3(A) \ .$$

Q.E.D.

Now we want to show that the formulae for $f_\mu, f_{\mu\nu}, f^\pm$ give the correct result for $\xi = \pi$ where the model is equivalent to the free fermion one. But first we would like to show what simplification appears in the formulae at the points $\xi = \frac{\pi}{\nu}$ for an arbitrary positive integer ν. At these points the reflection of solitons disappears and the soliton S-matrix becomes diagonal:

$$S_{\varepsilon_1,\varepsilon_2}^{\varepsilon_1',\varepsilon_2'}(\beta) = S_0(\beta)\delta_{\varepsilon_1}^{\varepsilon_1'}\delta_{\varepsilon_2}^{\varepsilon_2'}(-1)^{(\varepsilon_1-\varepsilon_2)(\nu-1)}$$

$$S_0(\beta) = \prod_{k=0}^{\nu-1} \frac{\operatorname{ch}\frac{1}{2}(\beta + \frac{\pi i}{\nu}k)}{\operatorname{ch}\frac{1}{2}(\beta - \frac{\pi i}{\nu}k)} \ .$$

Straightforward application of the formulae of Sec. 3 to this case is difficult. For example the proof of Lemma 5 of Sec. 3 becomes senseless. We shall obtain the necessary formulae from those for $\frac{\pi}{\nu} + \delta$, taking the limit $\delta \to 0$.

Consider for example the form factor

$$f^{+}(\beta_1,\dots,\beta_{2n})_{-\frac{1}{2},\dots,-\frac{1}{2},\frac{1}{2},\dots,\frac{1}{2}} = d^n 2^{\frac{-(n-1)(n-2)}{2}}$$
$$\times \exp\left(\frac{\pi}{2\xi}\sum_{p=1}^{n}(\beta_{n+p}-\beta_p-\pi i)\right)$$
$$\times \prod_{i<j}\zeta(\beta_i-\beta_j)\prod_{p=1}^{n}\prod_{q=n+1}^{2n}\frac{1}{\operatorname{sh}\frac{\pi}{\xi}(\beta_p-\beta_q+\pi i)}\det\|\mathfrak{A}_{ij}\|\,, \quad (103)$$

where

$$\mathfrak{A}_{ij} = \frac{1}{2\pi i}⨍_{-\infty}^{\infty}d\alpha\prod\varphi(\alpha-\beta_j)\exp\left(-\frac{\pi}{\xi}(n-2)\alpha+(n-2j)\alpha\right)$$
$$\times A_i\left(\exp\frac{2\pi}{\xi}\alpha\middle|\exp\frac{2\pi}{\xi}\beta_1,\dots,\exp\frac{2\pi}{\xi}\beta_n\middle|\exp\frac{2\pi}{\xi}\beta_{n+1},\dots,\exp\frac{2\pi}{\xi}\beta_{2n}\right).$$

From the very definition of the functions $A_i(x|a_1,\dots,a_n|b_1,\dots,b_n)$ it follows that $A_i(x|a_1,\dots,a_n|b_1,\dots,b_n) = \mathrm{O}(\delta)$ for $\delta\to 0$. Consider the relation

$$\prod(x\tau^{-3}-e^{\frac{2\pi}{\xi}}\beta_j)A_i(x|\dots) = P_1^{(i)}(x)+\prod(x\tau^{-2}-\tau^{-1}e^{\frac{2\pi}{\xi}\beta_j})$$
$$\times P_2^{(i)}(x)-\tau^{-2(n-2)}\prod(x\tau^{-3}-e^{\frac{2\pi}{\xi}\beta_j})P_2^{(i)}(x\tau^4)\,,$$

which occurs in the definition of the regularization $⨍$. Let us choose for

$$P_2^{(i)}(x) = 2^{-n-i}\sum_{k=0}^{n+i-1}x^{n-k+i-1}\tau^{3k}\sigma_k\left(\exp\frac{2\pi}{\xi}\beta_1,\dots,\exp\frac{2\pi}{\xi}\beta_{2n}\right).$$

Evidently $P_1^{(i)} = \mathrm{O}(\delta)$ for $\delta\to 0$; this is why

$$\frac{1}{2\pi i}⨍_{-\infty}^{\infty}\prod\varphi(\alpha-\beta_j)\exp\left(-\frac{\pi}{\xi}(n-2)\alpha+(n-2j)\alpha\right)$$
$$\times A_i\left(\exp\frac{2\pi}{\xi}\alpha\middle|\dots\right)d\alpha\overset{\delta\to 0}{\to}$$
$$\frac{1}{2\pi i}\int_\Gamma\prod\varphi(\alpha-\beta_j)\exp\left(-\frac{\pi}{\xi}(n-2)\alpha+(n-2j)\alpha\right)P_2\left(\exp\frac{2\pi\alpha}{\xi}\right)d\alpha\,. \quad (104)$$

Notice now that for $\xi = \frac{\pi}{\nu}$

$$\varphi(\alpha) = \psi(\alpha)\mathrm{sh}^{-1}\nu\left(\alpha - \frac{\pi i}{2}\right) ,$$

$$\psi(\alpha) = 2^{\nu-1}\prod_{j=1}^{\nu-1}\mathrm{sh}\,\frac{1}{2}\left(\alpha - i\frac{\pi j}{\nu}\right) , \quad \nu \geq 2;\ \psi(\alpha) = 1 , \quad \nu = 1 .$$

The integrand in (104) becomes a periodic function of α with the period $2\pi i$. Hence the contour of integration (see Fig. 5a) can be deformed to the joint of two segments $(\Lambda_1, \Lambda_1 - 2\pi i)$ and $(\Lambda_2, \Lambda_2 - 2\pi i)$, with $\Lambda_1 < \beta_j, \Lambda_2 > \beta_j$ for $\forall j$ (see Fig. 5b).

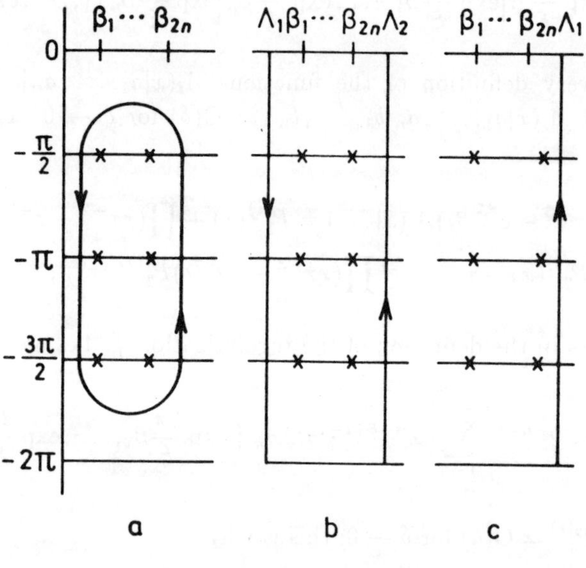

Fig. 5.

Notice that the integrand decreases for $\alpha \to -\infty$; consequently the integral over $(\Lambda_1, \Lambda_1 - 2\pi i)$ can be omitted. Present the integrand as follows:

$$2^{n-i}\frac{1}{2\pi i}\int_{\Lambda_2-2\pi i}^{\Lambda_2}\prod\psi(\alpha-\beta_j)\exp((n-2j)\alpha+\nu(n-2i))R_i(e^{\nu\alpha})d\alpha , \quad (105)$$

where $R_i(x)$ is a rational function of x possessing the following expansion for $x \to \infty$:

$$R_i(x) \doteq 1 + d_i^{(1)} x^{-1} + d_i^{(2)} x^{-2} + \dots \,. \tag{106}$$

The function $\psi(\alpha)$ behaves as $\exp\left(\frac{\nu}{2}\alpha - \frac{\alpha}{2}\right)$ for $\alpha \to \infty$, hence one can retain only a part of the series (106) in the integral (105):

$$\tilde{R}_i(x) = 1 + d_i^{(1)} x^{-1} + \dots + d_i^{(i-1)} x^{-i+1} \,.$$

Now substitute these formulae into the determinant (103). Adding the first row with coefficient $-d_2^{(1)}$ to the second, the second with coefficient $-d_3^{(1)}$ and the first with coefficient $-d_3^{(2)}$ to the third, etc., one obtains

$$f^+(\beta_1, \dots, \beta_{2n})_{-\frac12, \dots, -\frac12, \frac12, \dots, \frac12} = d^n \exp\left(\frac{\nu}{2} \sum_{p=1}^{n} (\beta_{n+p} - \beta_p - \beta_p - \pi i)\right)$$

$$\times \prod_{i<j} \zeta(\beta_i - \beta_j) \prod_{p=1}^{n} \prod_{q=n+1}^{2\nu} \mathrm{sh}^{-1} \nu(\beta_p - \beta_q) \det \|\mathfrak{A}_{ij}^{(0)}\|_{(n-1)\times(n-1)} \,,$$

where

$$\mathfrak{A}_{ij}^{(0)} = \frac{1}{2\pi i} \int_{\Lambda - 2\pi i}^{\Lambda} \prod \psi(\alpha - \beta_j) \exp\left((n - 2j)\alpha + \nu(n - 2i)\alpha\right) d\alpha 2^{n-i} \,.$$

Notice that the value of Λ is of no importance now. Evidently $\mathfrak{A}_{ij}^{(0)}$ is a symmetric function of $\beta_1 \dots \beta_{2n}$. The S-matrix is diagonal, hence

$$f^+(\beta_1, \dots, \beta_{2n})_{\varepsilon_1, \dots, \varepsilon_{2n}} = \prod_{j=1}^{2n} (2\varepsilon_j)^j d^n \prod_{i<j} \zeta(\beta_1 - \beta_j) \det \|\mathfrak{A}_{ij}^{(0)}\|$$

$$\times \prod_{p:\varepsilon_p = \frac12} \prod_{q:\varepsilon_q = -\frac12} \mathrm{sh}^{-1} \nu(\beta_p - \beta_q) \exp\left(\nu \sum \varepsilon_j \beta_j\right) \,.$$

Similarly one can consider $f_\mu, f_{\mu\nu}$, expressing them in terms of $\det \|\mathfrak{A}^{(\pm)}\|$:

$$\mathfrak{A}_{ij}^{(\pm)} = \frac{2^{n-i}}{2\pi i} \int_{\Lambda - 2\pi i}^{\Lambda} \prod \psi(\alpha - \beta_j) \exp(n - 2j + 1)\alpha + (n - 2i)\nu\alpha) d\alpha \,.$$

For $\nu = 1$ one has $\psi(\alpha) \equiv 1, \zeta(\beta) = \mathrm{sh}\, \frac{\beta}{2}$,

$$\mathfrak{A}_{ij}^{(0)} = \delta_{i,n-i}, \mathfrak{A}_{ij}^{(+)} = \mathfrak{A}_{ij}^{(-)} = 0 \,, \quad n \ge 2 \,.$$

Hence

$$f^{\pm}(\beta_1,\dots,\beta_{2n})_{-\frac{1}{2},\dots,-\frac{1}{2},\frac{1}{2},\dots,\frac{1}{2}} = \left(\frac{1}{2\pi i}\right)^n 2^{-n^2}\exp\left(\pm\frac{1}{2}\sum_{p=1}^{n}(\beta_{n+p}-\beta_p)\right) \times \prod_{i<j}\operatorname{sh}\frac{1}{2}(\beta_j-\beta_j)\prod_{p=1}^{n}\prod_{q=n+1}^{2n}\operatorname{sh}^{-1}(\beta_p-\beta_q)\,, \tag{107}$$

$$f_\mu(\beta_1,\beta_2)_{-\frac{1}{2},\frac{1}{2}} = \frac{1}{4\pi i}\left(\exp\left(\frac{1}{2}(\beta_1+\beta_2)\right)-(-1)^\mu\exp\left(\frac{1}{2}(\beta_1+\beta_2)\right)\right)\,, \tag{108}$$

$$f_{\mu\nu}(\beta_1,\beta_2)_{-\frac{1}{2},\frac{1}{2}} = \frac{1}{8\pi i}\left(\exp\left(\frac{1}{2}(\beta_1+\beta_2)\right)-(-1)^\mu\exp\left(-\frac{1}{2}(\beta_1+\beta_2)\right)\right) \times\left(\exp\left(\frac{1}{2}(\beta_1+\beta_2)\right)-(-1)^\nu\exp\left(-\frac{1}{2}(\beta_1+\beta_2)\right)\right)\operatorname{sh}\frac{1}{2}(\beta_1-\beta_2)\,. \tag{109}$$

Let us check the agreement of these formulae with those corresponding to the massive free Fermi-field. The SG model for $\xi=\pi$ is identified with the massive free Fermi-field in accordance with the rule that $Z^*_\varepsilon(\beta), Z_\varepsilon(\beta)$ are the creation-annihilation operators of fermions. Hence

$$\psi(x) = \int\Bigg(e^{ip_\mu(\beta)x_\mu}\begin{pmatrix}e^{-\frac{\beta}{2}}\\ e^{\frac{\beta}{2}}\end{pmatrix}Z^{\frac{1}{2}}(\beta)+e^{-p_\mu(\beta)x_\mu}\begin{pmatrix}e^{\frac{\beta}{2}}\\ e^{-\frac{\beta}{2}}\end{pmatrix}Z^{-\frac{1}{2}}(\beta) + e^{ip_\mu(\beta)x_\mu}\begin{pmatrix}e^{\frac{\beta}{2}}\\ e^{-\frac{\beta}{2}}\end{pmatrix}Z^*_{\frac{1}{2}}(\beta)+e^{-p_\mu(\beta)x_\mu}\begin{pmatrix}e^{-\frac{\beta}{2}}\\ e^{\frac{\beta}{2}}\end{pmatrix}Z^*_{-\frac{1}{2}}(\beta)\Bigg)\,d\beta\,.$$

The operators $j_\mu, T_{\mu\nu}$ are constructed for the massive free fermion field in the usual way:

$$j_\mu =: \overline{\psi}\gamma_\nu\psi : \,, T_{\mu\nu} =: \frac{1}{2}(\overline{\psi}\gamma_\mu\partial_\nu\psi-\partial_\nu\overline{\psi}\gamma_\mu\psi) : \,,$$
$$\gamma_0=\sigma^1\,,\quad \gamma_1=i\sigma^2\,.$$

Evidently

$$\langle ph|j_\mu(0) = \langle ph| \int d\beta_1 d\beta_2 \left(e^{\frac{1}{2}(\beta_1+\beta_2)} - (-1)^\mu e^{-\frac{1}{2}(\beta_1+\beta_2)}\right) Z^{-\frac{1}{2}}(\beta_1) Z^{\frac{1}{2}}(\beta_1) ,$$

$$\langle ph|T_{\mu\nu}(0) = \langle ph| \int d\beta_1 d\beta_2 \left(e^{\frac{1}{2}(\beta_1+\beta_2)} - (-1)^\mu e^{-\frac{1}{2}(\beta_1+\beta_2)}\right) \times \left(e^{\frac{1}{2}(\beta_1+\beta_2)} - (-1)^\mu e^{-\frac{1}{2}(\beta_1+\beta_2)}\right) \operatorname{sh} \frac{1}{2}(\beta_1 - \beta_2) Z^{-\frac{1}{2}}(\beta_1) Z^{\frac{1}{2}}(\beta_2) ,$$

which coincide with (108), (109).

The operators $\exp(\pm i\frac{\beta u}{2})$ are identified with $\exp(\pm\pi i \int_{-\infty}^x j_0(x)dx)$. The operators $\int_{-\infty}^x j_0(x)dx$ are presented through the creation-annihilation operators as follows:

$$\int_{-\infty}^{x_0} j_0(x_0, x_1)dx_0 = \int d\beta_1 d\beta_2 \Big(\operatorname{ch}^{-1}\frac{1}{2}(\beta_1 - \beta_2)(Z^{-\frac{1}{2}}(\beta_1) Z^{\frac{1}{2}}(\beta_2) \times \exp(-ix_\mu(p^\mu(\beta_1) + p^\mu(\beta_2))) + Z^*_{-\frac{1}{2}}(\beta_1) Z^*_{\frac{1}{2}}(\beta_2) \times \exp(-ix_\mu(p^\mu(\beta_1) + p^\mu(\beta_2)))) + \operatorname{sh}^{-1}(\beta_1 - \beta_2 - i0)(Z^*_{\frac{1}{2}}(\beta_1) Z^{\frac{1}{2}}(\beta_2) \times \exp(ix_\mu(p^\mu(\beta_1) - p^\mu(\beta_2))) + Z^*_{-\frac{1}{2}}(\beta_1) Z^{-\frac{1}{2}}(\beta_2) \times \exp(ix_\mu(p^\mu(\beta_1) - p^\mu(\beta_2))))\Big) .$$

To calculate the matrix elements of the operators $\exp(i\frac{\beta u}{2})$ and $\exp(-i\frac{\beta u}{2})$ we have to rewrite these operators in the normal form. To this end one can use the well-known procedure of normal order ring of an exponent of a bilinear form. Using the corresponding formulae from the book [62] one obtains

$$\vdots \exp\left(\pm\pi i : \int_{-\infty}^{x_1} j_0(x_1)dx_1 :\right) \vdots =: \exp(J_\pm(x_1)) : ,$$

$$J_\pm(x) = \int d\beta_1 d\beta_2 \exp\left\{\pm\frac{1}{2}(\beta_1 - \beta_2)\right\} \Big\{\frac{1}{\operatorname{ch}\frac{1}{2}(\beta_1 - \beta_2)} \times (Z^{-\frac{1}{2}}(\beta_1) Z^{\frac{1}{2}}(\beta_2) \exp(-ix_\mu(p^\mu(\beta_1) + p^\mu(\beta_2))) + Z^*_{-\frac{1}{2}}(\beta_1) Z^*_{\frac{1}{2}}(\beta_2) \exp(ix_\mu(p^\mu(\beta_1) + p^\mu(\beta_2)))) + \frac{1}{\operatorname{sh}\frac{1}{2}(\beta_1 - \beta_2 - i0)}(Z^*_{\frac{1}{2}}(\beta_1) Z^{\frac{1}{2}}(\beta_2) \exp(ix_\mu(p^\mu(\beta_1) - p^\mu(\beta_2))) . + Z^*_{-\frac{1}{2}}(\beta_1) Z^{-\frac{1}{2}}(\beta_2) \exp(ix_\mu(p^\mu(\beta_1) - p^\mu(\beta_2))))\Big\} .$$

Hence

$$\langle ph|\exp\left(\pm i\frac{\beta u(0)}{2}\right) = \langle ph|\sum_{n=0}^{\infty}\int d\beta_1\ldots d_{\beta_{2n}} Z^{-\frac{1}{2}}(\beta_1)\ldots Z^{-\frac{1}{2}}(\beta_n)$$

$$\times Z^{\frac{1}{2}}(\beta_{n+1})\ldots Z^{\frac{1}{2}}(\beta_{2n})\exp\left(\pm\frac{1}{2}\sum_{p=1}^{n}(\beta_{n+p}-\beta_p)\right)$$

$$\times\prod \operatorname{sh}\frac{1}{2}(\beta_i-\beta_j)\prod_{\substack{p\leq n\\ q>n}}\frac{1}{2\operatorname{sh}(\beta_p-\beta_q)}\,,$$

which coincides with Eq. (107).

This section contains several arguments in favour of identifying the operators defined by $f_\mu, f_{\mu\nu}, f^{\pm}$ with $j_\mu, T_{\mu\nu}, \exp(\pm i\frac{\beta u}{2})$ respectively. For the currents j_μ one more argument will be obtained in Sec. 10 which is the calculation of the singularities of the commutators $[j_\mu(x), j_\nu(y)]$ at the point $x = y$.

7

FORM FACTORS IN SU(2)-INVARIANT THIRRING MODEL

As it has been said in the Introduction, ITM particles (kinks) possess unusual statistics: they are spin-$\frac{1}{4}$ particles. That is why the charge conjugation matrix $C = i\sigma^2$ is skew-symmetric and the reasonings of Sec. 2 cannot be applied to ITM directly. However we can pass from spin-$\frac{1}{4}$ physical particles to auxiliary particles with the usual statistics. Consider the base of ITM space of states which consists of the vectors

$$Z^*_{\varepsilon_n}(\beta_n)\dots Z^*_{\varepsilon_1}(\beta_1)|ph\rangle \ .$$

Let us introduce formally the operators $\tilde{Z}^*_\varepsilon, \tilde{Z}^\varepsilon$ which satisfy the Zamolodchikov-Faddeev algebra with the following S-matrix:

$$\tilde{S}^{\varepsilon'_1,\varepsilon'_2}_{\varepsilon_1,\varepsilon_2}(\beta) = (-1)^{\varepsilon_1+\varepsilon'_2} S^{\varepsilon'_1\varepsilon'_2}_{\varepsilon_1,\varepsilon_2}(\beta) \ ,$$

where S is the ITM S-matrix ($\varepsilon = 1, 2$). The matrix $\tilde{S}$ satisfies the Yang-Baxter equation, unitarity and crossing-symmetry; the matrix of charge conjugation $c = \sigma^1$, i.e., the associated particles satisfy the usual statistics. Let us identify the spaces of states generated by Z^* and $\tilde{Z}^*$ via

$$\tilde{Z}^*_{\varepsilon_n}(\beta_n)\dots\tilde{Z}^*_{\varepsilon_1}(\beta_1)|ph\rangle = \prod_{j=1}^{n}(-1)^{j\varepsilon_j} Z^*_{\varepsilon_n}(\beta_n)\dots Z^*_{\varepsilon_1}(\beta_1)|ph\rangle \ .$$

Evidently this prescription is in agreement with the commutation relations. Now one can construct local operators whose auxiliary-particle form factors satisfy the usual equations. Locality is guaranteed by the theorem of Sec. 2. In terms of the physical particles the main equations for the form factors are written as follows:

$$\begin{aligned}
&f(\beta_1,\dots,\beta_i,\beta_{i+1},\dots,\beta_n)S_{i,i+1}(\beta_i-\beta_{i+1})\\
&\quad = f(\beta_,\dots,\beta_{i+1},\beta_i,\dots,\beta_n)\ ,\\
&f(\beta_1,\dots,\beta_{n-1},\beta_n+2\pi i) = i^n f(\beta_n,\beta_1,\dots,\beta_{n-1})\ ,\\
&2\pi i\,\mathrm{res}\, f(\beta_1,\dots,\beta_n)_{\varepsilon_1,\dots,\varepsilon_n}\\
&\quad = S_{n,n-1}(I-i^{n-1}S_{n-1,1}(\beta_{n-1}-\beta_1)\dots S_{n-1,n-2}(\beta_{n-1}-\beta_{n-2}))\ ,\\
&\qquad\qquad\qquad\qquad\qquad\qquad\qquad\qquad \beta_n=\beta_{n-1}+\pi i\ ,
\end{aligned} \tag{110}$$

where

$$S_{n,n-1} = e_{n,1}\otimes e_{n-1,2} - e_{n,2}\otimes e_{n-1,1}\ ,$$

$e_{k,\varepsilon}$ being the base in the isotopic space associated with the k-th particles.

Notice now that the matrix $\tilde{S}$ coincides with the SG S-matrix in the limit $\xi\to\infty$. This is why one can construct the form factors of the operators $T_{\mu\nu}, j^3_\mu$ taking the corresponding limit of SG form factors. Then one has to show that $T_{\mu\nu}$ is SU(2)-invariant while j^3_μ is the component of the SU(2) triplet, and to present the form factos of j^1_μ, j^2_μ. This programme can be realized but we would not like going into the details of the limiting process $\xi\to\infty$. We present only the final results directly in the terms of ITM.

Recall the S-matrix of ITM

$$\begin{aligned}
S^{\varepsilon'_1\varepsilon'_2}_{\varepsilon_1,\varepsilon_2}(\beta) &= S_0(\beta)\frac{(\beta\delta^{\varepsilon'_1}_{\varepsilon_1}\delta^{\varepsilon'_2}_{\varepsilon_2} - \pi i\delta^{\varepsilon'_2}_{\varepsilon_1}\delta^{\varepsilon'_1}_{\varepsilon_2})}{(\beta-\pi i)} \equiv S_0(\beta)\hat{S}^{\varepsilon'_1,\varepsilon'_2}_{\varepsilon,\varepsilon_2}(\beta)\ ,\\
S_0(\beta) &= \frac{\Gamma(\frac12+\frac{\beta}{2\pi i})\Gamma(-\frac{\beta}{2\pi})}{\Gamma(\frac12-\frac{\beta}{2\pi i})\Gamma(\frac{\beta}{2\pi i})}\ .
\end{aligned}$$

Now we present several definitions similar to those given in Sec. 3. Consider a set of rapidities $\beta_1,\dots,\beta_{2n}$ connected with isotopic spaces $h_i\simeq \mathrm{C}^2$. Introduce an auxiliary space h_0 and a matrix

$$\begin{pmatrix} A(\sigma|\beta_1,\dots,\beta_{2n}) & B(\sigma|\beta_1,\dots,\beta_{2n})\\ C(\sigma|\beta_1,\dots,\beta_{2n}) & D(\sigma|\beta_1,\dots,\beta_{2n})\end{pmatrix} = \hat{S}_{0,1}(\sigma-\beta_1)\dots\hat{S}_{0,2n}(\sigma-\beta_{2n})\ , \tag{111}$$

where

$$\hat{S}_{i,j}(\sigma) = \frac{1}{\sigma - \pi i}\left(\sigma - \frac{\pi i}{2} - \frac{\pi i}{2}\sigma_i^a \sigma_j^a\right) ,$$

$\sigma_i^a (a = 1, 2, 3)$ are the Pauli matrices acting in the spacing h_i. The matrix $\hat{S}$ differs from S by the absence of the multiplier S_0. In every space h_i^* we consider a base $e_{i,\varepsilon}(\varepsilon = 1, 2)$, the covector $\Pi \otimes e_{i,1}$ being denoted by $\langle 0|$. Besides the natural base of the tensor product $e_{\varepsilon_1,\dots\varepsilon_{2n}} = \Pi \otimes e_{i,\varepsilon_i}$, we introduce a base consisting of the vectors

$$w_{\varepsilon_1,\dots,\varepsilon_{2n}}(\beta_1, \dots, \beta_{2n}) = \langle 0| \prod_{p:\varepsilon_p = 2} B(\beta_p|\beta_1, \dots, \beta_{2n}) .$$

In contrast with the SG model we do not impose any restriction on $\Sigma\varepsilon_j$. The base $w_{\varepsilon_1,\dots,\varepsilon_{2n}}$ is connected with $e_{\varepsilon_1,\dots,\varepsilon_{2n}}$ by a triangular transformation and satisfies several requirements similar to those presented in Sec. 3 (the correspondence is established via the rule: all sh $\frac{\pi}{\xi}\beta$ are to be changed to β). A dual base $\tilde{w}_{\varepsilon_1,\dots,\varepsilon_{2n}}(\beta_1, \dots, \beta_{2n})$ is constructed as in the SG model. Lemma 5 of Sec. 3 remains valid.

The base $w_{\varepsilon_1,\dots,\varepsilon_{2n}}(\beta_1, \dots, \beta_{2n})$ possesses some specific properties associated with the SU(2)-invariance of the model. The algebra SU(2) acts in the space $h_1 \otimes \dots \otimes h_{2n}$ whose generators are $\Sigma^a = \Sigma_i \sigma_i^a$. The following Lemma describes the properties of $w_{\varepsilon_1,\dots,\varepsilon_{2n}}(\beta_1, \dots, \beta_{2n})$ with respect to this action.

Lemma 1. The covectors $w_{\varepsilon_1,\dots\varepsilon_{2n}}(\beta_1, \dots, \beta_{2n})$ are transformed under the action of the algebra SU(2) in the following way:

$$w_{\varepsilon_1,\dots,\varepsilon_{2n}}(\beta_1, \dots, \beta_{2n})\Sigma^3 = \left(\sum_{j=1}^{2n}(2\varepsilon_j - 3)\right) w_{\varepsilon_1,\dots,\beta_{2n}}(\beta_1, \dots, \beta_{2n}) , \tag{112}$$

$$\begin{aligned} &w_{\varepsilon_1,\dots,\varepsilon_{2n}}(\beta_1, \dots, \beta_{2n})\Sigma^+ \\ &= \sum_{p:\varepsilon_p=2} w_{\varepsilon_1,\dots,\varepsilon_{p-1},1,\varepsilon_{p+1},\dots,\varepsilon_{2n}}(\beta_1, \dots, \beta_{2n}) \prod_{\substack{q:\varepsilon_q=2\\ q\neq p}} \frac{\beta_p - \beta_q + \pi i}{\beta_p - \beta_1} , \end{aligned} \tag{113}$$

$$w_{\varepsilon_1,\dots,\varepsilon_{2n}}(\beta_1,\dots,\beta_{2n})\Sigma^- = \sum_{p:\varepsilon_p=1} w_{\varepsilon_1,\dots,\varepsilon_{p-1},2,\varepsilon_{p+1},\dots\varepsilon_{2n}}(\beta_1,\dots,\beta_{2n}) \prod_{\substack{q:\varepsilon_q=1\\ q\neq p}} \frac{\beta_p-\beta_q-\pi i}{\beta_p-\beta_q} . \tag{114}$$

Proof. Eq. (112) is evident because $B(\sigma)$ is a lowering operator:

$$B(\sigma)\Sigma^3 = (\Sigma^3+1)\,B(\sigma) .$$

Let us prove Eq. (113). The covector $w_{\varepsilon_1,\dots,\varepsilon_{2n}}(\beta_1,\dots,\beta_{2n})\Sigma^+$ has simple poles at the points $\beta_j=\beta_i+\pi i, j>i, \varepsilon_j=2, \varepsilon_i=1$, with residues

$$\begin{aligned}\operatorname{res} w_{\varepsilon_1,\dots,\varepsilon_{2n}}(\beta_1,\dots,\varepsilon_{2n})\Sigma^+ &= (e_{i,1}\otimes e_{j,2}-e_{i,2}\otimes e_{j,1})\\ &\otimes w_{\varepsilon_1,\dots,\hat{\varepsilon}_i,\dots,\hat{\varepsilon}_j,\dots,\varepsilon_{2n}}(\beta_1,\dots,\hat{\beta}_i\dots,\hat{\beta}_j,\dots,\beta_{2n})\sum_{k\neq i,j}\sigma_k^+\\ &\times\prod_{k>j}\frac{\beta_k-\beta_i}{\beta_k-\beta_i-\pi i}\hat{S}_{j-1,i}(\beta_{j-1}-\beta_i)\dots\hat{S}_{i+1,i}(\beta_{i+1}-\beta_i) ,\end{aligned} \tag{115}$$

where we have rearranged Σ^+ and $\hat{S}_{k,i}$ and made use of the SU(2)-invariance of the S-matrix. The covector $w_{\varepsilon_1,\dots,\varepsilon_{2n}}(\beta_1,\dots,\beta_{2n})\Sigma^+$ can be reconstructed by means of Eqs. (115). The RHS of (113) has the same singularities (the poles at the points $\beta_p=\beta_q$ are obviously fictious). The residue of the RHS of (113) at the point $\beta_j=\beta_i+\pi i, \varepsilon_j=2, \varepsilon_i=1$, is given by

$$\begin{aligned}&\operatorname{res}\left(\sum_{p:\varepsilon_p=2} w_{\varepsilon_1,\dots,\varepsilon_{p-1},1,\varepsilon_{p+1},\dots,\varepsilon_{2n}}(\beta_1,\dots,\beta_{2n})\prod_{\substack{q:\varepsilon_q=2\\ q\neq p}}\left(\frac{\beta_p-\beta_q+\pi i}{\beta_p-\beta_q}\right)\right)\\ &=(e_{i,1}\otimes e_{j,2}-e_{j,1}\otimes e_{i,2})\otimes\sum_{\substack{p:\varepsilon_p=2\\ p\neq j}}\prod_{\substack{q:\varepsilon_q=2\\ q\neq j}}\frac{\beta_p-\beta_q+\pi i}{\beta_p-\beta_q}\\ &\times w_{\varepsilon_1,\dots,\hat{\varepsilon}_i,\dots,\varepsilon_{p-1},1,\varepsilon_p,\dots,\hat{\varepsilon}_j,\dots\varepsilon_{2n}}(\beta_1,\dots,\hat{\beta}_i\dots\hat{\beta}_j,\dots,\beta_{2n})\\ &\times\prod_{k>j}\frac{\beta_k-\beta_i}{\beta_k-\beta_i-\pi i}\hat{S}_{j-1,i}(\beta_{j-1}-\beta_i)\dots\hat{S}_{i+1,i}(\beta_{i+1}-\beta_i) .\end{aligned} \tag{116}$$

The RHS of (113) decreases for $\beta_j \to \infty$, hence it can be reconstructed by means of Eqs. (116). Comparing Eqs. (115) and Eqs. (116), we can see that they provide us with the possibility of an inductive proof of Eq. (113). The verification of Eq. (133) for $n = 1$ is an easy exercise. The identity (114) can be proven in a similar way.
Q.E.D.

Definition 1. Let $F(\lambda_1, \dots, \lambda_{n+c}|\mu_1, \dots, \mu_{n-c})$ be a function symmetric with respect to $\lambda_1, \dots, \lambda_{n+c}$ and $\mu_1, \dots, \mu_{n-1}$ separately, c being an integer $-n \le c \le n$. The operation associates with the function F a vector from $h_1^* \otimes \dots \otimes h_{2n}^*$ in the following way:

$$\begin{aligned} &\langle F\rangle_n(\beta_1, \dots, \beta_{2n}) \\ &= \sum_{\substack{\{1,\dots,2n\} \\ =\{i_1,\dots,i_{n+c}\}\cup\{j_1,\dots,j_{n-c}\}}} F(\beta_{i_1}, \dots, \beta_{i_{n+c}}|\beta_{j_n}, \dots, \beta_{j_{n-c}}) \\ &\times \prod_{p,q} \frac{1}{\beta_{i_p} - \beta_{j_q}} w_{\varepsilon_1,\dots,\varepsilon_{2n}}(\beta_1, \dots, \beta_{2n}) , \end{aligned} \tag{117}$$

where $\varepsilon_{i_p} = 1, \varepsilon_{j_q} = 2, p = 1, \dots, n+c, q = 1, \dots, n-c$.

This definition is quite similar to Definition 1 of Sec. 3, the only difference being that here we do not restrict ourselves to the consideration of the Q^3 neutral sector.

Definition 2. Consider an antisymmetric polynomial $P(\alpha_1, \dots, \alpha_{n-1})$. The integral transformation $\phi_{n,\sigma}(\sigma = 0, \pm 1)$ associates with this polynomial a function of $\beta_1, \dots, \beta_{2n}$ in the following way:

$$\begin{aligned} \phi_{n,\sigma}(P)(\beta_1, \dots, \beta_{2n}) = \frac{(2\pi i)^{-n+1}}{(n-1)!} \int_{-\infty}^{\infty} d\alpha_1 \dots \int_{-\infty}^{\infty} d\alpha_{n-1} \prod_{i,j} \varphi(\alpha_i - \beta_j) \\ \times \prod_{i<j} \mathrm{sh}(\alpha_i - \alpha_j) \exp\left(\sigma \sum \alpha_i\right) P(\alpha_1, \dots, \alpha_{n-1}) , \end{aligned}$$

where

$$\varphi(\alpha) = \frac{1}{2}\pi^{-\frac{3}{2}} \Gamma\left(\frac{1}{4} - \frac{\alpha}{2\pi i}\right) \Gamma\left(\frac{1}{4} + \frac{\alpha}{2\pi i}\right) .$$

In contrast with Definition 2 of Sec. 3, there is no necessity of regularization of integrals here because they are well-defined for an arbitrary degree of P.

Definition 3. Polynomials $\Delta_n^c(\alpha_1, \dots, \alpha_{n-1}|\lambda_1, \dots \lambda_{n+c}|\mu_1, \dots, \mu_{n-c})$, $c = 0, \pm 1$ are the determinants of $(n-1) \times (n-1)$ matrices A^c with the following matrix elements:

$$
\begin{aligned}
&A_{ij}^c = A_i^c(\alpha_j|\lambda_1,\dots,\lambda_{n+c}|\mu_1,\dots,\mu_{n-c})\,,\\
&A_i^c(\alpha|\lambda_1,\dots,\lambda_{n+c}|\mu_1,\dots,\mu_{n-c})\\
&\quad=\prod_{j=1}^{n-c}\left(\alpha-\lambda_j-\frac{\pi i}{2}\right)[Q_{i-c}(\alpha-\pi i|\mu_1-\pi i,\dots,\mu_{n-c}-\pi i)\\
&\quad-\delta_{c,-1}\sigma_i(\lambda_1,\dots,\lambda_{n+c})]+\prod_{j=1}^{n-c}\left(\alpha-\mu_j+\frac{\pi i}{2}\right)[Q_{i+c}(\alpha|\lambda_1,\dots,\lambda_{n+c})\\
&\quad-\delta_{c,1}\sigma_i(\mu_1-\pi i,\dots,\mu_{n-c}-\pi i)]
\end{aligned}
\tag{118}
$$

where

$$
\begin{aligned}
&Q_i(\alpha|\lambda_1,\dots,\lambda_k)\\
&\quad=\sum_{l=0}^{i}\left[\left(\alpha+\frac{\pi i}{2}\right)^l-\left(\alpha-\frac{\pi i}{2}\right)^l\right](-1)^{i-l}\sigma_{i-l}(\lambda_1,\dots,\lambda_k)\,.
\end{aligned}
\tag{119}
$$

This definition is similar to Definition 3 of Sec. 3. Let us formulate some Lemmas which describe the properties of Δ_n^c.

Lemma 2. The polynomials Δ_n^c satisfy the recurrent relation

$$
\begin{aligned}
&\Delta_n^c(\alpha_1,\dots,\alpha_{n-1}|\lambda_1,\dots,\lambda_{n+c}|\mu_1,\dots,\mu_{n-c})|_{\mu_{n-c}=\lambda_{n+c}+\pi i}\\
&\quad=\prod_{j=1}^{n-1}\left(\alpha_j-\lambda_{n+c}-\frac{\pi i}{2}\right)\sum_{k=0}^{n-1}(-1)^k\\
&\quad\times\left\{\prod_{p=1}^{n+c-1}\prod_{q=1}^{n-c-1}\left(\alpha_k-\lambda_p-\frac{\pi i}{2}\right)\left(\alpha_k-\mu_q-\frac{\pi i}{2}\right)\right.\\
&\quad\left.-\prod_{p=1}^{n+c-1}\prod_{q=1}^{n-c-1}\left(\alpha_k-\lambda_p+\frac{\pi i}{2}\right)\left(\alpha_k-\mu_q+\frac{\pi i}{2}\right)\right\}\\
&\quad\times\Delta_n^c(\alpha_1,\dots,\hat{\alpha}_k,\dots,\alpha_{n-1}|\lambda_1,\dots,\lambda_{n+c-1}|\mu,\dots,\mu_{n-c-1})\,.
\end{aligned}
\tag{120}
$$

Proof. The proof follows immediately from two simple identities (see the proof of similar lemma of Sec. 4):

$$Q_i(\alpha|\lambda_1,\ldots,\lambda_k) = Q_i(\alpha|\lambda_1,\ldots,\lambda_{k-1}) - \lambda_k Q_{i-1}(\lambda_1,\ldots,\lambda_{k-1}) \ , \tag{121}$$

$$Q_k(\alpha|\lambda_1,\ldots,\lambda_k) = \prod_{j=1}^{k}\left(\alpha-\lambda-\frac{\pi i}{2}\right) - \prod_{j=1}^{k}\left(\alpha-\lambda_j+\frac{\pi i}{2}\right) \ . \tag{122}$$

Q.E.D.

Remark. The polynomial $\Delta_n^c(\alpha_1,\ldots,\alpha_{n-1}|\lambda_1,\ldots,\lambda_{n+c}|\mu_1,\ldots,\mu_{n-c})$ has a total degree $n(n-1)+\frac{1}{2}(n-1)(n-2)$. Being antisymmetric with respect to $\alpha_1,\ldots,\alpha_{n-1}$ it should contain $\Pi_{i<j}(\alpha_i-\alpha_j)$ as multiplier. Lemma 2 gives the values of this polynomial at n^2-c^2 points, defining it uniquely.

Lemma 3. Consider the functions $\langle\Delta_n^0\rangle_n, \langle\Delta_n^1\rangle_n, \langle\Delta_n^{-1}\rangle_n, \langle\tilde{\Delta}_n^0\rangle_n$ whose range of values is $h_1^*\otimes\ldots\otimes h_{2n}^*$ ($\tilde{\Delta}^0$ is defined as $\tilde{\Delta}_n^0 = \Sigma(\mu_j-\lambda_j-\pi i)\Delta_n^0$). This statement means that the first of these functions belongs to a singlet subspace of $h_1^*\otimes\ldots\otimes h_{2n}^*$ while the others belong to triplet subspaces; more precisely

$$\begin{aligned}
&\langle\Delta_n^0\rangle_n\Sigma^3 = \langle\Delta_n^0\rangle_n\Sigma^+ = \langle\Delta_n^0\rangle_n\Sigma^- = 0 \ ;\\
&\langle\Delta_n^c\rangle_n\Sigma^3 = c\langle\Delta_n^c\rangle_n \ , \quad c=\pm 1 \ ; \quad \langle\tilde{\Delta}_n^0\rangle\Sigma^3 = 0 \ ;\\
&\langle\Delta_n^\pm\rangle_n\Sigma^\pm = 0 \ ; \quad \langle\Delta_n^\pm\rangle_n\Sigma^\mp = \langle\tilde{\Delta}_n^0\rangle_n \ ;\\
&\langle\tilde{\Delta}_n^0\rangle_n\Sigma^\pm = \langle\Delta_n^\pm\rangle_n \ .
\end{aligned}$$

Proof. The equation $\langle\Delta_n^0\rangle\Sigma^3 = 0$ is obvious. Let us prove that $\langle\Delta_n^0\rangle_n\Sigma^+ = 0$. Equations (113), (117) imply

$$\langle\Delta_n^0\rangle_n\Sigma^+ = \langle X_n^{(1)}\rangle_n \ ,$$

where

$$\begin{aligned}
X_n^{(1)}(\alpha_1,\ldots\alpha_{n-1}|\lambda,\ldots,\lambda_{n+1}|\mu,\ldots,\mu_{n-1}) &= \sum_{k=1}^{n+1}\prod_{l\neq k}\frac{1}{\lambda_k-\lambda_l}\\
\times\prod_{l=1}^{n-1}(\alpha_k-\mu_l+\pi i)\Delta_n^0(\alpha_1,\ldots,\alpha_{n-1}|\lambda_1,\\
\ldots,\hat{\lambda}_k,\ldots,\lambda_{n+1}|\mu_1,\ldots,\mu_{n-1},\lambda_k) \ .
\end{aligned} \tag{123}$$

The poles at the points $\lambda_k = \lambda_l$ in RHS of Eq. (123) are obviously fictitious, hence $X_n^{(1)}$ is a polynomial of degree $n(n-1)+\frac{(n-1)(n-2)}{2}-1$. It can be shown that $X_n^{(1)}$ satisfy the same recurrent relation as Δ_n^1, moreover $X_1^{(1)}=0$. Consequently $X_n^{(1)}=0$ for every n. Equation $\langle\Delta_n^0\rangle_n\Sigma^-=0$ can be proven in a similar way.

Equations $\langle\Delta_n^c\rangle_n\Sigma^3=c\langle\Delta_n^c\rangle_n$ are obvious. Let us prove any one of the remaining equations, for example, $\langle\tilde{\Delta}_n^0\rangle_n\Sigma^+=\langle\Delta_n^+\rangle_n$. From Eqs. (113), (117) it follows that

$$\langle\tilde{\Delta}_n^0\rangle_n\Sigma^+=\langle X_n^{(2)}\rangle_n\ ,$$

where

$$\begin{aligned}
&X_n^{(2)}(\alpha_1,\dots,\alpha_{n-1}|\lambda,\dots,\lambda_{n+1}|\mu_1,\dots,\mu_{n-1})\\
&=\sum_{k=1}^{n+1}\prod_{l\neq k}\frac{1}{\lambda_k-\lambda_l}\prod_l(\lambda_k-\mu_l+\pi i)\\
&\times\left[\left(\sum_{j=1}^{n+1}\lambda_j-\sum_{j=1}^{n-1}\mu_j+\pi in\right)-2\lambda_k\right]\\
&\times\Delta_n^0(\alpha_1,\dots,\alpha_{n-2}|\lambda_1,\dots,\hat{\lambda}_k,\dots,\lambda_{n+1}|\mu_1,\dots,\mu_{n-1},\lambda_k)\ .
\end{aligned}$$

It can be shown that $X_n^{(2)}$ satisfies the same recurrent relation as Δ_n^+ and its degree is equal to $\frac{(n-1)(n-2)}{2}+n(n-1)$. These observations together with $X_1^{(2)}=1$ imply $X_n^{(2)}=\Delta_n^+$ for every n. Proofs of the other equations are quite similar.

Q.E.D.

Let us present explicit formulae for the form factors:

$$\begin{aligned}
f_{\mu\nu}(\beta_1,\dots\beta_{2n})&=M^2d^n\prod_{i<j}\zeta(\beta_i-\beta_j)\left[\sum_{j=1}^{2n}(e^{\beta_j}-(-1)^\nu e^{-\beta_j})\right]\\
&\times\left[\phi_{n,1}(\langle\Delta_n^0\rangle_n)\exp\left(-\frac{1}{2}\sum\beta_j\right)-(-1)^\mu\phi_{n,-1}(\langle\Delta_n^0\rangle_n)\right.\\
&\left.\times\exp\left(\frac{1}{2}\sum\beta_j\right)\right]
\end{aligned}$$

$$f^3_\mu(\beta_1,\dots,\beta_{2n}) = Md^n\prod_{i<j}\zeta(\beta_i-\beta_j)$$
$$\times\left[\phi_{n,1}(\langle\tilde{\Delta}^0_n\rangle_n)\exp\left(-\frac{1}{2}\sum\beta_j\right)-(-1)^\mu\phi_{n,-1}(\langle\tilde{\Delta}^0_n\rangle_n)\right.$$
$$\left.\times\exp\left(\frac{1}{2}\sum\beta_j\right)\right]$$
$$f^a_\mu(\beta_1,\dots,\beta_{2n}) = Md^n\prod_{i<j}\zeta(\beta_i-\beta_j)\quad a=1,2$$
$$\times\left\{\left[\phi_{n,1}(\langle\Delta^+_n\rangle_n)\exp\left(-\frac{1}{2}\sum\beta_j\right)-(-1)^\mu\phi_{n,-1}(\langle\Delta^+_n\rangle_n)\right.\right.$$
$$\left.\times\exp\left(\frac{1}{2}\sum\beta_j\right)\right]+(-1)^a\left[\phi_{n,1}(\langle\Delta^-_n\rangle_n)\exp\left(-\frac{1}{2}\sum\beta_j\right)\right.$$
$$\left.\left.-(-1)^\mu\phi_{n,-1}(\langle\Delta^{-1}_n\rangle_n)\times\exp\left(\frac{1}{2}\sum\beta_j\right)\right]\right\}, \tag{124}$$

where

$$\zeta(\beta)=c\,\mathrm{sh}\frac{\beta}{2}\exp\left\{-\int_0^\infty\frac{\sin^2\frac{1}{2}(\beta+\pi i)k\exp(-\frac{\pi k}{2})}{k\,\mathrm{sh}\,\pi k\,\mathrm{ch}\,\frac{\pi k}{2}}dk\right\},$$
$$c=2\pi^{\frac{3}{4}}\Gamma^{-1}\left(\frac{1}{4}\right)\exp\left\{-\int_0^\infty\frac{\mathrm{sh}^2\frac{\pi k}{4}\exp\left(-\frac{\pi k}{2}\right)}{k\,\mathrm{sh}\,\pi k\,\mathrm{ch}\,\frac{\pi k}{2}}dk\right\},$$
$$d=(2\pi c)^{-1},,\quad \zeta(\beta)\zeta(\beta-\pi i)=\varphi^{-1}\left(\beta+\frac{\pi i}{2}\right).$$

Let us list the main properties of these form factors. The form factors $f_{\mu\nu}, f^a_\mu$ satisfy Eqs. (13), (14), (16). This fact is a corollary of the connection with SG model for $\xi\to\infty$. Form factors f^1_μ, f^2_μ also satisfy Eqs. (13), (14), (16) since $f^a_\mu = f^3_\mu\Sigma^a, a=1,2$ and the S-matrix is SU(2)-invariant. Lemma 3 implies that $f_{\mu\nu}(\beta_1,\dots,\beta_{2\mu})$ creates a singlet while $f^a_\mu(\beta_1,\dots,\beta_{2n})$ creates a triplet. From the connection with SG for $\xi\to\infty$ it follows that

$$f_{\mu\nu}=f_{\nu\mu},\quad f^a_0\left(\sum\mathrm{ch}\,\beta_j\right)=f^a_1\left(\sum\mathrm{sh}\,\beta_j\right).$$

Operators $P_\mu = \int T_{\mu 0}dx, Q^a = \int j^a_0 dx_1$ are the operators of energy-momentum and charges:

$$\begin{aligned}
&P_\mu Z^*_{\varepsilon_n}(\beta_n)\dots Z^*_{\varepsilon_1}(\beta_1)|ph\rangle \\
&\quad = M\sum_{j=1}^{n}(e^{\beta_j}+(-1)^\mu e^{-\beta_j})Z^*_{\varepsilon_n}(\beta_n)\dots Z^*_{\varepsilon_1}(\beta_1)|ph\rangle\ , \\
&Q^a Z^*_{\varepsilon_n}(\beta_n)\dots Z^*_{\varepsilon_1}(\beta_1)|ph\rangle \\
&\quad = \sum_{j=1}^{n}(\sigma^a)^{\varepsilon'_j}_{\varepsilon_j} Z^*_{\varepsilon_n}(\beta_n)\dots Z^*_{\varepsilon'_j}(\beta_j)\dots Z^*_{\varepsilon_1}(\beta_1)|ph\rangle\ .
\end{aligned}$$

Formulae for norms similar to those given in Sec. 4 can also be obtained, for example

$$\begin{aligned}
&(\langle\Delta^0_n\rangle_n, \langle\Delta^0_n\rangle_n) \\
&\quad = \sum_{\substack{\{1\dots 2n\} \\ =\{i_1,\dots,i_n\}\cup\{j_1,\dots,j_n\}}} (\Delta^0_n(\alpha_1,\dots,\alpha_{n-1}|\beta_{i_n},\dots,\beta_{in}|\beta_{j_1},\dots,\beta_{j_n}))^2 \\
&\quad \times \prod_{p,q}\frac{1}{(\beta_{i_p}-\beta_{j_q})(\beta_{i_p}-\beta_{j_q}+\pi i)}\ .
\end{aligned} \tag{125}$$

8
FORM FACTORS IN O(3)-NONLINEAR σ-MODEL

In this section we present the formulae for form factors in NLS. It will be convenient to realize the group O(3) such that it preserves the bilinear form

$$\xi_{-1}\eta_1 + \xi_1\eta_{-1} - \xi_0\eta_0 \ ,$$

the corresponding indices being denoted as $\omega = 0, \pm 1$. The connection with the usual realization of O(3) is established by similitude with the matrix

$$U = \frac{1}{\sqrt{2}}\begin{pmatrix} 1 & 0 & i \\ 0 & -\sqrt{2} & 0 \\ 1 & 0 & -i \end{pmatrix} \ .$$

Operators Z^*_w, Z^w are defined by

$$Z^*_w = U^\varepsilon_w Z^*_\varepsilon \ , \quad Z^w = U^w_\varepsilon Z^\varepsilon \ .$$

The S-matrix in this realization is given by

$$S^{w'_1,w'_2}_{w_1,w_2}(\beta) = \frac{1}{(\beta+\pi i)(\beta-2\pi i)}\Big\{(\beta-\pi i)\beta\delta^{w'_1}_{w_1}\delta^{w'_2}_{w_2}$$
$$-2\pi i(\beta-\pi i)\delta^{w'_2}_{w_1}\delta^{w'_1}_{w_2} + 2\pi i\beta(-1)^{w_1+w_2}\delta_{w_1,-w_2}\delta^{w'_1,-w'_2}\Big\} \ . \tag{126}$$

It is well-known that the NLS S-matrix can be obtained from the ITM one via the fusion procedure [50, 51]. To avoid confusion of terms, we denote in this section the ITM S-matrix by cursive letters $\mathcal{S}_{\varepsilon_1,\varepsilon_2}^{\varepsilon'_1\varepsilon'_2}$. The equation for fusion procedure reads

$$S_{w_1,w_2}^{w'_1,w'_2}(\beta) = \tilde{P}_{\varepsilon_1,\varepsilon_2}^{w'_1}\tilde{P}_{\varepsilon_3,\varepsilon_4}^{w'_2}\mathcal{S}_{\varepsilon''_1,\varepsilon''_4}^{\varepsilon_1,\varepsilon_4}(\beta-\pi i)\mathcal{S}_{\varepsilon''_2,\varepsilon'_4}^{\varepsilon_2,\varepsilon''_4}(\beta)$$
$$\times\, \mathcal{S}_{\varepsilon'_1,\varepsilon''_3}^{\varepsilon''_1,\varepsilon_3}(\beta)\mathcal{S}_{\varepsilon'_2,\varepsilon'_3}^{\varepsilon''_2,\varepsilon''_3}(\beta+\pi i)\tilde{\tilde{P}}_{w_1}^{\varepsilon'_1,\varepsilon'_2}\tilde{\tilde{P}}_{w_2}^{\varepsilon'_3,\varepsilon'_4}\,, \qquad (127)$$

where $\tilde{P} = EP, P$ being the projection operator onto the symmetric subspace of $\mathbb{C}^2 \otimes \mathbb{C}^2$, and E identifying this space with $\mathbb{C}^3, \tilde{\tilde{P}} = \tilde{P}^t$. The following consequence of the Yang-Baxter equation is of great importance for the fusion procedure:

$$\mathcal{S}_{\varepsilon''_2,\varepsilon''_3}^{\varepsilon_2,\varepsilon_3}(\beta+\pi i)\mathcal{S}_{\varepsilon''_1,\varepsilon_3}^{\varepsilon_1,\varepsilon''_3}(\beta)P_{\varepsilon'_1,\varepsilon'_2}^{\varepsilon''_1,\varepsilon''_2} = P_{\varepsilon''_1,\varepsilon''_2}^{\varepsilon_1,\varepsilon_2}\mathcal{S}_{\varepsilon'_1,\varepsilon''_3}^{\varepsilon''_1,\varepsilon_3}(\beta)\mathcal{S}_{\varepsilon'_2,\varepsilon'_3}^{\varepsilon''_2,\varepsilon''_3}(\beta+\pi i)\,. \qquad (128)$$

Consider the form factor of any operator in ITM, $f(\beta_1\ldots\beta_{2n})_{\varepsilon_1\ldots\varepsilon_{2n}}$. It satisfies the equations

$$f(\beta_1,\ldots,\beta_{2n})_{\varepsilon_1,\ldots,\varepsilon_{2n}}\mathcal{S}_{\varepsilon'_i,\varepsilon'_{i+1}}^{\varepsilon_1,\varepsilon_{i+1}}(\beta_i-\beta_{i+1})$$
$$= f(\beta_1,\ldots,\beta_{i+1},\beta_i,\ldots,\beta_{2n})_{\varepsilon_1,\ldots,\varepsilon'_{i+1},\varepsilon'_i,\ldots,\varepsilon_{2n}}\,,$$
$$f(\beta_1,\ldots,\beta_{2n}+2\pi)_{\varepsilon_1,\ldots,\varepsilon_{2n}} = (-1)^n f(\beta_{2n},\beta_1,\ldots)_{\varepsilon_{2n},\varepsilon_1,\ldots}\,.$$

Equation (127) means that the NLS particle can be interpreted as a triple part of a "bound state" of two ITM kinks whose rapidities are $\beta-\frac{\pi i}{2},\beta+\frac{\pi i}{2}$. Clearly the term "bound state" must be understood here very carefully because the difference in the kinks' rapidities corresponds to their annihilation. Consider the following object:

$$f\left(\beta_1-\frac{\pi i}{2}+i\delta,\beta_1+\frac{\pi i}{2}-i\delta,\beta_2-\frac{\pi i}{2}+i\delta,\beta_2+\frac{\pi i}{2}-i\delta,\ldots\right.$$
$$\left.\beta_n-\frac{\pi i}{2}+i\delta,\beta_n+\frac{\pi i}{2}-i\delta\right)_{\varepsilon_1,\varepsilon_2,\varepsilon_3,\varepsilon_4,\ldots,\varepsilon_{2n-1},\varepsilon_{2n}}\,. \qquad (129)$$

At this point the hope appears that the limit of (129) for $\delta\to 0$ is related to the NLS form factor. This limit is singular because ITM form factors have simple poles at the point $\beta_{2j} = \beta_{2j-1}+\frac{\pi i}{2}$. As will be clear from what follows, the expression (129) increases as δ^{-n+1} for $\delta\to 0$. Consider the function

$$
\begin{aligned}
&F(\beta_1, \beta_2, \dots, \beta_n)_{w_1, \dots, w_n} \\
&= \lim_{\delta \to 0} \delta^{n-1} f \left(\beta_1 - \frac{\pi i}{2} + i\delta \, , \beta_1 + \frac{\pi i}{2} - i\delta, \beta_2 - \frac{\pi i}{2} + i\delta, \beta_2 + \frac{\pi i}{2} \right. \\
&\left. -i\delta, \dots, \beta_n - \frac{\pi i}{2} + i\delta, \beta_n + \frac{\pi i}{2} - i\delta \right)_{\varepsilon_1, \varepsilon_2, \varepsilon_3, \varepsilon_4, \dots, \varepsilon_{2n-1}, \varepsilon_{2n}} \\
&\times \tilde{\tilde{P}}^{\varepsilon_1, \varepsilon_2}_{w_1} \tilde{\tilde{P}}^{\varepsilon_3, \varepsilon_4}_{w_2} \dots \tilde{\tilde{P}}^{\varepsilon_{2n-1}, \varepsilon_{2n}}_{w_n} \, .
\end{aligned}
\tag{130}
$$

Equations (127), (128) and $P^2 = P$ imply

$$
\begin{aligned}
&F(\dots, \beta_i, \beta_{i+1}, \dots)_{\dots, w_i, w_{i+1}, \dots} \mathcal{S}^{w_i, w_{i+1}}_{w'_i, w'_{i+1}}(\beta_i - \beta_{i+1}) \\
&= \lim_{\delta \to 0} \delta^{n-1} f \left(\dots, \beta_i - \frac{\pi i}{2} + i\delta, \beta_i + \frac{\pi i}{2} - i\delta, \beta_{i+1} - \frac{\pi i}{2} + i\delta \, , \right. \\
&\left. \beta_{i+1} + \frac{\pi i}{2} - i\delta, \dots \right)_{\dots, \varepsilon_{2i-1}, \varepsilon_{2i}, \varepsilon_{2i+1}, \varepsilon_{2i+2}, \dots} \tilde{\tilde{P}}^{\varepsilon_1, \varepsilon_2}_{w_1} \dots P^{\varepsilon_{2i-1}, \varepsilon_{2i}}_{\varepsilon'_{2i-1}, \varepsilon'_{2i}} \\
&\times P^{\varepsilon_{2i+1} \varepsilon_{2i+2}}_{\varepsilon'_{2i+1}, \varepsilon'_{2i+2}} \dots \tilde{\tilde{P}}^{\varepsilon_{2n-1}, \varepsilon_{2n}}_{w_n} \mathcal{S}^{\varepsilon'_{2i-1}, \varepsilon'_{2i+2}}_{\varepsilon'''_{2i-1}, \varepsilon'''_{2i+2}}(\beta_i - \beta_{i+1} - \pi i) \\
&\times \mathcal{S}^{\varepsilon'_{2i}, \varepsilon'''_{2i+2}}_{\varepsilon''_{2i}, \varepsilon'''_{2i+2}}(\beta_i - \beta_{i+1}) \mathcal{S}^{\varepsilon'''_{2i-1}, \varepsilon'_{2i+1}}_{\varepsilon''_{2i-1}, \varepsilon'''_{2i+1}}(\beta_i - \beta_{i+1}) \mathcal{S}^{\varepsilon'''_{2i}, \varepsilon'''_{2i+1}}_{\varepsilon''_{2i}, \varepsilon''_{2i+1}}(\beta_i - \beta_{i+1} + \pi i) \\
&\times \tilde{\tilde{P}}^{\varepsilon''_{2i}, \varepsilon''_{2i-1}}_{w'_i} \tilde{\tilde{P}}^{\varepsilon''_{2i+2}, \varepsilon''_{2i+1}}_{w'_{i+1}} \\
&= \lim_{\delta \to 0} \delta^{n-1} f \left(\dots, \beta_i - \frac{\pi i}{2} + i\delta, \beta_i + \frac{\pi i}{2} - i\delta, \right. \\
&\left. \beta_{i+1} - \frac{\pi i}{2} + i\delta, \beta_{i+1} + \frac{\pi i}{2} - i\delta, \dots \right)_{\dots, \varepsilon_{2i-1}, \varepsilon_{2i}, \varepsilon_{2i+1}, \varepsilon_{2i+2}, \dots} \\
&\times \mathcal{S}^{\varepsilon_{2i}, \varepsilon_{2i+1}}_{\varepsilon''_{2i}, \varepsilon''_{2i+1}}(\beta_i - \beta_{i+1} + \pi i) \mathcal{S}^{\varepsilon_{2i-1}, \varepsilon''_{2i+1}}_{\varepsilon''_{2i-1}, \varepsilon'_{2i+1}}(\beta_i - \beta_{i+1}) \\
&\times \mathcal{S}^{\varepsilon''_{2i}, \varepsilon_{2i+2}}_{\varepsilon'_{2i}, \varepsilon''_{2i+2}}(\beta_i - \beta_{i+1}) \mathcal{S}^{\varepsilon''_{2i-1}, \varepsilon''_{2i+2}}_{\varepsilon_{2i-1}, \varepsilon'_{2i+2}}(\beta_i - \beta_{i+1} - \pi i) \\
&\times \tilde{\tilde{P}}^{\varepsilon_1, \varepsilon_2}_{w_1} \dots \tilde{\tilde{P}}^{\varepsilon_{2i-1}, \varepsilon_{2i}}_{w'_i} \tilde{\tilde{P}}^{\varepsilon'_{2i+1}, \varepsilon'_{2i+2}}_{w'_{i+1}} \dots \tilde{\tilde{P}}^{\varepsilon_{2n-1}, \varepsilon_{2n}}_{w_n} \\
&= F(\dots, \beta_{i+1}, \beta_i, \dots)_{w_1, \dots, w'_{i+1}, w'_i, \dots w_n} \, .
\end{aligned}
$$

It is evident from the very definition that

$$
F(\beta_1, \dots, \beta_n + 2\pi i)_{w_1, \dots, w_n} = F(\beta_n, \beta_1, \dots)_{w_n, w_1, \dots} \, .
$$

Thus F satisfies the two principal requirements imposed on the form factors.

Our further programme is to calculate explicitly the functions F and prove that they satisfy the third necessary requirement which is Eq. (16).

We are going to calculate F explicitly. Let us consider as an example the ITM form factor $f_+^3 = \frac{1}{2}(f_0^3 + f_1^3)$. According to Eqs. (124) this form factor can be written as

$$f_+^3(\beta_1', \ldots, \beta_{2n}') = d^n \prod_{i<j} \zeta(\beta_i' - \beta_j') \exp\left(\frac{1}{2}\sum \beta_j'\right) \langle \phi_{n,-1}(\tilde{\Delta}_n^0)\rangle_n \, .$$

To avoid confusion we indicate the ITM kinks' rapidities by primes. Our first purpose is to consider the limit of the function $\phi_{n,-1}(\tilde{\Delta}_n^0)$ when $\beta_1', \ldots, \beta_{2n}'$ are partitioned into pairs $\beta_{2i}' = \beta_i + \frac{\pi i}{2} - i\delta$, $\beta_{2i-1}' = \beta_i - \frac{\pi i}{2} + i\delta$ and $\delta \to 0$. Having in mind the following application of the operation $\langle\;\rangle_n$ we are interested in all possible partitions of $\beta_i', \ldots, \beta_{2n}'$ in the function

$$\tilde{\Delta}_n^0(\alpha_1, \ldots, \alpha_{n-1} | \beta_{i_1}', \ldots, \beta_{i_n}' | \beta_j', \ldots, \beta_{j_n}') \, .$$

The function $\phi_{n,-1}(\tilde{\Delta})$ can be presented as follows:

$$\begin{aligned}\phi_{n,-1}(\tilde{\Delta}_n^0(\alpha_1, \ldots, \alpha_{n-1} | \beta_{i_1}', \ldots, \beta_{i_n}' | \beta_{j_1}', \ldots, \beta_{j_n}'))\\ = \sum_{p=1}^{n} (\beta_{j_p}' - \beta_{i_p}' - \pi i) \det \|\mathfrak{A}_{ij}\| \, ,\end{aligned}$$

where

$$\begin{aligned}\mathfrak{A}_{ij} = \frac{1}{2\pi i}\int_{-\infty}^{\infty} \prod_p \varphi(\alpha - \beta_p')\\ \times \exp((n-2j-1)\alpha) A_i(\alpha | \beta_{i_1}', \ldots, \beta_{i_n}' | \beta_{j_1}', \ldots, \beta_{j_n}') d\alpha \, .\end{aligned}$$

The function $\varphi(\alpha)$ has simple poles at the points $\alpha = \pm\frac{\pi i}{2}$; that is why when $\beta_{2i}' \to \beta_i + \frac{\pi i}{2}, \beta_{2i-1}' \to \beta_i - \frac{\pi i}{2}$ the singularities of the integrand tend to the contour of integration. Extract these singularities explicitly:

$$\varphi\left(\alpha - \frac{\pi i}{2} + i\delta\right)\varphi\left(\alpha + \frac{\pi i}{2} - i\delta\right) = \frac{1}{\alpha^2 + \delta^2}\tilde{\varphi}_\delta(\alpha) \, , \qquad \tilde{\varphi}_\delta(\alpha) \to \tilde{\varphi}_0(\alpha) \quad \delta \to 0$$

where $\tilde{\varphi}_0(x)$ is the function

$$\varphi_0(\alpha) = \frac{1}{4\pi^3}\Gamma\left(\frac{1}{2}-\frac{\alpha}{2\pi i}\right)\Gamma\left(-\frac{\alpha}{2\pi i}\right)\Gamma\left(\frac{\alpha}{2\pi i}\right)\Gamma\left(\frac{1}{2}+\frac{\alpha}{2\pi i}\right) = \frac{\alpha}{\operatorname{sh}\alpha}\,,$$

which has no singularities on real axis. Hence

$$\begin{aligned}\prod_{p=1}^{2n}\varphi(\alpha-\beta'_p) &= \prod_{j=1}^{n}\frac{1}{(\alpha-\beta_j)^2+\delta^2}\tilde{\varphi}_\delta(\alpha-\beta_j)\\ &= \prod_{j=1}^{n}\varphi_\delta(\alpha-\beta_j)\sum_{k=1}^{n}\frac{1}{(\alpha-\beta_k)^2+\delta^2}\prod_{j\neq k}\frac{1}{(\beta_k-\beta_j)(2\alpha-\beta_j-\beta_k)}\,.\end{aligned}$$

Thus

$$\begin{aligned}\mathfrak{A}_{ij} &= \sum_{k=1}^{n}\frac{1}{2\pi i}\int_{-\infty}^{\infty}d\alpha\prod_{p=1}^{n}\varphi_\delta(\alpha-\beta_p)\frac{1}{(\alpha-\beta_k)^2+\delta^2}\exp(n-2j-1)\alpha\\ &\times\prod_{j\neq k}\frac{1}{(\beta_k-\beta_j)(2\alpha-\beta_k-\beta_j)}A_i(\alpha|\beta'_{i_1},\ldots,\beta'_{i_n}|\beta'_{j_1},\ldots,\beta'_{j_n})\,.\end{aligned} \tag{131}$$

The poles at the points $\alpha=\frac{1}{2}(\beta_k+\beta_j)$ are made artificially; as usual in such cases it makes no difference how to understand them. We can suppose, for example, that all β_k are slightly moved to the upper half-plane. Integrals in (131) behave as δ^{-1} for $\delta\to 0$. Using the formula

$$\lim_{\delta\to 0}\frac{\delta}{\alpha^2+\delta^2} = \pi\delta(\alpha)\,,$$

one obtains

$$\begin{aligned}\mathfrak{A}_{ij}\xrightarrow[\delta\to 0]{}\delta^{-1}\sum_{k=1}^{n}\prod_{p\neq k}\frac{1}{\operatorname{sh}(\beta_k-\beta_p)(\beta_k-\beta_p)}\exp(n-2j-1)\beta_k\\ \times A_i(\beta_k|\beta'_{i_1},\ldots,\beta'_{i_n}|\beta'_{j_1},\ldots,\beta'_{j_n})\,.\end{aligned}$$

Hence

$$\det\|\mathfrak{A}_{ij}\| = \delta^{-n+1}\det\|(\mathfrak{A}^{(1)}\mathfrak{A}^{(2)})_{ij}\|\,, \tag{132}$$

where $\mathfrak{A}^{(2)}$ is a $n\times(n-1)$ matrix with the following matrix elements:

$$\mathfrak{A}^{(2)}_{ij} = A_i(\beta_j|\beta'_{i_1},\ldots,\beta'_{i_n}|\beta'_{j_1},\ldots,\beta'_{j_n})\,.$$

$\mathfrak{A}^{(1)}_{ij}$ is a $(n-1)\times n$ matrix with the following matrix elements:

$$\mathfrak{A}^{(1)}_{ij} = \exp((n-2j-1)\beta_j)\prod_{p\neq i}\frac{1}{(\beta_i-\beta_p)\mathrm{sh}\,(\beta_i-\beta_p)}\ .$$

Recall that the set $\{\beta'_1,\beta'_2,\ldots,\beta'_{2n}\}$ coincides with the set $\{\beta_1-\frac{\pi i}{2},\beta_1+\frac{\pi i}{2},\ldots,\beta_n+\frac{\pi i}{2}\}$. Due to a well-known theorem a determinant of the kind (132) can be expanded into the sum

$$\det\|\mathfrak{A}^{(1)}\mathfrak{A}^{(2)}\| = \sum_{k=1}^{n}\det\|\mathfrak{A}^{(1)}_k\|\det\|\mathfrak{A}^{(2)}_k\|\ ,$$

where $\mathfrak{A}^{(1)}_k$ is a minor of $\mathfrak{A}^{(1)}$ with the k-th column omitted and $\mathfrak{A}^{(2)}_k$ is a minor of $\mathfrak{A}^{(2)}$ with the k-th row omitted. The determinant of $\mathfrak{A}^{(1)}_k$ can be evaluated explicitly:

$$\det\|\mathfrak{A}^{(1)}\mathfrak{A}^{(2)}\| = \frac{1}{\prod\limits_{i<j}\mathrm{sh}(\beta_i-\beta_j)}\exp\left(-\frac{1}{2}\sum\beta'_i\right)\frac{1}{\prod\limits_{i<j}(\beta_i-\beta_j)}$$
$$\times\left\{\sum_{k=1}^{n}\frac{1}{\prod\limits_{\substack{i<j\\ j\neq k}}(\beta_i-\beta_j)}\det\|\mathfrak{A}^{(2)}_k\|e^{\beta_k}\right\}\ .$$

Notice now that if we consider the form factor $f^3_- = \frac{1}{2}(f^3_0-f^3_1)$ the transformation $\phi_{n,-1}$ would change to $\phi_{n,1}$ and for $\phi_{n,1}(\tilde{\Delta}^0_n)$ we would undoubtedly obtain

$$\phi_{n,1}(\tilde{\Delta}^0_n)\to\delta^{-n+1}\left(\sum(\beta'_{j_l}-\beta'_{i_l}-\pi i)\right)\frac{1}{\prod\limits_{i<j}\mathrm{sh}(\beta_i-\beta_j)}\exp\left(\frac{1}{2}\sum\beta_j\right)$$
$$\times\frac{1}{\prod\limits_{i<j}(\beta_i-\beta_j)}\left\{\sum_{k=1}^{n}\frac{1}{\prod\limits_{\substack{i<j\\ i,j\neq k}}(\beta_i-\beta_j)}\det\|\mathfrak{A}^{(2)}_k\|e^{-\beta_k}\right\}\ .$$

According to Theorem 3 of Sec. 6

$$\exp\left(-\frac{1}{2}\sum\beta'_j\right)\left(\sum e^{\beta'_j}\right)\phi_{n,1}(\tilde{\Delta}^0_n)$$
$$=\exp\left(\frac{1}{2}\sum\beta'_j\right)\left(\sum e^{-\beta'_j}\right)\phi_{n,-1}(\tilde{\Delta}^0_n)\ ,$$

hence for $\delta \to 0$

$$\left\{ \sum_{k=1}^{n} \frac{1}{\prod\limits_{\substack{i<j \\ i,j\neq k}} (\beta_i - \beta_j)} \det \|\mathfrak{A}_k^{(2)}\| e^{\beta_k} \right\} \left(\sum e^{-\beta_j} \right)$$

$$= \left\{ \sum_{k=1}^{n} \frac{1}{\prod\limits_{\substack{i<j \\ i,j\neq k}} (\beta_i - \beta_j)} \det \|\mathfrak{A}_k^{(2)}\| e^{-\beta_k} \right\} \left(\sum e^{\beta_j} \right) . \tag{133}$$

Notice now that $\det \|\mathfrak{A}_k^{(2)}\|$ are polynomials, and the exponent $\exp(\beta_j)$ are linearly independent over the field of rational functions. Hence the following statement is valid.

Lemma 1. Consider a set $\beta_1, \dots, \beta_n$. Let $\{\beta'_{i_1}, \dots, \beta'_{i_n}\} \cup \{\beta'_{j_1}, \dots, \beta'_{j_n}\}$ be an arbitrary partition of the set $\{\beta_1 - \frac{\pi i}{2}, \beta_1 + \frac{\pi i}{2}, \beta_2 - \frac{\pi i}{2}, \beta_2 + \frac{\pi i}{2}, \dots, \beta_n + \frac{\pi i}{2}\}$. Consider now $(n-1) \times (n-1)$ matrices $\| \;_k^{(2)}\|_{ij}$ with the following matrix elements:

$$\|\mathfrak{A}_k^{(2)}\|_{ij} = A_i(\beta_j | \beta'_1, \dots, \beta'_{i_n} | \beta'_{j_1}, \dots, \beta'_{j_n}) , \quad j \neq k$$

where A_i are given by (118). The following identity holds for every k, l:

$$\det \|\mathfrak{A}_k^{(2)}\| \frac{1}{\prod\limits_{\substack{i,j\neq k \\ i<j}} (\beta_i - \beta_j)} = \det \|\mathfrak{A}_l^{(2)}\| \frac{1}{\prod\limits_{\substack{i,j\neq l \\ k<j}} (\beta_i - \beta_j)} \equiv X .$$

The proof of Lemma 1 follows immediately from (133). It is also possible to give an algebraic proof but it is very bulky.

Let us denote the set $\beta_1, \dots, \beta_n$ by B and the sets $\beta'_{i_1}, \dots \beta'_{i_n}; \beta'_{j_1}, \dots \beta'_{j_n}$ by $B'_1, B'_2, B'_1 \cup B'_2 \equiv B' = (B - \frac{\pi i}{2}) \cup (B + \frac{\pi i}{2}), (B + \alpha \equiv \{\beta_i + \alpha\})$. Fix a partition of B' into B'_1 and B'_2; it induces the partition of B into four subsets $B = B^+ \cup B^- \cup B^0 \cup \overline{B}^0$ according to the rule $\beta_j \in B^+$ if $\beta_j + \frac{\pi i}{2}, \beta_j - \frac{\pi i}{2} \in B'_2$; $\beta_j \in B^-$ if $\beta_j + \frac{\pi i}{2}, \beta_j - \frac{\pi i}{2} \in B'_1$; $\beta_j \in B^0$ if $\beta_j + \frac{\pi i}{2} \in B'_1$, $\beta_j - \frac{\pi i}{2} \in B'_2$; $\beta_j \in \overline{B^0}$ if $\beta_j + \frac{\pi i}{2} \in B'_2, \beta_j - \frac{\pi i}{2} \in B'_1$. The function X introduced in Lemma 1 is a function of $B^-, B^0, B^+, \overline{B}^0$: $X(B^- | B^0 | \overline{B}^0 | B^+)$. Evidently $n(B^+) = n(B^-)$. The following statement holds.

Lemma 2. The function $X(B^- | B^0 | \overline{B}^0 | B^+)$ vanishes if $\overline{B^0} \neq \emptyset$. The function $X(B^- | B^0 | \emptyset | B^+)$ contains the multiplier (except for the case $B^0 =$

$\emptyset, n(B^+) = n(B^-) = 1)$:

$$\prod_{\beta_i \in B^+} \prod_{\beta_j \in B^-} (\beta_i - \beta_j)^2 (\beta_i - \beta_j - \pi i)^2 \prod_{\beta_i \in B^+} \prod_{\beta_j \in B^0} (\beta_i - \beta_j - \pi i)$$
$$\times \prod_{\beta_i \in B^-} \prod_{\beta_j \in B^0} (\beta_i - \beta_j + \pi i) \,.$$

Proof. According to (118)

$$A_i(\alpha|B_1'|B_2') = \prod_{\beta' \in B_1'} \left(\alpha - \beta' - \frac{\pi i}{2}\right) Q_i(\alpha - \pi i|B_2' - \pi i)$$
$$+ \prod_{\beta' \in B_2'} \left(\alpha - \beta' + \frac{\pi i}{2}\right) Q_i(\alpha|B_1') \,. \tag{134}$$

According to Lemma 1, one of n realizations can be chosen for $X(B^-|B^0|\overline{B}^0|B^+)$:

$$X(B^-|B^0|\overline{B}^0|B^+) = \det \|\mathfrak{A}_k^{(2)}\| \frac{1}{\prod\limits_{\substack{i<j \\ i,j \neq k}} (\beta_i - \beta_j)} \,.$$

Suppose $\overline{B}^0 \neq \emptyset$, i.e. $\overline{B}^0$ contains at least one element β_j. Take for $X(B^-|B^-|\overline{B}^0|B^+)$ a realization with $k \neq j$ then $\mathfrak{A}_k^{(2)}$ contains the column $A(\beta_j|B_1'|B_2')$. As it follows from (134) this column vanishes.

Now let $\beta_i \in B^- \beta_j \in B^+$. Consider the realization of X with $k \neq i, j$. For the column corresponding to β_i one has

$$A_l(\beta_i|B_1'|B_2') = \prod(\beta_i - \beta^-)(\beta_i - \beta^- - \pi i)(\beta_i - \beta^0 - \pi i)$$
$$\times Q_l\left(\beta_i - \pi i|B^+ - \frac{\pi i}{2}, B^+ - \frac{3\pi i}{2}, B^0 - \frac{3\pi i}{2}\right)$$
$$+ \prod(\beta_i - \beta^+)(\beta_i - \beta^+ + \pi i)(\beta_i - \beta^0 + \pi i) \,,$$
$$\times Q_l\left(\beta_i|B^- + \frac{\pi i}{2}, B^- - \frac{\pi i}{2}, B^0 + \frac{\pi i}{2}\right) ,$$

where $\Pi(\beta_i - \beta^-) \equiv \prod\limits_{\beta_k \in B^-} (\beta_i - \beta_k)$, etc. Clearly, the first term disappears (there is β_i among β^-), and hence $\Pi(\beta_i - \beta^+)(\beta_i - \beta^+ + \pi i)(\beta_i - \beta^0 + \pi i)$ can

be extracted from $A_l(\beta_i|B_1'|B_2')$. Similarly $\Pi(\beta_j-\beta^-)(\beta_j-\beta^--\pi i)(\beta_j-\beta^0-\pi i)$ can be extracted from the columns corresponding to β_j. Thus X contains as a multiplier $(\beta_i-\beta_j)^2(\beta_i-\beta_j-\pi i)^2$. The cases $\beta_i\in B^-,\beta_j\in B^0$ and $\beta_i\in B^+,\beta_j\in B^0$ can be treated in a similar ways.
Q.E.D.

Lemma 2 implies the possibility of introducing the polynomial (except for the case $n(B^0)=0, n(B^+)=n(B^-)=1$)

$$\phi(B^-|B^0|B^+)=X(B^-|B^0|\emptyset|B^+)$$
$$\times\prod\frac{1}{(\beta^+-\beta^--\pi i)^2(\beta^+-\beta^-)^2(\beta^+-\beta^0-\pi i)(\beta^--\beta^0-\pi i)}\,.$$

For $n(B^+)=n(B^-)=1, n(B^0)=0$, the function ϕ is equal to

$$\phi(\beta^-|\emptyset|\beta^+)=\frac{\pi i}{(\beta^+-\beta^--\pi i)}\,.$$

For the constructive definition of ϕ see Definition 2 to be given later.

Let us return to formulae (130). We have to understand what the operation $\langle\ \rangle_n$ turns into after the partition of B' into $B-\frac{\pi i}{2}, B+\frac{\pi i}{2}$ and the projection onto spaces symmetric with respect to pairs of particles. The operator $B(\sigma)$ defined by

$$\begin{pmatrix}A(\sigma) & B(\sigma)\\ C(\sigma) & D(\sigma)\end{pmatrix}=\hat{\mathcal{S}}_{0,1}(\sigma-\beta_1')\dots\hat{\mathcal{S}}_{0,2n}(\sigma-\beta_{2n}') \tag{135}$$

is of fundamental importance for the definition of the operation $\langle\ \rangle_n$. We have to calculate $B(\sigma|\beta_1-\frac{\pi i}{2},\beta_1+\frac{\pi i}{2},\dots\beta_n-\frac{\pi i}{2},\beta_n+\frac{\pi i}{2})\times\approx{P}_{1,2}\dots\approx{P}_{2n-1,2n}$, where $\approx{P}_{2i-1,2i}$ is the operator $\approx{P}$ acting from $H_i\simeq\mathbb{C}^3$ to $h_{2i-1}\otimes h_{2i}\simeq\mathbb{C}^2\otimes\mathbb{C}^2$. Multiply (135) from the left by $\Pi\approx{P}_{2i-1,2i}$ and use the usual trick: present $\approx{P}_{2i-1,2i}$ as $P_{2i,2i-1}\approx{P}_{2i,2i-1}$ and move all $P_{2i-1,2i}$ to the left using the equation (see (128))

$$\hat{\mathcal{S}}_{0,2i-1}\left(\sigma-\beta_i+\frac{\pi i}{2}\right)\hat{\mathcal{S}}_{0,2i}\left(\sigma-\beta_i-\frac{\pi i}{2}\right)P_{2i-1,2i}$$
$$=P_{2i,2i-1}\hat{\mathcal{S}}_{0,2i}\left(\sigma-\beta_i-\frac{\pi i}{2}\right)\hat{\mathcal{S}}_{0,2i-1}\left(\sigma-\beta_i+\frac{\pi i}{2}\right)$$
$$=\approx{P}_{2i,2i-1}\tilde{P}_{2i,2i-1}\hat{\mathcal{S}}_{0,2i}\left(\sigma-\beta_i-\frac{\pi i}{2}\right)\hat{\mathcal{S}}_{0,2i-1}\left(\sigma-\beta_i+\frac{\pi i}{2}\right)\,.$$

Hence

$$\hat{S}_{0,1}\left(\sigma-\beta_1+\frac{\pi i}{2}\right)\hat{S}_{0,2}\left(\sigma-\beta_1-\frac{\pi i}{2}\right)\ldots$$
$$\hat{S}_{0,2n-1}\left(\sigma-\beta_n+\frac{\pi i}{2}\right)\hat{S}_{0,2n}\left(\sigma-\beta_n-\frac{\pi i}{2}\right)$$
$$\times\prod\tilde{\tilde{P}}_{2i-1,2i}=\prod\tilde{\tilde{P}}_{2i-1,2i}\hat{\hat{S}}_{0,1}\left(\sigma-\beta_1-\frac{\pi i}{2}\right)\ldots$$
$$\hat{\hat{S}}_{0,n}\left(\sigma-\beta_n-\frac{\pi i}{2}\right),$$

where $\hat{\hat{S}}_{0,i}$ is an operator acting in $h_0\otimes H_i\simeq \mathrm{C}^2\otimes\mathrm{C}^3$ in the following way:

$$\hat{\hat{S}}_{0,i}(\sigma)=\frac{1}{\sigma-2\pi i}(\sigma-\pi i+\pi i\sigma^a S_i^a)\ ,$$

where σ^a are the Pauli matrices acting in h_0, S_i^a are the spin-1 generators of SU(2) acting in H_i:

$$S^3\sim\begin{pmatrix}1&0&0\\0&0&0\\0&0&-1\end{pmatrix},\quad S^1\sim\frac{1}{\sqrt{2}}\begin{pmatrix}0&1&0\\1&0&1\\0&1&0\end{pmatrix},\quad S^2\sim\frac{1}{\sqrt{2}}\begin{pmatrix}0&i&0\\-i&0&i\\0&-i&0\end{pmatrix}.$$

Define the operators

$$\begin{pmatrix}\mathcal{A}(\sigma)&\mathcal{B}(\sigma)\\\mathcal{C}(\sigma)&\mathcal{D}(\sigma)\end{pmatrix}\equiv\tau(\sigma)=\hat{\hat{S}}_{0,1}(\sigma-\beta_1)\ldots\hat{\hat{S}}_{0,n}(\sigma-\beta_n)\ .$$

The operators satisfy the same commutation relations as $A(\sigma), B(\sigma), C(\sigma), D(\sigma)$, which follows from

$$\hat{\mathcal{S}}_{a,b}(\sigma_1-\sigma_2)\hat{\hat{\mathcal{S}}}_{a,i}(\sigma_1)\hat{\hat{\mathcal{S}}}_{b,i}(\sigma_2)=\hat{\hat{\mathcal{S}}}_{b,i}(\sigma_2)\hat{\hat{\mathcal{S}}}_{a,i}(\sigma_1)\hat{\mathcal{S}}_{a,b}(\sigma_1-\sigma_2)\ .$$

The following relation holds:

$$S_{i,j}(\beta_i-\beta_j)\hat{\hat{\mathcal{S}}}_{a,j}(\sigma-\beta_j)\hat{\hat{\mathcal{S}}}_{a,i}(\sigma-\beta_i)=\hat{\hat{\mathcal{S}}}_{a,i}(\sigma-\beta_i)\hat{\hat{\mathcal{S}}}_{a,j}(\sigma-\beta_j)S_{i,j}(\beta_i-\beta_j)\ ,\tag{136}$$

where S is the NLS S-matrix (126). It can be concluded from (136) that

$$\mathcal{B}(\sigma|\beta_1,\ldots,\beta_i,\beta_{i+1},\ldots,\beta_n)S_{i,i+1}(\beta_i-\beta_{i+1})$$
$$=S_{i,i+1}(\beta_i-\beta_{i+1})\mathcal{B}(\sigma|\beta_1,\ldots,\beta_{i+1},\beta_i,\ldots,\beta_n)\ .$$

As before, we suppose H_i and H_{i+1} to change places if β_i and β_{i+1} change places.

Consider some function $F(B_1'|B_2')$ which is invariant under independent permutations of the rapidities comprising B_1' and B_2'. The operation $\langle\ \rangle_n$ was defined as follows:

$$\langle F\rangle_n(\beta_1',\ldots,\beta_{2n}') = \sum_{B'=B_1'\cup B_2'} F(B_1'|B_2') \prod_{\substack{\beta_i'\in B_1' \\ \beta_j'\in B_2'}} \frac{1}{\beta_i'-\beta_j'} \times \langle 0| \prod_{\beta_j'\in B_2'} B(\beta_j') \,. \tag{137}$$

Partition rapidities into pairs $\beta_{2j}' = \beta_j + \frac{\pi i}{2}, \beta_{2j-1}' = \beta_j - \frac{\pi i}{2}$. This partition induces, as was explained above, the partition of B into $B^-, B^0, \overline{B}^0, B^+$, so we can define a function $F(B^-|B^0|\overline{B}^0|B^+)$ through $F(B_1'|B_2')$. After B' is partitioned into pairs and projected onto symmetric subspaces, Eq. (137) transforms to

$$\begin{aligned} &\sum_{B=B^-\cup B^0\cup\overline{B}^0\cup B^+} F(B^-|B^0|\overline{B}^0|B^+) \\ &\times \prod \frac{1}{(\beta^+-\beta^--\pi i)(\beta^+-\beta^-+\pi i)(\beta^+-\beta^-)^2} \\ &\times \prod \frac{1}{(\beta^+-\beta^0-\pi i)(\beta^+-\beta^0)(\beta^+-\overline{\beta}^0)(\beta^+-\overline{\beta}^0+\pi i)} \\ &\times \frac{1}{(\beta^--\beta^0-\pi i)(\beta^--\beta^0)} \\ &\times \prod \frac{1}{(\beta^--\beta^0)(\beta^--\overline{\beta}^0-\pi i)(\beta^0-\beta^0-\pi i)} \\ &\times \frac{1}{(\overline{\beta}^0-\overline{\beta}^0-\pi i)(\beta^0-\overline{\beta}^0)^2} \\ &\times \langle\langle 0|| \prod \mathcal{B}(\beta^0)\mathcal{B}(\beta^+)\mathcal{B}(\beta^++\pi i)\mathcal{B}(\overline{\beta}^0+\pi i) \,, \end{aligned}$$

where $\langle\langle 0|| = \Pi \otimes e_{i,-1}$, and $e_{i,w}(w=0,\pm 1)$ generate the base in H_i. The fact may give rise to misunderstanding that no singularities appear. As a matter of fact, the operation $\langle\ \rangle\rangle_n$ for ITM does have singularities at the points $\beta_{2j} = \beta_{2j-1} + \pi i$. What has become of them? This contradiction has a simple resolution: the residues of $\langle\ \rangle_n$ at its poles belong to skew-symmetric subspaces, but we have projected it onto the tensor product of symmetric subspaces.

Now we have to substitute the polynomial $\phi(B^-|B^0|B^+)$ for F:

$$\phi_{n-1}(\langle\tilde{\Delta}_n^0\rangle)_n \overset{\delta\to 0}{\longrightarrow} \delta^{-n+1}\Bigg[\sum_{B=B^-\cup B^0\cup B^+} \phi(B^-|B^0|B^+)$$

$$\times\left\{\sum(\beta^+-\beta^-)-\pi i(n(B^+)+n(B^0))\right\}$$

$$\times\prod\frac{(\beta^+-\beta^--\pi i)}{(\beta^+-\beta^-+\pi i)(\beta^+-\beta^0)(\beta^--\beta^0)(\beta^0-\beta^0-\pi i)(\beta^+-\beta^-)}$$

$$\times\langle\langle 0||\prod\mathcal{B}(\beta^0)\mathcal{B}(\beta^+)\mathcal{B}(\beta^++\pi i)\Bigg]$$

$$\times\left(\sum e^{\beta_j}\right)\exp\left(-\sum\beta_j\right)\prod_{i<j}\frac{1}{\mathrm{sh}(\beta_i-\beta_j)(\beta_i-\beta_j)},$$

the terms with $\overline{B}^0\neq\emptyset$ disappearing in agreement with Lemma 2.

Now that we have some idea about the degeneration of ITM form factors to the NLS ones, let us pass on to formal definitions.

First, let us define a base in $H_1^*\otimes\ldots\otimes H_n^*$ consisting of covectors

$$w_{w_1,\ldots,w_n}(\beta_1,\ldots,\beta_n)=\langle\langle 0||\prod_{j:w_j=0}\mathcal{B}(\beta_j)\prod_{j:w_j=1}\mathcal{B}(\beta_j)\mathcal{B}(\beta_j+\pi i)\,,$$

$w_j=0,\pm 1$. This base is an eigenbasis for the operator $\mathcal{A}(\sigma)$ with the eigenvalues

$$w_{w_1,\ldots,w_n}(\beta_1,\ldots,\beta_n)\mathcal{A}(\sigma)=\prod_{j:w_j=-1}\frac{\sigma-\beta_j}{\sigma-\beta_j-2\pi i}$$

$$\times\prod_{j:w_j=0}\frac{\sigma-\beta_i-\pi i}{\sigma-\beta_j-2\pi i}w_{w_1,\ldots,w_n}(\beta_1,\ldots,\beta_n)\,. \tag{138}$$

The base $w_{w_1,\ldots,w_n}$ is connected with the natural base of tensor product $e_{w_1,\ldots,w_n}$ via a triangular transformation which means that in the expansion

$$w_{w_1,\ldots,w_n}(\beta_1,\ldots,\beta_n)=\sum_{\Sigma w_i=\Sigma w_1'}C_{w_1,\ldots,w_n}^{w_1',\ldots,w_n'}(\beta_1,\ldots,\beta_n)e_{w_1',\ldots,w_n'}$$

only those $C^{w'_1,\dots,w'_n}_{w_1,\dots,w_n}$ differ from zero for which $(w'_1,\dots,w'_n) < (w_1,\dots,w_n)$, the multiindices being ordered as ternary numbers (for example, $(0,-1) < (0,0) < (0,1) < (1-1) < (10) < (11)$). The covector $w_{w_1,\dots,w_n}(\beta_1,\dots,\beta_n)$ can possess simple poles at the points $\beta_j = \beta_i + 2\pi i, \beta_j = \beta_i + \pi i$; the poles at the points $\beta_j = \beta_i + \pi i$ exist only for $j > i, w_j = 1, w_i = -1$. All above statements can be proven similarly to the Lemmas of Sec. 3. A base dual to $w_{w_1,\dots,w_n}(\beta_1,\dots,\beta_n)$ can be defined:

$$\begin{aligned}&\tilde{w}_{w_1,\dots,w_n}(\beta_1,\dots,\beta_n)\\&= \langle\langle 0\| \prod_{j:w_j=0} \mathcal{B}'(\beta_j + 2\pi i) \prod_{j:w_j=1} \mathcal{B}(\beta_j + \pi i)\mathcal{B}'(\beta_j + 2\pi i)\ ,\end{aligned}$$

where $\mathcal{B}'(\beta_p + 2\pi i) = \lim_{\sigma\to\beta_p} \mathcal{B}(\sigma + 2\pi i)(\sigma - \beta_p)$. The following equation similar to (80) holds

$$(w_{w_1,\dots,w_n}(\beta_1,\dots,\beta_n), \tilde{w}_{w'_1,\dots,w'_m}(\beta_1,\dots,\beta_m)) = \delta_{w_1,w'_1}\cdots\delta_{w_m,w'_m}\ .$$

Definition 1. Consider a set of functions $F_l(\lambda_1,\dots\lambda_{l+c}|\mu_1,\dots,\mu_{n-2l+c}|\nu_1,\dots,\nu_l), c = 0, \pm 1, l = \min(0,-c),\dots,[\frac{n-c}{2}]$. Operation $\langle\langle\ \rangle\rangle_n$ associates with these functions the following vector from $H_1^*\otimes\dots\otimes H_n^*$:

$$\begin{aligned}\langle\langle\{F_l\}\rangle\rangle_n = &\sum_{B=B^-\cup B^0\cup B^+} F_{n(B^+)}(B^-|B^0|B^+)\\&\times\prod\frac{(\beta^+-\beta^--\pi i)}{(\beta^+-\beta^--\pi i)(\beta^+-\beta^0)(\beta^--\beta^0)(\beta^+-\beta^-)(\beta^0-\beta^0-\pi i)}\\&\times w_{w_1,\dots,w_n}(B)\ ,\end{aligned}\tag{139}$$

where $w_j = -1, 0, 1$ for $\beta_j \in B^-, B^0, B^+$ respectively.

Definition 2. Polynomials $\phi^c(B^-|B^0|B^+)\,(n(B^+) - n(B^-) = c)$ are defined as follows:

$$\phi^c(B^-|B^0|B^+) = \prod_{\substack{i<j\\ i,j\neq k}}\prod(\beta^+-\beta^-)f_k(B^-|B^0|B^+)\det M_k^c$$

$$\begin{aligned}&f_k(B^-|B^0|B^+)\\&= \begin{cases}1\ , \beta_k\in B^0\ ,\\ (\Pi(\beta_k-\beta^-)(\beta_k-\beta^--\pi i)(\beta_k-\beta^0-\pi i))^{-1}\ , & \beta_k\in B^+\ ,\\ (\Pi(\beta_k-\beta^+)(\beta_k-\beta^++\pi i)(\beta_k-\beta^0+\pi i))^{-1}\ , & \beta_k\in B^-\ ,\end{cases}\end{aligned}$$

where M_k^c is a $(n-1)\times(n-1)$ matrix obtained by omitting the k-th column from the $n\times(n-1)$ matrix M^c with the following matrix elements:

$$M_{ij}^c = Q_{i+c}\left(\beta_j|B^- + \frac{\pi i}{2}, B^- - \frac{\pi i}{2}, B^0 - \frac{\pi i}{2}\right) - \delta_{c,1}\sigma_i\left(B^+ - \frac{\pi i}{2}, B^+ - \frac{3\pi i}{2}, B^0 - \frac{3\pi i}{2}\right), \tag{140}$$

for $i = 1,\dots,n-1, \beta_j \in B^-$;

$$\begin{aligned}M_{ij}^c &= \prod(\beta_j - \beta^-)(\beta_j - \beta^- - \pi i)(\beta_j - \beta^0 - \pi i)\\ &\times\left\{Q_{i-c}\left(\beta_j - \pi i|B^+ - \frac{\pi i}{2}, B^+ - \frac{3\pi i}{2}, B^0 - \frac{3\pi i}{2}\right)\right.\\ &\left.-\delta_{c,-1}\sigma_i\left(B^- - \frac{\pi i}{2}, B^- + \frac{\pi i}{2}, B^0 + \frac{\pi i}{2}\right)\right\} + \prod(\beta_j - \beta^+)\\ &\times(\beta_j - \beta^+ + \pi i)(\beta_j - \beta^0 + \pi i)\left\{Q_{i+c}\left(\beta_j|B^- + \frac{\pi i}{2},\right.\right.\\ &\left.\left.B^- - \frac{\pi i}{2}, B^0 - \frac{\pi i}{2}\right) - \delta_{c,1}\sigma_i\left(B^+ - \frac{\pi i}{2}, B^+ - \frac{3\pi i}{2}, B^0 - \frac{3\pi i}{2}\right)\right\},\end{aligned}$$

for $i = 1,\dots,n-1, \beta_j \in B^0$;

$$M_{ij} = Q_{i-c}\left(\beta_j - \pi i|B^+ - \frac{\pi i}{2}, B^+ - \frac{3\pi i}{2}, B^0 - \frac{\pi i}{2}\right) - \delta_{c,-1}\sigma_i\left(B^- - \frac{\pi i}{2}, B^- + \frac{\pi i}{2}, B^0 + \frac{\pi i}{2}\right)$$

for $1 \le i \le n-1$, $\beta_j \in B^+$.

This definition does not depend on choice of k due to Lemma 1.

Remark. In the one case $n(B^-) = n(B^+) = 1, n(B^0) = 0$, the function ϕ^0 is not a polynomial but is equal to

$$\phi^0(\beta^-|\emptyset|\beta^+) = \frac{\pi i}{\beta^+ - \beta^- - \pi i}.$$

Let us discuss an important point. ITM form factors depend on even numbers of rapidities. NLS form factors obtained from the ITM ones depend on arbitrary numbers of rapidities since an NLS particle is created by two ITM kinks. In the space of states of NLS one can introduce the operator of charge conjugation, postulating its properties as follows:

$$C|ph\rangle = |ph\rangle \ ,$$
$$CZ^*_w(\beta)C^{-1} = -Z^*_w(\beta) \ .$$

Section 2 implies that even and odd parts of an operator are mutually independent which means that both parts are local operators. That is why the form factors of one local operator in ITM create the form factors of two operators in NLS, one of which is provided with even particle form factors, while the other with odd ones. Hence the form factors $f^a_\mu, f_{\mu\nu}$ create four sets of form factors in NLS:

$$f^a_\mu(\beta_1, \dots, \beta_n) = \lim_{\delta\to 0} \delta^{-n+1} f^a_\mu(\beta'_1, \dots \beta'_{2n})|_{\substack{\beta'_{2i}=\beta_i+\frac{\pi i}{2}-i\delta \\ \beta'_{2i-1}=\beta_i-\frac{\pi i}{2}+i\delta}}$$
$$\times \overset{\approx}{P}_{1,2}\dots\overset{\approx}{P}_{2n-1,2n} \prod_{i<j} \mathrm{th}^2(\beta_i-\beta_j) \ , \quad n\equiv 0 (\mathrm{mod}\ 2)$$
$$f^a(\beta_1, \dots, \beta_n) = \lim_{\delta\to 0} \delta^{-n+1} f^a_1(\beta'_1, \dots, \beta'_{2n})|_{\substack{\beta'_{2i}=\beta_i+\frac{\pi i}{2}-i\delta \\ \beta'_{2i-1}=\beta_i-\frac{\pi i}{2}+i\delta}}$$
$$\times \overset{\approx}{P}_{1,2}\dots\overset{\approx}{P}_{2n-1,2n} \left(\sum \mathrm{sh}\beta_j\right)^{-1} \prod_{i<j} \mathrm{th}^2\frac{1}{2}(\beta_i-\beta_j) \ , \quad n\equiv 1 (\mathrm{mod}\ 2) \ ,$$
$$f_{\mu\nu}(\beta_1, \dots, \beta_n) = \lim_{\delta\to 0} \delta^{-n+1} f_{\mu\nu}(\beta'_1, \dots, \beta'_{2n})|_{\substack{\beta'_{2i}=\beta_i+\frac{\pi i}{2}-i\delta \\ \beta'_{2i-1}=\beta_i-\frac{\pi i}{2}+i\delta}}$$
$$\times \overset{\approx}{P}_{1,2}\dots\overset{\approx}{P}_{2n-1,2n} \prod_{i<j} \mathrm{th}^2\frac{1}{2}(\beta_i-\beta_j) \ , \quad n\equiv 0 (\mathrm{mod}\ 2) \ ,$$
$$f(\beta_1, \dots, \beta_n) = \lim_{\delta\to 0} \delta^{-n+1} f_{00}(\beta'_1, \dots, \beta'_{2n})|_{\substack{\beta'_{2i}=\beta_i+\frac{\pi i}{2}-i\delta \\ \beta'_{2i-1}=\beta_i-\frac{\pi i}{2}+i\delta}}$$
$$\times \overset{\approx}{P}_{1,2}\dots\overset{\approx}{P}_{2n-1,2n} \prod_{i<j} \mathrm{th}^2\frac{1}{2}(\beta_i-\beta_j) \frac{\Sigma(\mathrm{ch}\,\beta_j)^2 - (\Sigma \mathrm{sh}\,\beta_j)^2 - 1}{(\Sigma \mathrm{sh}\,\beta_j)^2} \ ,$$
$$n\equiv 1 (\mathrm{mod}\ 2) \ .$$

We hope that the use of the same notations for the NLS form factors in the RHS's and for the ITM form factors in the LHS's will not cause any misunderstanding. Additional multipliers $\Pi \mathrm{th}^2(\beta_i-\beta_j)$ and $(\Sigma \mathrm{ch}\,\beta_j)^2 - (\Sigma \mathrm{sh}\,\beta_j)^2 - 1$ are introduced in order that form factors possess proper structure of poles; obviously they do not influence the validity of Eqs. (141), (142).

The constructive definitions of $f^a_\mu, f^a, f_{\mu\nu}, f$ are

$$f^3(\beta_1,\dots\beta_n) = 2^{-\frac{3n}{2}} M \prod_{i<j} \mathrm{th}^2 \frac{1}{2}(\beta_i-\beta_j)\left(\frac{\beta_i-\beta_j+\pi i}{\beta_i-\beta_j}\right)$$
$$\times \sum (e^{\beta_j} - (-1)^\mu e^{-\beta_j})\langle\langle\{\tilde{\phi}^0\}\rangle\rangle_n\ , \quad n \equiv 0 (\mathrm{mod}\ 2)\ ,$$

$$\tilde{\phi}^0(B^-|B^0|B^+) = \phi^0(B^-|B^0|B^+)\left(\sum(\beta^+-\beta^-) - \pi i(n(B^+)+n(B^0))\right)\ ,$$
$$f^a_\mu(\beta_1,\dots,\beta_n) = 2^{-\frac{3n}{2}} M \prod_{i<j} \mathrm{th}^2 \frac{1}{2}(\beta_i-\beta_j)\left(\frac{\beta_i-\beta_j+\pi i}{\beta_i-\beta_j}\right)$$
$$\times \{\langle\langle\phi^1\rangle\rangle_n + (-1)^a\langle\langle\{\phi^{-1}\}\rangle\rangle_n\}\ , \quad n \equiv 0(\mathrm{mod}\ 2),\ a=1,2,$$

$$f^3(\beta_1,\dots\beta_n) = 2^{-\frac{3n}{2}} \prod_{i<j} \mathrm{th}^2(\beta_i-\beta_j)\left(\frac{\beta_i-\beta_j+\pi i}{\beta_i-\beta_j}\right)\langle\langle\{\tilde{\phi}^0\}\rangle\rangle_n\ ,$$
$$n \equiv 1(\mathrm{mod}\ 2)\ ,$$

$$f^a(\beta_1,\dots,\beta_n) = 2^{-\frac{3n}{2}} \prod_{i<j} \mathrm{th}^2 \frac{1}{2}(\beta_i-\beta_j)\left(\frac{\beta_i-\beta_j+\pi i}{\beta_i-\beta_j}\right)$$
$$\times \{\langle\langle\{\phi^1\}\rangle\rangle_n + (-1)^a\langle\langle\{\phi^{-1}\}\rangle\rangle_n\}\ , \quad n \equiv 1(\mathrm{mod}\ 2)\ , \quad a=1,2\ ,$$

$$f_{\mu\nu}(\beta_1,\dots,\beta_n) = 2^{-\frac{3n}{2}} M^2 \prod_{i<j} \mathrm{th}^2 \frac{1}{2}(\beta_i-\beta_j)\left(\frac{\beta_i-\beta_j+\pi i}{\beta_i-\beta_j}\right)$$
$$\times \left(\sum(e^{\beta_i} - (-1)^\mu e^{-\beta_j})\right)\left(\sum(e^{\beta_j} - (-1)^\nu e^{-\beta_j})\right)\langle\langle\{\phi^0\}\rangle\rangle_n\ ,$$
$$n \equiv 0(\mathrm{mod}\ 2)\ ,$$

$$f(\beta_1,\dots,\beta_n) = 2^{-\frac{3n}{2}} M^2 \prod_{i<j} \mathrm{th}^2 \frac{1}{2}(\beta_i-\beta_j)\left(\frac{\beta_i-\beta_j+\pi i}{\beta_i-\beta_j}\right)\langle\langle\{\phi^0\}\rangle\rangle_n$$
$$\times \left[\left(\sum \mathrm{ch}\,\beta_j\right)^2 - \left(\sum \mathrm{sh}\,\beta_j\right)^2 - 1\right]\ , \quad n \equiv 1(\mathrm{mod}\ 2), n \geq 3\ .$$

The reasonings presented at the beginning of the present section implies the validity of the following statement.

Theorem 1. Form factors $f^a_\mu, f^a, f_{\mu\nu} f$ satisfy the requirements

$$f(\beta_1, \dots, \beta_i, \beta_{i+1}, \dots, \beta_n)_{w_1,\dots w_i, w_{i+1}, \dots, w_n} S^{w_i, w_{i+1}}_{w'_i, w'_{i+1}}(\beta_i - \beta_{i+1})$$

$$= f(\beta_1, \dots, \beta_{i+1}, \beta_i, \dots, \beta_n)_{w_1, \dots, w'_{i+1}, w'_i, \dots, w_n}, \tag{141}$$

$$f(\beta_1, \dots, \beta_{n-1}, \beta_n + 2\pi i)_{w_1, \dots, w_{n-1}, w_n}$$

$$= f(\beta_n, \beta_1, \dots, \beta_{n-1})_{w_n, w_1, \dots, w_{n-1}}. \tag{142}$$

The connection with ITM form factors implies that f^a_μ, f^a generate triplets and $f_{\mu\nu}, f$ generate singlets with respect to the action of SU(2) whose generators are $\Sigma^a = \Sigma S^a_i$. This fact clarifies the equality of the minimal number of arguments of $f(\beta_1, \dots, \beta_n)$ to 3: one-particle space has no singlet subspace.

Now we proceed to the most difficult problem which is the calculation of residues of form factors. Unfortunately, in contrast with Eqs. (141), (142) it seems to be impossible to obtain formulae for residues in connection with ITM. First, let us elucidate what singularities form factors as functions of β_n have in the strip $0 \le \operatorname{Im} \beta_n \le 2\pi$. The only singularities are simple poles at the points $\beta_n = \beta_j + \pi i$. Let us explain this fact. The operation $\langle\langle\ \rangle\rangle_n$ is regular in the strip besides simple poles at the points $\beta_n = \beta_j + 2\pi i$. The covector $w_{w_1, \dots, w_n}(\beta_1, \dots, \beta_n)$ has also simple poles at the points $\beta_n - \beta_j + \pi i$ for $w_n = 1, w_j = -1$, but they are cancelled by the multiplier $\Pi(\beta^+ - \beta^- - \pi i)$. The multiplier $\Pi_{i<j} \operatorname{th}^2 \frac{1}{2}(\beta_i - \beta_j)(\beta_i - \beta_j + \pi i)(\beta_i - \beta_j)^{-1}$ cancels the poles at the points $\beta_n = \beta_j + 2\pi i$ and produces simple poles at the points $\beta_n = \beta_j + \pi i$. As usual, the multipliers $(\Sigma(\exp(\beta_j) - (-1)^\mu \exp(-\beta_j))$ in f^a_μ and $(\Sigma(\exp(\beta_j) - (-1)^\mu \exp(-\beta_j))) \times (\Sigma(\exp(\beta_j) - (-1)^\nu \exp(-\beta_j)))$ in $f_{\mu\nu}$ cancel the poles in minimal (two-particles) form factors; $(\Sigma \operatorname{ch} p_j)^2 - (\Sigma \operatorname{sh} \beta_i)^2 - 1$ in f serves the same purpose for the three-particle form factors since

$$\left(\sum_{j=1}^{3} \operatorname{ch} \beta_j\right)^2 - \left(\sum_{j=1}^{3} \operatorname{sh} \beta_j\right)^2 - 1 = 2 + 2\sum_{i<j}^{3} \operatorname{ch}(\beta_i - \beta_j)$$

$$= 8 \prod_{i<j}^{3} \operatorname{ch} \frac{1}{2}(\beta_i - \beta_j).$$

For the calculation of residues of the form factors we need the following statement.

Lemma 3. The following equations hold:

$$\phi(B^- \cup \beta|B^0|B^+ \cup \beta) = \phi(B^-|B^0 \cup \beta|B^+)\,, \tag{143}$$

$$\begin{aligned}
&\phi(B^- \cup \beta|B^0|B^+ \cup (\beta+\pi i)) = \pi i \phi(B^-|B^0|B^+) \\
&\times \Big[\prod(\beta-\beta^0+\pi i)(\beta-\beta^0)(\beta-\beta^-+\pi i)(\beta-\beta^+) \\
&-\prod(\beta-\beta^0-\pi i)(\beta-\beta^0+2\pi i)(\beta-\beta^- -\pi i)(\beta-\beta^+ +2\pi i)\Big] \\
&B^+ \cup B^- \cup B^0 \neq \emptyset\,.
\end{aligned} \tag{144}$$

To avoid piling up of indices we have denoted ϕ^0 by ϕ.

Proof. Consider, at first, Eq. (143). Recall that

$$\begin{aligned}
\phi(B^- \cup \beta|B^0|B^+ \cup \beta) &= \frac{1}{\prod\limits_{\substack{i<j\\ i,j\neq k}} (\beta_i-\beta_j)} \prod_{\substack{\beta^+\in B^+\cup\beta\\ \beta^-\in B^-\cup\beta}} (\beta^+-\beta^-) \\
&\times \det M_k(B^- \cup \beta|B^0|B^+ \cup \beta) \prod_{i\neq k} \frac{1}{(\beta-\beta_i)^2} f_k(B^- \cup \beta|B^0|B^+ \cup \beta)\,,
\end{aligned} \tag{145}$$

where M_k is an $(n-1)\times(n-1)$ matrix, $n = n(B^+)+n(B^-)+n(B^0) = 2n(B^+)+n(B^0)$, with matrix elements given by (140), $\beta_k \in B^- \cup B^0 \cup B^+$. The two columns which correspond to $\beta \in B^- \cup \beta$ and $\beta \in B^+ \cup \beta$ are placed last. Use Eqs. (112) which imply

$$\begin{aligned}
&Q_i\left(\alpha|B^+ - \frac{\pi i}{2}, B^+ - \frac{3\pi i}{2}, B^0 - \frac{3\pi i}{2}, \beta - \frac{\pi i}{2}, \beta - \frac{3\pi i}{2}\right) \\
&= Q_i\left(\alpha|B^+ - \frac{\pi i}{2}, B^+ - \frac{3\pi i}{2}, B^0 - \frac{3\pi i}{2}, \beta - \frac{3\pi i}{2}\right) \\
&- \left(\beta - \frac{\pi i}{2}\right) Q_{i-1}\left(\alpha|B^+ - \frac{\pi i}{2}, B^+ - \frac{3\pi i}{2}, B^0 - \frac{3\pi i}{2}, \beta - \frac{3\pi i}{2}\right)\,, \\
&Q_i\left(\alpha|B^- - \frac{\pi i}{2}, B^- + \frac{\pi i}{2}, B^0 + \frac{\pi i}{2}, \beta - \frac{\pi i}{2}, \beta + \frac{\pi i}{2}\right) \\
&= Q_i\left(\alpha|B^- - \frac{\pi i}{2}, B^- + \frac{\pi i}{2}, B^0 + \frac{\pi i}{2}, \beta + \frac{\pi i}{2}\right)\,. \\
&- \left(\beta - \frac{\pi i}{2}\right) Q_{i-1}\left(\alpha|B^- - \frac{\pi i}{2}, B^- + \frac{\pi i}{2}, B^0 + \frac{\pi i}{2}, \beta + \frac{\pi i}{2}\right)\,.
\end{aligned} \tag{146}$$

Equation (146) allows us to rewrite (145) as follows:

$$\phi(B^- \cup \beta|B^0|B^+ \cup \beta) = \frac{1}{\prod\limits_{i<j}(\beta_i - \beta_j)} \prod_{\substack{\beta^+ \in B^+ \\ \beta^- \in B^-}} (\beta^+ - \beta^-) \frac{1}{\prod\limits_{i \neq k}(\beta - \beta_i)}$$
$$\times \det \tilde{M}_k f_k(B^-|B^0 \cup \beta|B^+) ,$$

where $\tilde{M}_k$ is a $(n+1) \times (n+1)$ matrix whose left-upper corner coincides with the matrix $M_k(B^-|B^0 \cup \beta|B^+)$.

Two last columns are equal to

$$Q_i\left(\beta|B^- + \frac{\pi i}{2}, B^- - \frac{\pi i}{2}, B^0 - \frac{\pi i}{2}, \beta + \frac{\pi i}{2}\right) ,$$
$$Q_i\left(\beta - \pi i|B^+ - \frac{3\pi i}{2}, B^+ - \frac{\pi i}{2}, B^0 - \frac{3\pi i}{2}, \beta - \frac{3\pi i}{2}\right) .$$

Elements of the last row and the first $(n-1)$ columns are equal to

$$Q_{n+1}\left(\beta_i - \pi i|B^+ - \frac{3\pi i}{2}, B^+ - \frac{\pi i}{2}; B^0 - \frac{3\pi i}{2}, \beta - \frac{3\pi i}{2}\right) , \quad \beta_i \in B^+;$$
$$Q_{n+1}\left(\beta_i|B^- + \frac{\pi i}{2}, B^- - \frac{\pi i}{2}, B^0 + \frac{\pi i}{2}, \beta + \frac{\pi i}{2}\right) , \quad \beta_i \in B^-;$$
$$\prod(\beta_i - \beta^-)(\beta_i - \beta^- - \pi i)(\beta_i - \beta^0 - \pi i)(\beta_i - \beta - \pi i)$$
$$\times Q_{n+1}\left(\beta_i - \pi i|B^+ - \frac{3\pi i}{2}, B^+ - \frac{\pi i}{2}, B^0 - \frac{3\pi i}{2}, \beta - \frac{3\pi i}{2}\right)$$
$$+ \prod(\beta_i - \beta^+)(\beta_i - \beta^+ + \pi i)(\beta_i - \beta^0 + \pi i)(\beta_i - \beta + \pi i)$$
$$\times Q_{n+1}\left(\beta_i|B^- - \frac{\pi i}{2}, B^- - \frac{\pi i}{2}, B^0 + \frac{\pi i}{2}, \beta + \frac{\pi i}{2}\right) , \quad \beta_i \in B^0 . \tag{147}$$

Notice that the number of arguments in Q_{n+1} in (147) is equal to $n+1$; hence one can use Eq. (129):

$$Q_{n(B)}(\alpha|B) = \prod_{\beta \in B}\left(\alpha - \beta - \frac{\pi i}{2}\right) - \prod_{\beta \in B}\left(\alpha - \beta + \frac{\pi i}{2}\right) . \tag{148}$$

Substituting (148) into (147) one proves that all the functions in (147) are equal to zero. For the last two elements of the last row one has

$$\pi i \prod(\beta - \beta^+)(\beta - \beta^+ + \pi i)(\beta - \beta^0 + \pi i) ,$$
$$- \pi i \prod(\beta - \beta^-)(\beta - \beta^- - \pi i)(\beta - \beta^0 - \pi i) .$$

Now expanding the determinant along the last row one obtains a sum of two $n \times n$ determinants with identical first $n-1$ columns. This sum can be rewritten as an $n \times n$ determinant with the same first $n-1$ columns and the last column being

$$\begin{aligned}
&\pi i \prod (\beta - \beta^+)(\beta - \beta^+ + \pi i)(\beta - \beta^0 + \pi i) \\
&\times Q_i \left(\beta | B^- + \frac{\pi i}{2}, B^- - \frac{\pi i}{2}, B^0 + \frac{\pi i}{2}, \beta + \frac{\pi i}{2} \right) \\
&+ (-\pi i) \prod (\beta - \beta^-)(\beta - \beta^- - \pi i)(\beta - \beta^0 - \pi i) \\
&\times Q_i \left(\beta - \pi i | B^+ - \frac{3\pi i}{2}, B^+ - \frac{\pi i}{2}, B^0 - \frac{3\pi i}{2}, \beta - \frac{3\pi i}{2} \right) .
\end{aligned}$$

Evidently we have obtained $\det M_k(B^-|B^0 \cup \beta|B^+)$.

Let us pass on to the proof of Eqs. (144). Applying (121) twice, one obtains

$$\begin{aligned}
Q_i \left(\alpha | B, \beta - \frac{\pi i}{2}, \beta + \frac{\pi i}{2} \right) &= Q_i(\alpha|B) - 2\beta Q_{i-1}(\alpha|B) \\
&+ \left(\beta - \frac{\pi i}{2} \right) \left(\beta + \frac{\pi i}{2} \right) Q_{i-2}(\alpha|B) .
\end{aligned}$$

Further using the identities

$$\begin{aligned}
Q_{n(B)}(\alpha|B) &= \prod_{\beta \in B} \left(\alpha - \beta - \frac{\pi i}{2} \right) - \prod_{\beta \in B} \left(\alpha - \beta + \frac{\pi i}{2} \right) , \\
Q_{n(B)+1}(\alpha|B) &= \left(\alpha - \frac{\pi i}{2} \right) \prod_{\beta \in B} \left(\alpha - \beta - \frac{\pi i}{2} \right) - \left(\alpha + \frac{\pi i}{2} \right) \\
&\times \prod \left(\alpha - \beta + \frac{\pi i}{2} \right)
\end{aligned}$$

we transform $\det M_k(B^- \cup \beta | B^0 | B^+ \cup (\beta + \pi i))$ to the following form: its left upper $(n-1) \times (n-1)$ corner coincides with $M_k(B^-|B^0|B^+)$, its two last rows of the first $n-1$ columns are equal to zero, its right-lower corner is a 2×2 matrix

$$\begin{bmatrix}
\begin{array}{l} \Pi(\beta - \beta^-)(\beta - \beta^- - \pi i)(\beta - \beta^0 - \pi i) \\ -\Pi(\beta - \beta^- + \pi i)(\beta - \beta^-)(\beta - \beta^0) \end{array} & \begin{array}{l} \Pi(\beta - \beta^+ + \pi i)(\beta - \beta^+ + 2\pi i)(\beta - \beta^0 + 2\pi i) \\ -\Pi(\beta - \beta^+)(\beta - \beta^+ + \pi i)(\beta - \beta^0 + \pi i) \end{array} \\
\\
\begin{array}{l} \left(\beta - \frac{\pi i}{2}\right) \Pi(\beta - \beta^-)(\beta - \beta^- - \pi i) \\ \times(\beta - \beta^0 - \pi i) - \left(\beta + \frac{\pi i}{2}\right) \Pi(\beta - \beta^0) \\ \times(\beta - \beta^-)(\beta - \beta^- + \pi i) \end{array} & \begin{array}{l} \left(\beta + \frac{\pi i}{2}\right) \Pi(\beta - \beta^+ + 2\pi i)(\beta - \beta^0 + 2\pi i) \\ \times(\beta - \beta^+ + \pi i) - \left(\beta - \frac{\pi i}{2}\right) \Pi(\beta - \beta^+) \\ \times(\beta - \beta^+ + \pi i)(\beta - \beta^0 + \pi i) \end{array}
\end{bmatrix} .$$

The determinant of the last matrix is equal to

$$\pi i\prod(\beta-\beta^-)(\beta-\beta^++\pi i)\Big\{\prod(\beta-\beta^0+\pi i)(\beta-\beta^0)(\beta-\beta^-+\pi i)$$
$$(\beta-\beta^+)-\prod(\beta-\beta^0-\pi i)(\beta-\beta^0+2\pi i)(\beta-\beta^--\pi i)(\beta-\beta^++2\pi i)\Big\}$$

Q.E.D.

Remark. It should be noted that the recurrent relation (144) has appeared in another branch of the theory of completely integrable models, which is Korepin's approach to Green's functions (see, for example review [24]). We hope that the explicit solution of these relations as provided by function ϕ will lead to progress in the framework of this method.

We approach the calculation of residues of the form factors at the points $\beta_n=\beta_j+\pi i$. The only proof we know of is rather cumbersome, so the reader who is interested in the result rather than the techniques may omit it.

Theorem 2. Form factors $f^a_\mu, f^a, f_{\mu\nu}, f$ considered as functions of β_n have in the strip $0\le \operatorname{Im}\beta_n\le 2\pi$ only simple poles at the points $\beta_n=\beta_j+\pi i, j\le n-1$, the corresponding residues being

$$2\pi i\operatorname{res} f(\beta_1,\dots,\beta_n)_{w_1,\dots,w_n}=(-1)^{w_n}\delta_{w_n,-w'_j}$$
$$\times f(\beta_1,\dots,\hat\beta_j,\dots,\beta_{n-1})_{w'_1,\dots,\hat w_j,\dots,w'_{n-1}}$$
$$\times\Big\{\delta^{w'_1}_{w_1}\dots\delta^{w'_{j-1}}_{w_{n-1}}S^{w'_{n-1},w'_j}_{w_{n-1},\tau_1}(\beta_{n-1}-\beta_j)\dots S^{w'_{j+1},\tau_{n-j-2}}_{w_{j+1},w_j}(\beta_{j+1}-\beta_j)$$
$$-S^{w'_j,w'_1}_{\tau_1,w_1}(\beta_j-\beta_1)\dots S^{\tau_{j-2},w'_{j-1}}_{w_j,w_{j-1}}(\beta_j-\beta_{j-1})\delta^{w'_{j+1}}_{w_{j+1}}\dots\delta^{w'_{n-1}}_{w_{n-1}}\Big\},$$
$$\beta_n=\beta_j+\pi i\,. \qquad (149)$$

In this formula f can be replaced by $f^a_\mu, f^a, f_{\mu\nu}$.

Proof. Consider any form factor, for example, $f^3(\beta_1,\dots,\beta_n)$. Evidently it is enough to prove (149) for $j=1$; other j can be considered using the symmetry (141). Let us introduce the notation

$$F(\beta_1,\dots,\beta_n)=\left(\frac{4}{\pi}\right)^{\frac{n}{2}}\prod_{i<j}\operatorname{cth}\frac{1}{2}(\beta_i-\beta_j)f^3(\beta_1,\dots,\beta_n)\,. \qquad (150)$$

The function F is regular at the point $\beta_n=\beta_1+\pi i$. Eq. (150) in conformity to F is equivalent to

$$F(\beta_1,\dots,\beta_n)|_{\beta_n=\beta_1+\pi i}=F(\beta_2,\dots,\beta_{n-1})\operatorname{sing}_{n,1}$$
$$\times\{U(\beta_1|\beta_2,\beta_3,\dots,\beta_{n-1})-1\}\,,$$

where

$$\text{sing}_{n,1} = e_{n,-1} \otimes e_{1,1} + e_{n,1} \otimes e_{1,-1} - e_{n,0} \otimes e_{1,0} \ ,$$
$$U(\beta_1|\beta_2, \dots, \beta_{n-1}) = S_{n-1,1}(\beta_{n-1} - \beta_1) \dots S_{2,1}(\beta_2 - \beta_1) \ .$$

Later on we shall denote $U(\beta_1|\beta_2, \dots, \beta_{n-1})$ by $U(\beta_1)$ if the context does not lead to misunderstanding. We shall extract explicitly the matrix structure of $U(\beta_1)$ with respect to H_1, writing as $U_{w_1}^{w_1'}(\beta_1)$. The proof is divided into two parts.

Part I. It is sufficient to prove the identity

$$F(\beta_1, \dots, \beta_n)_{w_1, w_n}|_{\beta_n = \beta_1 + \pi i} = (-1)^{w_n} \delta_{w_n', -w_1'} F(\beta_2, \dots, \beta_{n-1}) \times \{U_{w_1}^{w_1'}(\beta_1) - \delta_{w_1}^{w_1'}\} \tag{151}$$

for $(w_1, w_n) = (0, 1), (-1, 0), (-1, 1)$.

Proof of Part I. First, the formula

$$F(\beta_1, \dots, \beta_n)_{w_1, \dots, w_n} = F(\beta_1, \dots, \beta_n)_{-w_1, \dots, -w_n} \tag{152}$$

implies that if we prove Eq. (151) for $(w_1, w_n) = (0, 1), (-1, 0)$, it will also be proven automatically for $(w_1, w_n) = (0, -1), (1, 0)$. So only three components remain: $(w_1, w_n) = (-1, -1), (1, 1), (0, 0)$. Using once more Eq. (152) one makes sure that it is sufficient to consider $(w_1, w_n) = (1, 1), (0, 0)$. Thus our purpose is to prove Eq. (151) for $(w_1, w_n) = (1, 1), (0, 0)$ assuming its validity for

$$(w_1, w_n) = (0, 1), (-1, 0), (-1, 1), (0, -1), (1, 0), (1, -1) \ .$$

Consider the case $(w_1, w_n) = (1, 1)$. Rewrite the identity in question extracting explicitly the indices which correspond to the particle β_2:

$$F(\beta_1, \beta_2, \dots, \beta_n)_{1, w_2, \dots, 1}|_{\beta_n = \beta_1 + \pi i} = F(\beta_2, \dots)_{w_2'} \times \tilde{U}_{w_1'}^{-1}(\beta_1) S_{1, w_2}^{w_1', w_2'}(\beta_2 - \beta_1) \ ,$$

where $\tilde{U}(\beta_1) = U(\beta_1|\beta_3, \dots, \beta_{n-1})$, the second term in brackets has disappeared since $\delta_1^{-1} = 0$. Three possibilities exist: $w_2 = 0, -1, 1$. Let us consider these possibilities in turn.

1. Let $w_2 = 0$. The identity in question is rewritten as follows:

$$F(\beta_1, \beta_2, \ldots, \beta_n)_{1,0,\ldots,1}|_{\beta_n=\beta_1+\pi i} = F(\beta_2, \ldots)_0 S_{1,0}^{1,0}(\beta_2 - \beta_1) \times \tilde{U}_1^{-1}(\beta_1) + F(\beta_2, \ldots)_1 S_{1,0}^{0,1}(\beta_2 - \beta_1)\tilde{U}_0^{-1}(\beta_1) .$$

Using (141), (142) write a sequence of equalities:

$$\begin{aligned} F(\beta_1, \beta_2, \ldots, \beta_n)_{1,0,\ldots,1} &= [S_{0,1}^{1,0}(\beta_1 - \beta_2)]^{-1}\{F(\beta_2, \beta_1, \ldots, \beta_n)_{1,0,\ldots,1} \\ &- S_{0,1}^{0,1}(\beta_1 - \beta_2)F(\beta_1, \beta_2, \ldots, \beta_n)_{0,1,\ldots,1}\} = [S_{0,1}^{1,0}(\beta_1 - \beta_2)]^{-1} \\ &\times \{F(\beta_1, \ldots, \beta_n, \beta_2 + 2\pi i)_{0,\ldots,1,1} \\ &- S_{0,1}^{1,0}(\beta_1 - \beta_2)F(\beta_1, \beta_2, \ldots, \beta_n)_{0,1,\ldots,1}\} \\ &= [S_{0,1}^{1,0}(\beta_1 - \beta_2)]^{-1}\{F(\beta_1, \ldots, \beta_2 + 2\pi i, \beta_n)_{0,\ldots,1,1} S_{1,1}^{1,1}(\beta_2 - \beta_n + 2\pi i) \\ &- S_{0,1}^{0,1}(\beta_1 - \beta_2)F(\beta_1, \beta_2, \ldots, \beta_n)_{0,1,\ldots,1}\} . \end{aligned}$$

We suppose the value of $F(\beta_1, \ldots, \beta_n)_{0,\ldots,1}$ at the point is known and is given by Eq. (151). Hence

$$\begin{aligned} F(\beta_1, \beta_2, \ldots, \beta_n)_{1,0,\ldots,1}|_{\beta_n=\beta_1+\pi i} &= [S_{0,1}^{1,0}(\beta_1 - \beta_2)]^{-1} \\ &\times \{S_{1,1}^{1,1}(\beta_2 - \beta_1 + \pi i)F(\ldots, \beta_2 + 2\pi i)_{w_2'} S_{w_1',1}^{-1,w_2'}(\beta_2 - \beta_1 + 2\pi i)U_1^{w_1'}(\beta_1) \\ &- S_{0,1}^{0,1}(\beta_1 - \beta_2)F(\beta_2, \ldots)_{w_2'} U_{w_1'}^{-1} S_{0,1}^{w_1',w_2'}(\beta_2 - \beta_1)\} \\ &= F(\beta_2, \beta_3, \ldots, \beta_{n-1})_{w_2'}[S_{0,1}^{1,0}(\beta_1 - \beta_2)]^{-1}\{S_{1,1}^{1,1}(\beta_2 - \beta_1 + \pi i) \\ &\times S_{-1,1}^{1,1}(\beta_2 - \beta_1 + 2\pi i)\tilde{U}_1^{-1}(\beta_1)\delta_1^{w_2'} - S_{0,1}^{0,1}(\beta_1 - \beta_2)S_{0,1}^{1,0}(\beta_2 - \beta_1) \\ &\times U_1^{-1}(\beta_1)\delta_1^{w_2'} - S_{0,1}^{0,1}(\beta_1 - \beta_2)S_{0,1}^{0,1}(\beta_2 - \beta_1)\tilde{U}_0^{-1}(\beta_1)\delta_0^{w_2'}\} . \end{aligned}$$

Using the unitarity and crossing symmetry properties, one obtains the required result.

2. Let $w_2 = -1$. Using Eqs. (141), (142) one obtains

$$\begin{aligned} F(\beta_1, \beta_2, \ldots, \beta_n)_{1,-1,\ldots,1} &= [S_{-1,1}^{1,-1}(\beta_1 - \beta_2)]^{-1} \\ &\times \{-F(\beta_1, \beta_2, \ldots, \beta_n)_{0,0,\ldots,1} S_{-1,1}^{0,0}(\beta_1 - \beta_2) - S_{-1,1}^{-1,1}(\beta_1 - \beta_2) \\ &\times F(\beta_1, \beta_2, \ldots, \beta_n)_{-1,1,\ldots,1} + S_{1,1}^{1,1}(\beta_2 - \beta_n + 2\pi i) \\ &\times F(\beta_1, \ldots, \beta_2 + 2\pi i, \beta_n)_{-1,\ldots,1,1}\} . \end{aligned} \tag{153}$$

Using (153) and supposing that $F(\beta_1, \ldots, \beta_n)_{0,\ldots,1}$ and $F(\beta_1, \ldots, \beta_n)_{-1,\ldots,1}$ for $\beta_n = \beta_1 + \pi i$ are known, one obtains Eq. (151).

3. Let $w_2 = 1$. The following identity holds:

$$F(\beta_1, \beta_2, \dots, \beta_n)_{1,1,\dots,1} = S^{1,1}_{1,1}(\beta_2 - \beta_1) S^{1,1}_{1,1}(\beta_2 - \beta_1 + 2\pi i) \times F(\beta_1, \beta_3, \dots, \beta_2 + 2\pi i, \beta_n)_{1,w_3,\dots,1,1} . \quad (154)$$

If $w_3 = 0, -1$ we calculate (154) at the point $\beta_n = \beta_1 + \pi i$ using the results of the previous subsections. If $w_3 = 1$ we use (154) one more time, etc., until we get $w = 0$ or -1 for the first time (we shall reach $w = 0$ or -1 without fail because $\Sigma w_j = 0$).

Let us outline the proof for $(w_1, w_n) = (0,0)$. We assume Eq. (151) to be proven for $(w_1, w_n) = (0,1), (1,0), (-1,0), (0,-1), (1,-1), (-1,1)$ and consequently, according to the proof given above, $(w_1, w_n) = (1,1), (-1,-1)$. We consider again the three possibilities $w_2 = 0, 1, -1$ in turn.

1. Let $w_2 = 0$. Using Eqs. (141), (142) one obtains

$$\begin{aligned} F(\beta_1, \beta_2, \dots, \beta_n)_{0,0,\dots,0} &= [S^{0,0}_{-1,1}(\beta_1 - \beta_2)]^{-1} \{ -S^{-1,1}_{-1,1}(\beta_1 - \beta_2) \\ &\times F(\beta_1, \beta_2, \dots, \beta_n)_{-1,1,\dots,0} - S^{1,-1}_{-1,1}(\beta_1 - \beta_2) F(\beta_1, \beta_2, \dots, \beta_n)_{1,-1,\dots,0} \\ &+ S^{1,0}_{1,0}(\beta_2 - \beta_n + 2\pi i) F(\beta_1, \dots, \beta_2 + 2\pi i, \beta_n)_{-1,\dots,1,0} \\ &+ S^{0,1}_{1,0}(\beta_2 - \beta_n + 2\pi i) F(\beta_1, \dots, \beta_2 + 2\pi i, \beta_n)_{-1,\dots,0,1} \} . \end{aligned}$$

Substituting $F(\beta_1, \dots, \beta_n)_{w_1,\dots,w_n}$ for $\beta_n = \beta_1 + \pi i, (w_1, w_n) = (-1,0), (1,0), (-1,1)$, one obtains Eq. (151) for $w_1 = w_2 = w_n = 0$.

2. Let $w_2 = 1$. Using Eqs. (141), (142) one obtains

$$\begin{aligned} F(\beta_1, \beta_2, \dots, \beta_n)_{0,1,\dots,0} &= [S^{0,1}_{1,0}(\beta_1 - \beta_2)]^{-1} \{ -S^{1,0}_{0,1}(\beta_1 - \beta_2) \\ &\times F(\beta_1, \beta_2, \dots, \beta_n)_{1,0,\dots,0} + S^{0,0}_{0,0}(\beta_2 + \beta_n + 2\pi i) \\ &\times F(\beta_1, \dots, \beta_2 + 2\pi i, \beta_n)_{1,\dots,0,0} \\ &+ F(\beta_1, \dots, \beta_2 + 2\pi i, \beta_n)_{1,\dots,-1,1} S^{-1,1}_{0,0}(\beta_2 - \beta_n + 2\pi i) \\ &+ F(\beta_1, \dots, \beta_2 + 2\pi i, \beta_n)_{1,\dots,1,-1} S^{1,-1}_{0,0}(\beta_2 - \beta_n + 2\pi i) \} . \end{aligned}$$

Substituting $F(\beta_1, \dots, \beta_n)_{w_1,\dots,w_n}$ for $\beta_n = \beta_1 + \pi i, (w_1, w_n) = (1,0), (1,1), (1,-1)$, one obtains Eq. (151) for $w_2 = 1, w_1 = w_n = 0$.

3. The case $w_2 = -1$ is quite similar to the case $w_2 = 1$. The proof of Part I is finished.

Part 2. Function $F(\beta_1, \dots, \beta_n)_{w_1,\dots,w_n}$ satisfies Eq. (151) for $(w_1, w_n) = (-1,1), (0,1), (-1,0)$.

Proof of Part 2. At first, notice the following circumstance. Since

$$S(\beta_j-\beta_1)=c_1 S^{t_1}(\beta_1-\beta_j+\pi i)c_1\ ,$$
$$\mathrm{sing}_{1,n}A_1=\mathrm{sing}_{1,n}c_n A_n^{t_n}c_n\ ,$$

where c_1 is the crossing matrix c acting in H_1, t_1 is the transposition with respect to H_1, A is an arbitrary matrix, one can change in (151):

$$U(\beta_1|\beta_2,\dots,\beta_{n-1})\to c_1 V^{t_1}(\beta_1|\beta_2,\dots,\beta_{n-1})c_1$$

where

$$V(\beta_1|\beta_2,\dots,\beta_n)$$
$$=S_{n,2}(\beta_1-\beta_2+\pi i)\dots S_{n,n-1}(\beta_1-\beta_{n-1}+\pi i)\ .$$

The matrix $S(\beta_1-\beta_j+\pi i)$ can be presented as follows:

$$S^{w_1',w_2'}_{w_1,w_2}(\beta_1-\beta_j+\pi i)=\left(\frac{(\beta_1-\beta_j)}{(\beta_1-\beta_j+2\pi i)}\right)\tilde{P}^{w_1'}_{\varepsilon_1,\varepsilon_1'}\stackrel{\approx}{S}{}^{\varepsilon_1,w_2'}_{\varepsilon_1'',w_2''}(\beta_1-\beta_i+\pi i)$$
$$\times\stackrel{\approx}{S}{}^{\varepsilon_1,w_2''}_{\varepsilon_1''',w_2}(\beta_1-\beta_j+2\pi i)P^{\varepsilon_1'',\varepsilon_1'''}_{w_1}\ .$$

Hence

$$V^{w_1'}_{w_1}(\beta_1)=\prod_{j=2}^{n-1}\left(\frac{(\beta_1-\beta_j)}{(\beta_1-\beta_j+2\pi i)}\right)P^{w_1'}_{\varepsilon_1,\varepsilon_1'}\tilde{\tau}^{\varepsilon_1}_{\varepsilon_1''}(\beta_1+\pi i)$$
$$\times\tilde{\tau}^{\varepsilon_1}_{\varepsilon_1'''}(\beta_1+2\pi i)\stackrel{\approx}{P}{}^{\varepsilon_1'',\varepsilon_1''}_{w_1}\ . \qquad (155)$$

Here and later on $\tilde{\tau}(\beta)=\tau(\beta|\beta_2,\dots,\beta_{n-1})$. For example,

$$V^{-1}_{-1}(\beta_1)=\tilde{A}(\beta_1+\pi i)\tilde{A}(\beta_1+2\pi i)\prod\left(\frac{\beta_1-\beta_j}{\beta_1-\beta_j+2\pi i}\right)\ .$$

The following identity also holds [22]:

$$(P^-)^{\varepsilon_1,\varepsilon_1'}_{\varepsilon_1'',\varepsilon_1'''}\tilde{\tau}^{\varepsilon_1''}_{\sigma_1}(\beta_1+\pi i)\tilde{\tau}^{\varepsilon_1'''}_{\sigma_1'}(\beta_1+2\pi i)(P^-)^{\sigma_1,\sigma_1'}_{\sigma_1'',\sigma_1'''}$$
$$=\prod\left(\frac{\beta_1-\beta_j+2\pi i}{\beta_1-\beta_j}\right)\delta^{\varepsilon_1}_{\sigma_1''}\delta^{\varepsilon_1'}_{\sigma_1'''}\ , \qquad (156)$$

which implies for example

$$\tilde{\mathcal{A}}(\beta_1+\pi i)\tilde{\mathcal{B}}(\beta_1+2\pi i)-\tilde{\mathcal{B}}(\beta_1+\pi i)\tilde{\mathcal{A}}(\beta_1+2\pi i)=0\,.$$

Let us prove Eq. (151) for $(w_1, w_n) = (-1, 1)$. Eq. (151) can be written in the form

$$\begin{aligned}
&\Big\{\langle\langle 0\| \sum_{B=B^-\cup B^0\cup B^+} \prod \mathcal{B}(\beta^0)\mathcal{B}(\beta^+)\mathcal{B}(\beta^++\pi i)\tilde{\phi}(B^-|B^0|B^+)\\
&\times\prod\frac{(\beta^+-\beta^-+\pi i)}{(\beta^+-\beta^-+\pi i)(\beta^+-\beta^0)(\beta^--\beta^0)(\beta^+-\beta^-)(\beta^0-\beta^0-\pi i)}\Big\}\\
&\times P_{-1,1}|_{\beta_n=\beta_1+\pi i}\\
&=\langle\langle 0\|\sum_{(B\backslash\beta_1,\beta_n)=B^-\cup B^0\cup B^+}\prod\tilde{\mathcal{B}}(\beta^0)\tilde{\mathcal{B}}(\beta^+)\tilde{\mathcal{B}}(\beta^++\pi i)\phi(B^-|B^0|B^+)\\
&\times\frac{1}{2}(S_n^-)^2\\
&\times\prod\frac{(\beta^+-\beta^--\pi i)}{(\beta^+-\beta^-+\pi i)(\beta^+-\beta^0)(\beta^--\beta^0)(\beta^+-\beta^-)(\beta^0-\beta^0-\pi)}\\
&\times\left(\tilde{\mathcal{A}}(\beta_1+\pi i)\tilde{\mathcal{A}}(\beta_1+2\pi i)\prod\left(\frac{\beta_1-\beta_j}{\beta_1-\beta_j+2\pi i}\right)-1\right)\,,
\end{aligned}\tag{157}$$

where $P_{-1,1}$ is a projection operator onto $e_{1,-1}\otimes e_{n,1}\otimes H_2^*\otimes\dots,\otimes H_n^*$. What terms in LHS of Eq. (157) contain $\langle\langle 0\|(S_n^-)^2$, i.e., are not annihilated by the operator P_{-11}? The triangularity of the base $w_{w_1,\dots,w_n}(\beta_1,\dots,\beta_n)$ implies, at first, that these are the terms with $\beta_n\in B^+$. Present $\mathcal{B}(\sigma)$ in the form

$$\begin{aligned}
\mathcal{B}(\sigma)&=a_1(\sigma)\tilde{\mathcal{A}}(\sigma)b_n(\sigma)+a_1(\sigma)\tilde{\mathcal{B}}(\sigma)d_n(\sigma)\\
&+b_1(\sigma)\tilde{\mathcal{C}}(\sigma)b_n(\sigma)+b_1(\sigma)\tilde{\mathcal{D}}(\sigma)d_n(\sigma)
\end{aligned}\tag{158}$$

where

$$\begin{pmatrix} a_1(\sigma) & b_1(\sigma)\\ c_1(\sigma) & d_1(\sigma)\end{pmatrix}=\overset{\approx}{\mathcal{S}}_{0,1}(\sigma-\beta_1),\ \begin{pmatrix} a_n(\sigma) & b_n(\sigma)\\ c_n(\sigma) & d_n(\sigma)\end{pmatrix}=\overset{\approx}{\mathcal{S}}_{0,n}(\sigma-\beta_n)\,,$$

the partition into the blocks being made with respect to h_0. Note that

$$\langle\langle 0\|a_i(\sigma)=\left(\frac{\sigma-\beta_i}{\sigma-\beta_i-2\pi i}\right)\langle\langle 0\|,\quad \langle\langle 0\|d_i(\sigma)=\langle\langle 0\|$$

$$\langle\langle 0||b_i(\sigma) = \left(\frac{\pi i}{\sigma-\beta_i-2\pi i}\right)\langle\langle 0||S_i^-\,,\quad \langle\langle 0||c_i(\sigma)=0\,.$$

Consider a term with $\beta_1 \in B^0, B^+$ in LHS of Eq. (154). Move it to the left and apply it to the vacuum vector

$$\langle\langle 0||\mathcal{B}(\beta_1) = \frac{1}{2}\langle\langle 0||S_1^-\,. \tag{159}$$

The expansion (158) shows that $\mathcal{B}(\sigma)$ does not contain the raising operator S_1^+ and consequently further application of the operators cannot annihilate S_1^- in (159). Hence those terms in LHS of Eq. (157) for which $\beta_1 \in B^0$ or B^+ are annihilated by $P_{-1,1}$.

Thus only those terms in LHS of Eq. (157) are essential for which $\beta_1 \in B^-, \beta_n \in B^+$. To calculate

$$\begin{aligned}&\langle\langle 0|| \sum_{(B\backslash\beta_1,\beta_n)=B^-\cup B^0\cup B^+}\left(\frac{(\beta_n-\beta_1-\pi i)}{(\beta_n-\beta_1+\pi i)(\beta_n-\beta_1)}\right)\mathcal{B}(\beta_n)\mathcal{B}(\beta_n+\pi i)\\&\times\prod\mathcal{B}(\beta^0)\mathcal{B}(\beta^+)\mathcal{B}(\beta^++\pi i)\tilde{\Phi}(B\cup\beta_1|B^0|B^+\cup\beta_n)\\&\times\frac{(\beta_1-\beta^+)(\beta_1-\beta^++\pi i)}{(\beta_1-\beta^-+2\pi i)(\beta_1-\beta^+-\pi i)(\beta_1-\beta^0+\pi i)(\beta_1-\beta^0)(\beta_1-\beta^-+\pi i)(\beta_1-\beta^+)}\\&\times\prod\frac{(\beta^+-\beta^--\pi i)}{(\beta^+-\beta^-+\pi i)(\beta^+-\beta^0)(\beta^--\beta^0)(\beta^+-\beta^-)(\beta^0-\beta^0-\pi i)}\Bigg|_{\beta_n=\beta_1+\pi i} P_{-11}\end{aligned}$$

let us use the following trick: put before P_{-11} the product

$$S(\beta_{n-1}-\beta_n)\dots S(\beta_2-\beta_n)S(\beta_n-\beta_2)\dots S(\beta_n-\beta_{n-1}) = I$$

obtaining

$$\begin{aligned}&\langle\langle 0|| \sum_{(B\backslash\beta_1,\beta_2)=B^-\cup B^-\cup B^0\cup B^+}\left(\frac{(\beta_n-\beta_1-\pi i)}{(\beta_n-\beta_1+\pi i)(\beta_n-\beta_1)}\right)\hat{\mathcal{B}}(\beta_n)\hat{\mathcal{B}}(\beta_n+\pi i)\\&\times\prod\hat{\mathcal{B}}(\beta^0)\hat{\mathcal{B}}(\beta^+)\hat{\mathcal{B}}(\beta^++\pi i)\phi(B^-\cup\beta_1|B^0|B^+\cup(\beta_1+\pi i))\\&\times\prod\frac{(\beta^+-\beta^--\pi i)}{(\beta^+-\beta^-+\pi i)(\beta^+-\beta^0)(\beta^--\beta^0)(\beta^+-\beta^-)(\beta^0-\beta^0-\pi i)}\\&\times\prod\frac{(\beta_1-\beta^-)(\beta_1-\beta^++\pi i)}{(\beta_1-\beta^-+2\pi i)(\beta_1-\beta^+-\pi i)(\beta_1-\beta^0+\pi i)(\beta_1-\beta^0)(\beta_1-\beta^-+\pi i)(\beta_1-\beta^+)}\\&S_{n,2}(\beta_n-\beta_2)\dots S_{n,n-1}(\beta_n-\beta_{n-1})\,,\end{aligned} \tag{160}$$

where $\hat{\mathcal{B}}(\sigma)$ is constructed via $\hat{\tau}(\sigma) = \tau(\sigma|\beta_1, \beta_n, \beta_2, \ldots, \beta_{n-1})$, i.e.,

$$\begin{aligned}\hat{\mathcal{B}}(\sigma) = (a_1(\sigma)a_n(\sigma) + b_1(\sigma)c_n(\sigma)\tilde{\mathcal{B}}(\sigma) \\ + (a_1(\sigma)b_n(\sigma) + b_1(\sigma)d_n(\sigma))\tilde{\mathcal{D}}(\sigma) .\end{aligned}$$

Direct calculation gives

$$\begin{aligned}&\langle\langle 0\|\hat{\mathcal{B}}(\beta_n)\hat{\mathcal{B}}(\beta_n+\pi i)(\beta_n-\beta_1-\pi i)|_{\beta_n=\beta_1+\pi i} \\ &\quad = (-\pi i)\langle\langle 0\|((S_1^-)^2+(S_n^-)^2-S_1^-S_n^-) , \qquad (161)\\ &\langle\langle 0\|((S_1^-)^2+(S_n^-)^2-S_1^-S_n^-)\hat{\mathcal{B}}(\sigma) = \frac{(\sigma-\beta_1)(\sigma-\beta_1-\pi i)}{(\sigma-\beta_1-2\pi i)(\sigma-\beta_1-3\pi i)} \\ &\quad \times \langle\langle 0\|((S_1^-)^2+(S_n^-)^2-S_1^-S_n^-)\tilde{\mathcal{B}}(\sigma) . \qquad (162)\end{aligned}$$

Substituting these equations in Eq. (160) and using Eq. (155) one obtains

$$\begin{aligned}&\langle\langle 0\| \sum_{(B\backslash\beta_1,\beta_n)=B^-\cup B^0\cup B^+} \prod \tilde{\mathcal{B}}(\beta^0)\tilde{\mathcal{B}}(\beta^+)\tilde{\mathcal{B}}(\beta^++\pi i)\phi(B^-|B^0|B^+) \\ &\times \prod \frac{(\beta^+-\beta^--\pi i)}{(\beta^+-\beta^-+\pi i)(\beta^+-\beta^0)(\beta^--\beta^0)(\beta^+-\beta^-)(\beta^0-\beta^0-\pi i)} \\ &\times \left[-1-\prod \frac{(\beta_1-\beta^--\pi i)(\beta_1-\beta^++2\pi i)(\beta_1-\beta^0-\pi i)(\beta_1-\beta^0+2\pi i)}{(\beta_1-\beta^-+\pi i)(\beta_1-\beta^+)(\beta_1-\beta^0+\pi i)(\beta_1-\beta^0)}\right] \\ &\times \tilde{A}(\beta_1+\pi i)\tilde{A}(\beta_1+2\pi i)\frac{1}{2}(S_n^-)^2 \prod \left(\frac{(\beta_1-\beta)}{(\beta_1-\beta+2\pi i)}\right) .\end{aligned}$$

Recall now that the covectors $\langle\langle 0\|\Pi\tilde{\mathcal{B}}(\beta^0)\tilde{\mathcal{B}}(\beta^+)\tilde{\mathcal{B}}(\beta^++\pi i)$ are eigenvectors for $\tilde{A}(\sigma)$ with the eigenvalues

$$\begin{aligned}&\langle\langle 0\| \prod \tilde{\mathcal{B}}(\beta^0)\tilde{\mathcal{B}}(\beta^+)\tilde{\mathcal{B}}(\beta^++\pi i)\tilde{A}(\sigma) \\ &= \prod \frac{(\sigma-\beta^-)(\sigma-\beta^0-\pi i)}{(\sigma-\beta^--2\pi i)(\sigma-\beta^0-2\pi i)}\langle\langle 0\| \prod \tilde{\mathcal{B}}(\beta^0)\tilde{\mathcal{B}}(\beta^+)\tilde{\mathcal{B}}(\beta^++\pi i) . \qquad (163)\end{aligned}$$

This observation finishes the proof of (157) for $(w_1 w_n) = (-11)$.

Consider now $w_1 = -1, w_n = 0$. Eq. (157) reduces to

$$\Bigg\{ \langle\langle 0 \| \sum_{B=B^-\cup B^0\cup B^+} \prod \mathcal{B}(\beta^0)\mathcal{B}(\beta^+)\mathcal{B}(\beta^+ + \pi i)\tilde{\phi}(B^-|B^0|B^+)$$

$$\times \prod \frac{(\beta^+ - \beta^- - \pi i)}{(\beta^+ - \beta^- + \pi i)(\beta^+ - \beta^0)(\beta^- - \beta^0)(\beta^+ - \beta^-)(\beta^0 - \beta^0 - \pi i)} \Bigg\} P_{-1,0} \Bigg|_{\beta_n = \beta_1 + \pi i}$$

$$= -\frac{1}{\sqrt{2}} \langle\langle 0 \| \sum_{(B\backslash \beta_1, \beta_n)=B^-\cup B^0\cup B^+} \prod \tilde{\mathcal{B}}(\beta^0)\tilde{\mathcal{B}}(\beta^+)\tilde{\mathcal{B}}(\beta^+ + \pi i)\tilde{\phi}(B^-|B^0|B^+)$$

$$\times \prod \frac{1}{(\beta^+ - \beta^- + \pi i)(\beta^+ - \beta^0)(\beta^- - \beta^0)(\beta^+ - \beta^-)(\beta^0 - \beta^0 - \pi i)}$$

$$\times S_n^- \tilde{\mathcal{A}}(\beta_1 + \pi i)\tilde{\mathcal{B}}(\beta_1 + 2\pi i) \prod_{\beta \in B\backslash \beta_1, \beta_n} \left(\frac{\beta_1 - \beta}{\beta_1 - \beta + 2\pi i} \right), \tag{164}$$

where $P_{-1,0}$ is the projection operator onto $e_{1,-1} \otimes e_{n,0} \otimes H_2^* \dots H_{n-1}^*$. It can be shown that only those terms in LHS of Eq. (164) are not annihilated by $P_{-1,0}$, for which $\beta_n \in B^+, B^0, \beta_1 \in B^-$. For $\beta_n \in B, \beta_1 \in B^-$ the LHS of Eq. (164) can be calculated similarly to LHS of Eq. (157):

$$\langle\langle 0 | \sum_{(B\backslash \beta_1, \beta_n)=B^-\cup B^0\cup B^+} \prod \mathcal{B}(\beta^0)\mathcal{B}(\beta^+)\mathcal{B}(\beta^+ + \pi i)\mathcal{B}(\beta_n)\mathcal{B}(\beta_n + \pi i)$$

$$\times \left(\frac{(\beta_n - \beta_1 - \pi i)}{(\beta_n - \beta_1 + \pi i)(\beta_n - \beta_1)} \right) \tilde{\phi}(\beta_1, B^-|B^0|B^+, \beta_1 + \pi i)$$

$$\times \prod \frac{(\beta^+ - \beta^- - \pi i)}{(\beta^+ - \beta^- + \pi i)(\beta^+ - \beta^0)(\beta^- - \beta^0)(\beta^+ - \beta^-)(\beta^0 - \beta^0 - \pi i)}$$

$$\times \prod \frac{(\beta_1 - \beta^-)(\beta_1 - \beta^+ + \pi i)}{(\beta_1 - \beta^- + 2\pi i)(\beta_1 - \beta^+ - \pi i)(\beta_1 - \beta^0 + \pi i)(\beta_1 - \beta^0)(\beta_1 - \beta^- + \pi i)(\beta_1 - \beta^+)} P_{-1,0}$$

$$= \langle\langle 0 \| \sum_{(B\backslash \beta_1, \beta_2)=B^-\cup B^0\cup B^+} \prod \tilde{\mathcal{B}}(\beta^0)\tilde{\mathcal{B}}(\beta^+)\tilde{\mathcal{B}}(\beta^+ + \pi i)\tilde{\phi}(B^-|B^0|B^+)$$

$$\times \prod \frac{(\beta^+ - \beta^- - \pi i)}{(\beta^+ - \beta^- + \pi i)(\beta^+ - \beta^-)(\beta^- - \beta^0)(\beta^+ - \beta^-)(\beta^0 - \beta^0 - \pi i)} S_n^-$$

$$\times \prod_{j=2}^{n-1} \left(\frac{\beta_1 - \beta_j}{\beta_1 - \beta_j + 2\pi i} \right) (\tilde{\mathcal{A}}^{-1}(\beta_1 + 2\pi i)\tilde{\mathcal{B}}(\beta_1 + 2\pi i) - \tilde{\mathcal{A}}(\beta_1 + \pi i)\tilde{\mathcal{B}}(\beta_1 + 2\pi i)) .$$

Substitute this in Eq. (164). As a result, Eq. (164) is reduced to

$$
\begin{aligned}
&\langle\langle 0\| \sum_{(B\backslash\beta_1,\beta_n)=B^-\cup B^0\cup B^+} \mathcal{B}(\beta_n)\frac{1}{(-\pi i)(\beta_n-\beta_1)}\prod \mathcal{B}(\beta^0)\mathcal{B}(\beta^+)\mathcal{B}(\beta^++\pi i)\\
&\times \phi(B^-,\beta_1|B^0,\beta_1+\pi i|B^+)\prod\frac{(\beta^+-\beta^--\pi i)}{(\beta^+-\beta^-+\pi i)(\beta^+-\beta^0)(\beta^--\beta^0)}\\
&\times\prod\frac{1}{(\beta^+-\beta^-)(\beta^0-\beta^0-\pi i)}\prod\frac{1}{(\beta_1-\beta^+-\pi i)(\beta_1-\beta^0)(\beta_1-\beta^+)}\\
&\times\prod\frac{1}{(\beta_1-\beta^-+\pi i)(\beta_1-\beta^0)(\beta_1-\beta^0+2\pi i)}P_{-1,0}\Big|_{\beta_n=\beta_1+\pi i}\\
&=-\langle\langle 0\|\sum_{(B\backslash\beta_1,\beta_n)=B^-\cup B^0\cup B^+}\prod\tilde{\mathcal{B}}(\beta^0)\tilde{\mathcal{B}}(\beta^+)\tilde{\mathcal{B}}(\beta^++\pi i)\phi(B^-|B^0|B^+)\\
&\times\prod\frac{(\beta^+-\beta^--\pi i)}{(\beta^+-\beta^-+\pi i)(\beta^+-\beta^0)(\beta^+-\beta^-)(\beta^0-\beta^0-\pi i)(\beta^--\beta^0)}\\
&\times\prod\frac{(\beta_1-\beta^-)(\beta_1-\beta^0)}{(\beta_1-\beta^-+2\pi i)(\beta_1-\beta^0+\pi i)}S_n^-\tilde{\mathcal{B}}(\beta_1+2\pi i)\,,
\end{aligned}
\tag{165}
$$

where we have used the identity (156).

Calculate

$$
\langle\langle 0\|\prod\mathcal{B}(\beta^0)\mathcal{B}(\beta^+)\mathcal{B}(\beta^++\pi i)\mathcal{B}(\beta_n)P_{-1,0}
$$

for some partition of $(B\backslash\beta_1,\beta_n)$ into B^-,B^0,B^+. Substitute $\mathcal{B}(\sigma)$ in the form (158):

$$
\begin{aligned}
\mathcal{B}(\sigma)&=a_1(\sigma)\tilde{A}(\sigma)b_n(\sigma)+a_1(\sigma)\tilde{\mathcal{B}}(\sigma)d_n(\sigma)\\
&\quad+b_1(\sigma)\tilde{C}(\sigma)b_n(\sigma)+b_1(\sigma)\tilde{\mathcal{D}}(\sigma)d_n(\sigma)\,.
\end{aligned}
\tag{166}
$$

The last two terms can be omitted because they create S_1^-. We state that

$$
\begin{aligned}
&\langle\langle 0\|\prod\mathcal{B}(\beta^0)\mathcal{B}(\beta^+)\mathcal{B}(\beta^++\pi i)\\
&=\prod\frac{(\beta^0-\beta_1)(\beta^+-\beta_1)(\beta^+-\beta_1+\pi i)}{(\beta^0-\beta_1-2\pi i)(\beta^+-\beta_1-2\pi i)(\beta^+-\beta^--\pi i)}\\
&\times\langle\langle 0\|\prod\tilde{\mathcal{B}}(\beta^0)\tilde{\mathcal{B}}(\beta^+)\tilde{\mathcal{B}}(\beta^+-\pi i)\\
&+(\text{terms containing } S_1^-)\,.
\end{aligned}
\tag{167}
$$

Suppose we have proven (167) for some subsets of $\{\beta^0\}=B_1^0,\{\beta^+\}=B_1^+$. Let us apply one more operator $\mathcal{B}(\beta_j),\beta_j\in B^0$. The last two terms in (166) are omitted, the first term gives nothing because $\langle\langle 0\|\Pi\tilde{\mathcal{B}}(\beta^0)\tilde{\mathcal{B}}(\beta^+)\tilde{\mathcal{B}}(\beta^+ + \pi i)$ is the eigenvector for $\tilde{A}(\beta_j)$ with the eigenvalue (163) and $B_1^0 \cup B_1^+$ does not contain β_j. The case $\beta_i \in B^+$ is treated in a similar way. Now apply the operator $\mathcal{B}(\beta_n)P_{-1,0}$ to

$$\langle\langle 0\| \prod \tilde{\mathcal{B}}(\beta^0)\tilde{\mathcal{B}}(\beta^+)\tilde{\mathcal{B}}(\beta^+ + \pi i)\ .$$

The second term can be omitted because it does not contain S_n^-. As a result one has

$$\begin{aligned}&\langle\langle 0\| \prod \mathcal{B}(\beta^0)\mathcal{B}(\beta^+)\mathcal{B}(\beta^+ + \pi i)\mathcal{B}(\beta_n)P_{-1,0}\\ &= -\prod \frac{(\beta_1-\beta^0)(\beta_1-\beta^+)(\beta_1-\beta^+-\pi i)(\beta_1-\beta^-+\pi i)(\beta_1-\beta^0)}{(\beta_1-\beta^0+2\pi i)(\beta_1-\beta^++2\pi i)(\beta_1-\beta^++\pi i)(\beta_1-\beta^--\pi i)(\beta_1-\beta^0-\pi i)}\\ &\times \frac{1}{2}\langle\langle 0\| \prod \tilde{\mathcal{B}}(\beta_0)\tilde{\mathcal{B}}(\beta^+)\tilde{\mathcal{B}}(\beta^+ + \pi i)S_n^-\ .\end{aligned} \tag{168}$$

The following identities hold:

$$\begin{aligned}&\langle\langle 0\| \prod \tilde{\mathcal{B}}(\beta^0)\tilde{\mathcal{B}}(\beta^+)\tilde{\mathcal{B}}(\beta^+ + \pi i)\tilde{\mathcal{B}}(\beta_1 + 2\pi i)\\ &= (-\pi i)\prod \frac{(\beta_1-\beta^-+2\pi i)(\beta_1-\beta^0+\pi i)}{(\beta_1-\beta^-)(\beta_1-\beta^0)}\langle\langle 0\| \Bigg\{ 2\sum_{\beta\in B^-}\prod_{\beta\cup B^0}\tilde{\mathcal{B}}(\beta^0)\\ &\times \prod_{\beta^+\in B^+}\tilde{\mathcal{B}}(\beta^+)\tilde{\mathcal{B}}(\beta^+ + \pi i)\frac{1}{\beta_1-\beta+2\pi i}\prod_{\beta^0\in B^0}\left(\frac{\beta-\beta^0-2\pi i}{\beta-\beta^0-\pi i}\right),\\ &\times \prod_{B^-\backslash\beta}\left(\frac{\beta-\beta^--2\pi i}{\beta-\beta^-}\right) + \sum_{\beta\in B^0}\prod_{B^0\backslash\beta}\tilde{\mathcal{B}}(\beta^0)\prod_{B^+\cup\beta}\tilde{\mathcal{B}}(\beta^+)\tilde{\mathcal{B}}(\beta^+ + \pi i)\\ &\times \frac{1}{\beta_1-\beta+\pi i}\prod_{B^0\backslash\beta}\left(\frac{\beta-\beta^0-\pi i}{\beta-\beta^0}\right)\prod_{B^-}\left(\frac{\beta-\beta^--\pi i}{\beta-\beta^-+\pi i}\right),\end{aligned} \tag{169}$$

whose validity follows from the observations. The LHS and RHS are both rational functions of β_1 which decrease for $\beta_1 \to \infty$ and have $n(B^-)+n(B^0)$ simple poles and whose values at the points $\beta_1 = \beta^- - 2\pi i, \beta_1 = \beta^0 - \pi i$ coincide.

Transform LHS of Eq. (165) using (168) and RHS of (165) using (169) and use the identity

$$\begin{aligned}\tilde{\phi}(B^-,\beta_1|B^0,\beta_1+\pi i|B^+) &= -2(\pi i)^2\prod(\beta_1-\beta^--\pi i)(\beta_1-\beta^0-\pi i)\\ &\times(\beta_1-\beta^++2\pi i)(\beta_1-\beta^0+2\pi i)\tilde{\phi}(B_1,\beta_1+\pi i|B^0|B^+)\end{aligned}$$

following from Lemma 3. As a result the proof of Eqs. (165) reduces to the proof of the identity:

$$
\begin{aligned}
&\tilde{\phi}(B^-,\beta_1+\pi i|B^0|B^+) = \prod(\beta_1-\beta^+ +\pi i)(\beta_1-\beta^0+2\pi i) \\
&\times \left\{ -2(\pi i)^2 \sum_{\beta\in B^0} \tilde{\phi}(B^-,\beta|B^0\backslash\beta|B^+)\frac{1}{\beta_1-\beta+2\pi i} \right. \\
&\times \prod_{B^+,B^-,B^0\backslash\beta} \frac{(\beta-\beta^0-2\pi i)(\beta-\beta^0+\pi i)(\beta-\beta^- -2\pi i)(\beta-\beta^+ +\pi i)}{(\beta-\beta^0)(\beta-\beta^+ -\pi i)} \\
&+ \sum_{\beta\in B^+} \tilde{\phi}(B^-|B^0\cup\beta|B^+\backslash\beta)\frac{1}{\beta_1-\beta+\pi i} \\
&\times \prod \frac{1}{(\beta-\beta^0+\pi i)(\beta-\beta^+)} \,.
\end{aligned}
\tag{170}
$$

From the very definition of $\tilde{\phi}$ it follows that, considered as a function of β_1, $\tilde{\phi}(B^-\cup\beta_1|B^0|B^+)$ is a polynomial of degree $n(B^+)+n(B^0)-1$. This is why it is sufficient to prove that RHS and LHS of (170) coincide at the points $\beta_1=\beta^0-2\pi i$, $\beta_1=\beta^+-\pi i$. This fact follows immediately from Lemma 3.

Consider finally $w_1=0, w_n=1$. Eq. (157) reduces to

$$
\begin{aligned}
&\left\{ \langle\langle 0\| \sum_{B=B^-\cup B^0\cup B^+} \prod \mathcal{B}(\beta^0)\mathcal{B}(\beta^+)\mathcal{B}(\beta^+ +\pi i)\tilde{\phi}(B^-|B^0|B^+) \right. \\
&\left.\times \prod \frac{(\beta^+-\beta^- -\pi i)}{(\beta^+-\beta^-+\pi i)(\beta^+-\beta^0)(\beta^- -\beta^0)(\beta^+-\beta^-)(\beta^0-\beta^0-\pi i)} \right\} P_{0,1}\Bigg|_{\beta_n=\beta_1+\pi i} \\
&= \langle\langle 0\| \sum_{(B\backslash\beta_1,\beta_n)=B^-\cup B^0\cup B^+} \prod \tilde{\mathcal{B}}(\beta^0)\tilde{\mathcal{B}}(\beta^+)\tilde{\mathcal{B}}(\beta^+ +\pi i)\tilde{\phi}(B^-|B^0|B^+) \\
&\times \prod \frac{(\beta^+-\beta^- -\pi i)}{(\beta^+-\beta^-+\pi i)(\beta^+-\beta^0)(\beta^- -\beta^0)(\beta^+-\beta^-)(\beta^0-\beta^0-\pi i)} S_1^-(S_n^-)^2 \\
&\times \prod_{j=2}^{n-1} \frac{\beta_1-\beta_j}{\beta_1-\beta_j+2\pi i}\tilde{\mathcal{A}}(\beta_1+2\pi i)\tilde{\mathcal{C}}(\beta_1+\pi i) \,.
\end{aligned}
\tag{171}
$$

It can be shown that in LHS only those terms are essential in which $\beta_n\in B^+, \beta_1\in B^-$ or B^0. The accounting of terms with $\beta_n\in B^+, \beta_1\in B^-$ leads to the replacement in RHS of $\tilde{\mathcal{A}}(\beta_1+2\pi i)\tilde{\mathcal{C}}(\beta_1+\pi i)$ by $\tilde{\mathcal{A}}^{-1}(\beta_1+$

$\pi i)\tilde{\mathcal{C}}(\beta_1+\pi i)$. To transform the LHS for $\beta_n \in B^+, \beta_1 \in B^0$ one should use the identity

$$\lim_{\beta_n\to\beta_1+\pi i}\langle\langle 0\|\mathcal{B}(\beta_1)\prod \mathcal{B}(\beta^0)\mathcal{B}(\beta^+)\mathcal{B}(\beta^++\pi i)\mathcal{B}(\beta_n)\mathcal{B}(\beta_n+\pi i)P_{0,1}$$
$$=\langle\langle 0\|S_1^-(S_n^-)^2\prod \tilde{\mathcal{B}}(\beta^0)\tilde{\mathcal{B}}(\beta^+)\tilde{\mathcal{B}}(\beta^++\pi i)$$
$$\times\prod\frac{(\beta_1-\beta^0+\pi i)^2(\beta_1-\beta^-+\pi i)(\beta_1-\beta^++2\pi i)(\beta_1-\beta^+)}{(\beta_1-\beta^0+2\pi i)(\beta_1-\beta^0-\pi i)(\beta_1-\beta^--\pi i)(\beta_1-\beta^-)(\beta_1-\beta^++2\pi i)}, \tag{172}$$

whose proof is similar to that of Eq. (168): operators $\mathcal{B}$ should be applied to $\langle\langle 0\|$ in the order they are written in LHS of Eq. (172). Note that it is impossible to use Eqs. (161), (162) here since an indeterminacy appears: in Eq. (172) the multiplier $(\beta_n-\beta_1-\pi i)$ is absent, in Eq. (162) one has for $\sigma=\beta_1$ zero in the RHS. Formula (172) is the result of uncovering this indeterminacy.

To transform RHS of (171) one should use the identity

$$\langle 0\|\prod\tilde{\mathcal{B}}(\beta^0)\tilde{\mathcal{B}}(\beta^+)\tilde{\mathcal{B}}(\beta+\pi i)\tilde{\mathcal{C}}(\beta_1+\pi i)$$
$$=\prod\frac{(\beta_1-\beta^0)(\beta_1-\beta^-+\pi i)}{(\beta_1-\beta^0-\pi i)(\beta_1-\beta^--\pi i)}$$
$$\times\langle\langle 0\|\left\{2\sum_{\beta\in B^+}\prod_{B^0\cup\beta}\mathcal{B}(\beta^0)\prod_{B^+\backslash\beta}\mathcal{B}(\beta^+)\right.$$
$$\times\tilde{\mathcal{B}}(\beta^++\pi i)\frac{1}{\beta_1-\beta-\pi i}\prod\frac{(\beta-\beta^0+2\pi i)(\beta-\beta^++2\pi i)}{(\beta-\beta^0+\pi i)(\beta-\beta^+)}$$
$$+\sum_{\beta\in B^0}\prod_{B^0\backslash\beta}\tilde{\mathcal{B}}(\beta^0)\prod_{B^+}\tilde{\mathcal{B}}(\beta^++\pi i)\frac{1}{\beta_1-\beta}$$
$$\left.\times\prod\frac{(\beta-\beta^++\pi i)(\beta-\beta^0+\pi i)}{(\beta-\beta^+-\pi i)(\beta-\beta^0)}\right\}. \tag{173}$$

Equation (173) is not quite obvious; let us outline its proof. Both LHS and RHS are rational functions of β_1, decreasing for $\beta_1\to\infty$ and having $n(B^0)+n(B^-)+n(B^+)$ simple poles at the points $\beta_1=\beta+\pi i, \beta\in B^-\cup B^+\cup B^0$. That is why it is sufficient to prove Eq. (173) at $n(B^-)+n(B^0)+n(B^+)$ points. RHS has zeros at the point $\beta_1=\beta^--\pi i$. Among the commutation relations on elements of $\tilde{\tau}$ one can find the following:

$$\begin{aligned}&\tilde{\mathcal{B}}(\sigma_1)\tilde{\mathcal{C}}(\sigma_2)-\tilde{\mathcal{C}}(\sigma_2)\tilde{\mathcal{B}}(\sigma_1)\\&\quad=\left(\frac{1}{\sigma_1-\sigma_2}\right)(\tilde{\mathcal{A}}(\sigma_1)\tilde{\mathcal{D}}(\sigma_2)-\tilde{\mathcal{A}}(\sigma_2)\tilde{\mathcal{D}}(\sigma_1))\,.\end{aligned}\tag{174}$$

Let $\beta_k=\beta_1+\pi i, \beta_k\in B^-$, and move $\tilde{\mathcal{C}}(\beta_k)$ to the left, successively using (174). We obtain zero because

$$\begin{aligned}&\langle\langle 0\|\prod\tilde{\mathcal{B}}(\beta^0)\tilde{\mathcal{B}}(\beta^+)\tilde{\mathcal{B}}(\beta^++\pi i)\tilde{\mathcal{A}}(\beta)=0\,,\quad\beta\in B^-\\&\langle\langle 0\|\prod\tilde{\mathcal{B}}(\beta^0)\tilde{\mathcal{B}}(\beta^+)\tilde{\mathcal{B}}(\beta^++\pi i)\tilde{\mathcal{A}}(\beta+\pi i)=0\,,\quad\beta\in B^0\,.\end{aligned}\tag{175}$$

Consider now $\beta_k=\beta_1, \beta_k\in B^0$. In RHS one has

$$\langle\langle 0\|\prod_{B^0\backslash\beta_k}\tilde{\mathcal{B}}(\beta^0)\tilde{\mathcal{B}}(\beta^+)\tilde{\mathcal{B}}(\beta^++\pi i)\prod_{j\neq k}\frac{\beta_1-\beta_j+\pi i}{\beta_1-\beta_j-\pi i}\,.\tag{176}$$

In LHS one has

$$\langle\langle 0\|\prod_{B^0\backslash\beta_k}\tilde{\mathcal{B}}(\beta^0)\tilde{\mathcal{B}}(\beta^+)\tilde{\mathcal{B}}(\beta^++\pi i)\tilde{\mathcal{B}}(\beta_k)\tilde{\mathcal{C}}(\beta_k+\pi i)\,.\tag{177}$$

From Eq. (156) it follows that

$$\tilde{\mathcal{B}}(\beta_k)\tilde{\mathcal{C}}(\beta_k+\pi i)=\tilde{\mathcal{A}}(\beta_k)\tilde{\mathcal{D}}(\beta_k+\pi i)-\prod_{j\neq k}\left(\frac{\beta_k-\beta_j+\pi i}{\beta_k-\beta_j-\pi i}\right)\,.$$

Substitute this identity into Eq. (177). The first term disappears by virtue of Eq. (175) and the remaining term is equal to (176). The case of $\beta_k=\beta_1-\pi i, \beta_k\in B^+$ is treated in a similar way. This completes the proof of Eq. (173).

Let us return to Eq. (171). Transform LHS using Eq. (172) and RHS using Eq. (173) and exploit

$$\begin{aligned}&\tilde{\phi}(B^-|B^0\cup\beta_1|B^+\cup(\beta_1+\pi i))=2(\pi i)^2\tilde{\phi}(B^-|B^0|B^+\cup\beta_1)\\&\times\prod(\beta_1-\beta^++2\pi i)(\beta_1-\beta^0+2\pi i)(\beta_1-\beta^--\pi i)(\beta_1-\beta^0-\pi i)\,.\end{aligned}$$

As a result the proof of Eq. (171) reduces to the proof of the identity

$$
\begin{aligned}
\tilde{\phi}(B^-|B^0|B^+\cup\beta_1) &= \prod(\beta_1-\beta^0-\pi i)(\beta_1-\beta^-) \\
&\times\Bigg\{\sum_{\beta\in B^-}\tilde{\phi}(B^-|\beta|B^0\cup\beta|B^+)\frac{1}{\beta_1-\beta}\prod\frac{1}{(\beta-\beta^0-\pi i)(\beta-\beta^-)} \\
&+2\pi^2\sum_{\beta\in B^0}\tilde{\phi}(B^-|B^0\backslash\beta|B^+\cup\beta)\frac{1}{(\beta_1-\beta-\pi i)} \\
&\times\prod\frac{(\beta-\beta^0+2\pi i)(\beta-\beta^++2\pi i)(\beta-\beta^0-\pi i)(\beta-\beta^--\pi i)}{(\beta-\beta^0)(\beta-\beta^-+\pi i)}\Bigg\}\,,
\end{aligned}
\tag{178}
$$

which follows from Lemma 3 and the fact that $\tilde{\phi}(B^-|B^0|B^+\cup\beta_1)$ is a polynomial of degree of $n(B^0)+n(B^+)-1$ with respect to β_1.

Q.E.D.

We have shown that form factors $f_\mu^a, f^a, f_{\mu\nu}, f$ satisfy Axioms 1, 2, 3 and thus defined some local operators. What are the properties of these operators? With respect to the Lorentz group the operators defined by f_μ^a and $f_{\mu\nu}$ are a vector and an (1, 1) tensor respectively, the operators defined by f^a, f are scalars. With respect to the isotopic group the operators defined by f_μ^a, f^a are vectors, the operators defined by $f_{\mu\nu}, f$ are scalars. The possibility of introducing the nontrivial operator of charge conjugation C in the space of states is connected, of course, with the possibility of introducing it in terms of initial Lagrangian formalism: $Cn^aC^{-1}=-n^a$. Evidently the Lagrangian is invariant under this transformation. The operators defined by $f_{\mu\nu}, f_\mu^a$ are even with respect to the charge conjugation operators f^a; f is odd.

The above facts show that it is plausible to identify the operators defined by f_μ^a and $f_{\mu\nu}$ with $j_\mu^a=\varepsilon^{abc}\partial_\mu n^b n^c$ and $T_{\mu\nu}$. This identification is confirmed by the following theorem.

Theorem 3. If operators $j_\mu^a, T_{\mu\nu}$ are defined by $f_\mu^a, f_{\mu\nu}$ then

$$
\partial_\mu j_\mu^a=0\,,\quad \partial_\mu T_{\mu\nu}=0\,,\quad T_{\mu\nu}=T_{\nu\mu} \tag{179}
$$

and the operators $P_\mu=\int_{-\infty}^{\infty}T_{\mu 0}(x_1)dx_1, Q^a=\int_{-\infty}^{\infty}j_0^a(x_1)dx_1$ are the operators of energy-momentum and charge, which means

$$\begin{aligned}
&P_\mu Z^*_{w_1}(\beta_1)\dots Z^*_{w_n}(\beta_n)|ph\rangle \\
&\quad = M\sum_{j=1}^{n}(e^{\beta_j}+(-1)^\mu e^{-\beta_j})Z^*_{w_1}(\beta_1)\dots Z^*_{w_n}(\beta_n)|ph\rangle\ , \\
&Q^a Z^*_{w_1}(\beta_1)\dots Z^*_{w_n}(\beta_n)|ph\rangle \\
&\quad = \sum_{j=1}^{n}(S^a)^{w_j'}_{w_j} Z^*_{w_1}(\beta_1)\dots Z^*_{w_j'}(\beta_j)\dots Z^*_{w_n}(\beta_n)|ph\rangle\ .
\end{aligned} \tag{180}$$

Proof. Eqs. (179) follow from the very definition of form factors. The proof of Eq. (180) is quite similar to those of Theorem 5 of Sec. 6; it is based on the explicit formulae for the two-particle form factors presented above.
Q.E.D.

The operator defined by f^a is C-odd Lorentz scalar transformed under vector representation of the isotopic group. It is natural to identify it with the field n^a itself. Additional arguments in favour of this identification will be presented in Sec. 10 by calculating the singularities of commutators at the origin of coordinates. At last, the operator defined by f is C-odd Lorentz and isotopic. It is natural to identify it with the operator of the euclidean topological charge $q=\varepsilon^{abc}\varepsilon_{\mu\nu}\partial_\mu n^a\partial_\nu n^b n^c$ which is the simplest operator of the kind.

The norms of the form factors are calculated by means of formulae similar to (80), (125). For example,

$$\begin{aligned}
(\langle\langle\{\phi^0\}\rangle\rangle_n, \langle\langle\{\phi^0\}\rangle\rangle_n) &= \sum_{B=B^-\cup B^0\cup B^+} (\phi^0(B^-|B^0|B^+))^2 \\
&\times \frac{(\beta^+-\beta^- -\pi i)}{(\beta^+-\beta^-)(\beta^+-\beta^0)(\beta^0-\beta^-)(\beta^0-\beta^0-\pi i)(\beta^+-\beta^-+\pi i)} \\
&\times \frac{1}{(\beta^0-\beta^0-2\pi i)(\beta^0-\beta^- -2\pi i)(\beta^+-\beta^- -2\pi i)(\beta^+-\beta^0-2\pi i)}\ .
\end{aligned} \tag{181}$$

9

ASYMPTOTICS OF FORM FACTORS

The local commutativity theorem (Theorem 1 of Sec. 2) requires besides the validity of Axioms 1, 2, 3, 4 special asymptotic behaviors of $f(\beta_1, \ldots, \beta_k, \beta_{k+1} + \sigma, \ldots, \beta_n + \sigma)$ for $\sigma \to \infty$, which will be studied in the present section. The asymptotic for ITM will be considered in detail; for SG model only statements will be given; the asymptotic for NLS will be obtained from the ITM ones. We consider in detail ITM asymptotics, but not the SG ones, because of two reasons: first, proofs are more descriptive for ITM; second, for ITM more refined formulae can be obtained which will be used in the next section.

Before we calculate the asymptotic of form factors in ITM let us prove some subsidiary statements. Theorem 4 of Sec. 6 implies for the limit $\xi \to \infty$ the following Lemma.

Lemma 1. Form factors $f_\mu^a, f_{\mu\nu}$ in ITM can be presented in the form

$$
\begin{aligned}
f_\mu^3(\beta_1, \ldots, \beta_{2n}) &= \left[\sum (e^{\beta_j} - (-1)^\mu e^{-\beta_j})\right] \prod_{i<j} \zeta(\beta_i - \beta_j) \langle F_n^3 \rangle_n \ , \\
f_\mu^a(\beta_1, \ldots, \beta_{2n}) &= \left[\sum (e^{\beta_j} - (-1)^\mu e^{-\beta_j})\right] \prod_{i<j} \zeta(\beta_i - \beta_j) \\
&\quad \times (\langle F_n^+ \rangle_n + (-1)^a \langle F_n^- \rangle_n) \ , \quad a = 1, 2 \\
f_{\mu\nu}(\beta_1, \ldots, \beta_{2n}) &= \left[\sum (e^{\beta_j} - (-1)^\mu e^{-\beta_j}) \sum (e^{\beta_j} - (-1)^\nu e^{-\beta_j})\right] \\
&\quad \times \prod_{i<j} \zeta(\beta_i - \beta_j) \langle F_n \rangle_n \ , \qquad (182)
\end{aligned}
$$

where

$$F_n(\beta_1,\dots,\beta_n|\beta_{n+1},\dots,\beta_{2n}) = \frac{(2\pi i)^{-n}}{n!}\int d\alpha_1\dots d\alpha_n \prod \varphi(\alpha_i-\beta_j)$$

$$\prod_{i<j}\mathrm{sh}(\alpha_i-\alpha_j)\Pi^0(\alpha_1,\dots,\alpha_n|\beta_1,\dots,\beta_n|\beta_{n+1},\dots,\beta_{2n}),$$

$$F_n^3(\beta_1,\dots,\beta_n|\beta_{n+1},\dots\beta_{2n})$$

$$= \left(\sum_{j=1}^{n}(\beta_{n+j}-\beta_j-\pi i)\right)F(\beta_1,\dots,\beta_n|\beta_{n+1},\dots,\beta_{2n}),$$

$$F_n^{\pm}(\beta_1,\dots,\beta_{n\pm1}|\beta_{n+1\mp1},\dots,\beta_{2n})$$

$$= \frac{(2\pi i)^{-n}}{n!}\int d\alpha_1\dots,d\alpha_n\prod\varphi(\alpha_i-\beta_j)$$

$$\times\prod_{i<j}\mathrm{sh}(\alpha_i-\alpha_j)\Pi^{\pm}(\alpha_1,\dots,\alpha_n|\beta_1,\dots,\beta_{n+1}|\beta_{n+1\mp1},\dots,\beta_{2n}).$$

Polynomials $\Pi^c(\alpha_1,\dots,\alpha_n|\lambda_1,\dots,\lambda_{n+c}|\mu_1,\dots,\mu_{n-c})$ are the determinants of the $n\times n$ matrices $\mathcal{B}_{ij}^c(c=0,\pm)$ with matrix elements

$$\mathcal{B}_{ij}^c = A_{i-1}^c(\alpha_j|\lambda_1,\dots,\lambda_{n+c}|\mu_1,\dots,\mu_{n-c}),(2\le i\le n),\mathcal{B}_{1j}^c=\phi^c(\alpha_j),$$

where the functions A_i^c are given in Sec. 7, $\phi^c(\alpha)$ are arbitrary polynomials of degree $(n-1)$ whose senior coefficient is equal to 1. Later on we shall take for $\phi^c(\alpha_j), c=\pm$, the following polynomials:

$$\phi^+(\alpha)=\prod_{q=1}^{n-1}\left(\alpha-\mu_q+\frac{\pi i}{2}\right),\quad \phi^-(\alpha)=\prod_{q=1}^{n-1}\left(\alpha-\lambda_q-\frac{\pi i}{2}\right).$$

In this case the terms containing $\delta_{c,-1}\sigma_i(\lambda_1,\dots,\lambda_{n-1}),\delta_{c,1}\sigma_i(\mu_1-\pi i,\dots,\mu_{n-1}-\pi i)$ can be omitted in the formulae for A_i^c (118), since their contributions to the 2-nd,..., n-th rows are proportional to the first row.

For further estimation we shall need the following representation of the polynomials Π^c.

Lemma 2. Polynomials $\Pi^c(\alpha_1,\dots,\alpha_n|\lambda_1,\dots,\lambda_{n+c}|\mu_1,\dots,\mu_{n-c})$ can be presented as follows:

$$\Pi^c(\alpha_1,\dots,\alpha_n|\lambda_1,\dots,\lambda_{n+c}|\mu_1,\dots,\mu_{n-c})=\frac{\det\mathcal{D}^c}{\pi i\prod_{i<j}(\alpha_i-\alpha_j)},\tag{183}$$

where $\mathcal{D}^c$ are $n \times n$ matrices with matrix elements

$$\mathcal{D}^c_{ij} = \prod_{q=1}^{n+c}\left(\alpha_j - \lambda_q - \frac{\pi i}{2}\right)\sum_{m=1}^{n-c-1}\left\{\prod_{q=m+2}^{n-c}(\alpha_i - \mu_q)\sum_{s=0}\frac{1}{(2s+1)!}\right.$$
$$\times\left(\frac{\pi i}{2}\frac{d}{d\alpha_j}\right)^{2s+1}\prod_{q=1}^{m}(\alpha_j - \mu_q)\prod_{q=1}^{n-c}\left(\alpha_j - \mu_q + \frac{\pi i}{2}\right)$$
$$\times\sum_{m=1}^{n+c-1}\prod_{q=m+2}^{n+c}(\alpha_i - \lambda_q - \pi i)\frac{1}{(2s+1)!}\left(\frac{\pi i}{2}\frac{d}{d\alpha_j}\right)^{2s+1}$$
$$\left.\times\prod_{q=1}^{m}(\alpha_j - \lambda_q)\right\} - \pi i \delta_{c,0}\phi^0(\alpha_j)\phi(\alpha_i) \tag{184}$$

where $\phi(\alpha_j)$ is an arbitrary polynomial of degree $(n-1)$ whose senior coefficient is equal to 1.

Proof. Consider, for example, $\Pi^+(\alpha_1,\dots,\alpha_n|\lambda_1,\dots,\lambda_{n+1}|\mu_1,\dots,\mu_{n-1})$. According to Lemma 1 it is the determinant of an $n \times n$ matrix with matrix elements

$$\mathcal{B}^+_{ij} = \prod_{q=1}^{n+1}\left(\alpha_j - \lambda_q - \frac{\pi i}{2}\right)Q_{i-2}(\alpha_j - \pi i|\mu_1 - \pi i,\dots,\mu_{n-1} - \pi i)$$
$$+\prod_{q=1}^{n-1}\left(\alpha_j - \mu_q + \frac{\pi i}{2}\right)Q_i(\alpha_j|\lambda_1,\dots,\lambda_{n+1}), \tag{185}$$

where the polynomials Q_i are defined in Sec. 7, Eq. (119)

$$Q_i(\alpha|\lambda_1,\dots,\lambda_k)$$
$$= \sum_{l=0}^{i}\left[\left(\alpha + \frac{\pi i}{2}\right)^l - \left(\alpha - \frac{\pi i}{2}\right)^l\right](-1)^{i-l}\sigma_{i-l}(\lambda_1,\dots,\lambda_k).$$

Let us multiply and divide Π^+ by the Vandermond determinant

$$\det\begin{vmatrix} (\alpha_1 - \pi i)^{n-1}, & (\alpha_1 - \pi i)^{n-2}, & \cdots\cdots, & 1 \\ \cdot & & & \cdot \\ \cdot & & & \cdot \\ \cdot & & & \cdot \\ \cdot & & & \cdot \\ \cdot & & & \cdot \\ (\alpha_n - \pi i)^{n-1}, & (\alpha_n - \pi i)^{n-2}, & \cdots\cdots, & 1 \end{vmatrix}.$$

We obtain

$$\Pi^+(\alpha_1,\dots,\alpha_n|\lambda_1,\dots,\lambda_{n+1}|\mu_1,\dots,\mu_{n-1}) = \frac{\det \mathcal{D}^+}{\pi i \prod\limits_{i<j}(\alpha_i-\alpha_j)} ,$$

where

$$\mathcal{D}^+_{ij} = \prod_{q=1}^{n+1}\left(\alpha_j-\lambda_q-\frac{\pi i}{2}\right)\sum_{m=1}^{n-2} Q_m(\alpha_j-\pi i|\mu_1-\pi i,\dots,$$
$$\times\ \mu_{n-1}-\pi i)(\alpha_i-\pi i)^{n-m-2} + \prod_{q=1}^{n-1}\left(\alpha_j-\mu_q+\frac{\pi}{2}\right)$$
$$\times\sum_{m=1}^{n} Q_m(\alpha_j|\lambda_1,\dots,\lambda_{n+1})(\alpha_i-\pi i)^{n-m} .$$

Thus we have to show that

$$\sum_{m=1}^{k-1} Q_m(\alpha_1|\nu_1,\dots,\nu_k)\alpha_2^{k-1-m} = \sum_{m=1}^{k-1}\prod_{q=m+2}^{k}(\alpha_2-\nu_q)$$
$$\times\sum_{s=0}\frac{1}{(2s+1)!}\left(\frac{\pi i}{2}\frac{d}{d\alpha_i}\right)^{2s+1}\prod_{q=1}^{m}(\alpha_1-\nu_q) . \tag{186}$$

Let us prove this identity. Notice that Q_m can be presented as follows:

$$Q_m(\alpha_1|\nu_1,\dots,\nu_k) = \sum_{l=0}^{m}\sum_{u=l}^{m}\left(\frac{\pi i}{2}\right)^l[1-(-1)^l]C_u^l\alpha_1^{u-l}$$
$$\times\ (-1)^{m-u}\sigma_{m-u}(\nu_1,\dots,\nu_k)$$
$$= \sum_{s=0}\left(\frac{\pi i}{2}\right)^{2s+1}\frac{1}{(2s+1)!}\left(\frac{d}{d\alpha_1}\right)^{2s+1}\sum_{u=0}^{m}\sigma_{m-u}(\nu_1,\dots,\nu_k) ,$$

hence

$$\sum_{m=1}^{k-1} Q_m(\alpha_1|\nu_1,\dots,\nu_k)\alpha_2^{k-m-1} = \sum_{s=0}\left(\frac{\pi i}{2}\right)^{2s+1}\frac{1}{(2s+1)!}$$
$$\times\left(\frac{d}{d\alpha_1}\right)^{2s+1}\sum_{p=1}^{k-1}\sum_{q=0}^{k-p-1}\alpha_1^p\alpha_2^q(-1)^{k-p-q-1}\sigma_{k-p-q-1}(\nu_1,\dots,\nu_k) .$$

Now we have only to use the identity

$$\sum_{p=1}^{k-1}\sum_{q=0}^{k-p-1} \alpha_q^p \alpha_2^q (-1)^{k-p-q-1}\sigma_{k-p-q-1}(\nu_1,\ldots,\nu_k)$$
$$= \sum_{m=0}^{n-1}\prod_{q=1}^{m}(\alpha_1-\nu_q)\prod_{q=m+2}^{k}(\alpha_2-\nu_q)\,.$$

The proof of the last identity is not very complicated but a beautiful combinatorial problem which we leave to the reader.

Thus the representation (183) for Π^+ is proven. Π^- can be considered in a similar way; to prove Eq. (183) for Π^0 one should replace the first column of the Vandermond determinant used by $\phi(\alpha_i), i = 1,\ldots,n$.

Remark 1. The series of the type $\sum\limits_{s=0}\frac{1}{(2s+1)!}\left(\frac{\pi i}{2}\frac{d}{d\alpha}\right)^{2s+1}\prod\limits_{q=1}^{m}(\alpha-\nu_q)$ encountered in Eq. (184) terminates. Evidently

$$\sum_{s=0}^{\infty}\frac{1}{(2s+1)!}\left(\frac{\pi i}{2}\frac{d}{d\alpha}\right)^{2s+1} = \frac{1}{2}\left(\exp\left(\frac{\pi i}{2}\frac{d}{d\alpha}\right) - \exp\left(-\frac{\pi i}{2}\frac{d}{d\alpha}\right)\right)\,,$$

and consequently

$$\sum_{s=0}^{\infty}\frac{1}{(2s+1)!}\left(\frac{\pi i}{2}\frac{d}{d\alpha}\right)^{2s+1}\prod_{q=1}^{m}(\alpha-\nu_q)$$
$$= \frac{1}{2}\left[\prod_{q=1}^{m}\left(\alpha-\nu_q+\frac{\pi i}{2}\right) - \prod_{q=1}^{m}\left(\alpha-\nu_q-\frac{\pi i}{2}\right)\right]\,.$$

Remark 2. From the proof of Lemma 2 it follows that ordering of the sets $\{\lambda\}$ and $\{\mu\}$ is of no importance. For example, in the expression

$$\prod_{q=m+2}^{n-c}(\alpha_i-\mu_q)\sum_{s=0}\left(\frac{\pi i}{2}\right)^{2s+1}\frac{1}{(2s+1)!}\left(\frac{d}{d\alpha_j}\right)^{2s+1}\prod_{q=1}^{m}(\alpha_j-\mu_q)$$

all μ_q can be replaced by $\mu_{\pi(q)}$, where π is an arbitrary permutation.

The asymptotic behaviour of $f(\beta_1,\ldots,\beta_k,\beta_{k+1}+\sigma,\ldots,\beta_{2n}+\sigma)$ is described by the following theorem.

Theorem 1. Consider the form factors $f_\mu^a, f_{\mu\nu}$ in ITM. Denote by g^a and g the following functions:

$$g^3(\beta_1, \dots, \beta_{2n}) = \prod_{i<j} \zeta(\beta_i - \beta_j) \langle F_n^3 \rangle_n \ ,$$

$$g^a(\beta_1, \dots, \beta_{2n}) = \prod_{i<j} \zeta(\beta_i - \beta_j)(\langle F_n^+ \rangle_n + (-1)^a \langle F_n^- \rangle_n) \ , \quad a = 1, 2 \ ,$$

$$g(\beta_1, \dots, \beta_{2n}) = \prod_{i<j} \zeta(\beta_i - \beta_j) \langle F_n \rangle_n$$

which are connected with $f_\mu^a, f_{\mu\nu}$ by Eq. (182). The following asymptotic holds: let $\beta_1, \dots, \beta_k$ be finite while β_j for $j \geq k+1$ are equal to $\tilde{\beta}_j + \sigma$, $\tilde{\beta}_j$ being finite quantities and $\sigma \to \infty$, then

$$\begin{aligned} g^a(\beta_1, \dots, \beta_{2n}) &= \mathrm{O}\left(\sigma^{\frac{1}{4}} \exp\left(-\frac{1}{4}|\sigma|\right)\right) \ , & k \equiv 1 (\mathrm{mod}\ 2) \ , \\ g(\beta_1, \dots, \beta_{2n}) &= \mathrm{O}\left(\sigma^{-\frac{3}{4}} \exp\left(-\frac{1}{4}|\sigma|\right)\right) \ , & k \equiv 1 (\mathrm{mod}\ 2) \ , \\ g(\beta_1, \dots, \beta_{2n}) &= \mathrm{O}(\sigma^{-2}) \ , & k \equiv 0 (\mathrm{mod}\ 2) \ , \\ g^a(\beta_1, \dots, \beta_{2n}) &= 2\pi\sigma^{-1}\varepsilon^{abc} g^b(\beta_1, \dots, \beta_k) & \\ &\quad \times g^c(\tilde{\beta}_{k+1}, \dots, \beta_{2n})(1 + \mathrm{O}(\sigma^{-1})) \ , & k \equiv 0 (\mathrm{mod}\ 2) \ . \end{aligned} \tag{187}$$

We have presented in explicit expression the principal term of the asymptotics in the last formula as it will be important later.

Proof. Notice first that we can limit ourselves to the consideration of the case $k = n$. In fact, suppose we have proven Eqs. (187) for $k = n$. Then using Eq. (110), we get

$$\begin{aligned} &\mathrm{res}\, g(\beta_1, \beta_2, \dots \beta_{2n}) = g(\beta_3, \dots, \beta_{2n}) \otimes (e_{1,1} \otimes e_{2,2} - e_{2,1} \otimes e_{1,2}) \\ &\times (S(\beta_{2n} - \beta_1) \dots S(\beta_3 - \beta_1) - 1) = g(\beta_3, \dots, \beta_{2n}) \\ &\otimes (e_{1,1} \otimes e_{2,2} - e_{1,2} \otimes e_{2,1})(S(\beta_n - \beta_1) \dots S(\beta_3 - \beta_1) - 1 + \mathrm{O}(\sigma^{-1})) \ , \\ &\qquad \beta_2 = \beta_1 + \pi i \ . \end{aligned}$$

Here we have used the asymptotics $S(\sigma) = 1 + \mathrm{O}(\sigma^{-1})$. The asymptotics of the LHS have been assumed known, so that $g(\beta_3, \dots, \beta_{2n})$ also has the required asymptotics (the operator $1-S(\beta_n-\beta_1)\ldots S(\beta_3-\beta_1)$ has no kernel independent of β_1, the latter being arbitrary). The case $\beta_{2n}=\beta_{2n-1}+\pi i$ can be treated in a similar way. Thus we can obtain from Eq. (187) for $k=n$ Eqs. (187) for $k>n$ and $k<n$.

Let us suppose for the sake of definiteness $\sigma>0$ and consider the function g^3:

$$g^3(\beta_1,\ldots,\beta_{2n})=\prod_{i<j}\zeta(\beta_i-\beta_j)\langle F^3\rangle_n\,. \tag{188}$$

For the form factors in ITM the combination of exponential and power behaviour is characteristic. Consider, at first, the exponents. The function $\zeta(\beta)$ has the following asymptotics:

$$\zeta(\beta)\underset{\beta\to\infty}{=}2^{-\frac{1}{4}}\exp\left(\frac{\beta}{4}\right)\beta^{\frac{1}{4}}(1+\mathrm{O}(\beta^{-1}))\,,$$

hence

$$\prod_{i<j}^{2n}\zeta(\beta_i-\beta_j)\sim\left(\frac{\sigma}{2}\right)^{\frac{n^2}{4}}e^{\frac{n^2}{4}\sigma}\prod_{i<j}^{n}\zeta(\beta_i-\beta_j)\prod_{i<j\leq n+1}^{2n}\zeta(\tilde\beta_i-\tilde\beta_j)\,. \tag{189}$$

Let us now turn to $\langle F_n^3\rangle_n$. From the very definition it follows that

$$\langle F_n^3\rangle_n=\sum_{\{1\ldots 2n\}=\{i_1\ldots i_n\}\cup\{j_1\ldots j_n\}}\langle 0|\prod B(\beta_{jp})F(\beta_{i_1},\ldots,\beta_{i_n}|_{j_1},\ldots,\beta_{j_n})$$
$$\times\prod\frac{1}{\beta_{i_p}-\beta_{j_q}}\,. \tag{190}$$

The factors $\Pi B(\beta_{j_p})\Pi(\beta_{i_p}-\beta_{j_q})^{-1}$ give only power contribution; that is why we have only to estimate the exponential behaviour of

$$F^3(\beta_{i_1},\ldots,\beta_{i_n}|\beta_{j_1},\ldots,\beta_{j_n})$$
$$=\left(\sum(\beta_{j_p}-\beta_{i_p}-\pi i)\right)F(\beta_{i_1},\ldots,\beta_{i_n}|\beta_{j_1}\ldots,\beta_{j_n})\,,$$

where F is the determinant of an $n\times n$ matrix with matrix elements

$$\mathcal{F}_{ij}=\frac{2^{-\frac{n+1}{2}}}{\pi i}\int d\alpha\prod_{j=1}^{2n}\varphi(\alpha-\beta_j)A_{i-1}(\alpha|\beta_{i_1},\ldots,\beta_{i_n}|\beta_{j_1},\ldots,\beta_{j_n})$$
$$\times\exp(n+1-2j)\alpha\,,\qquad i>j\,,$$
$$\mathcal{F}_{1j}=\frac{2^{-\frac{n+1}{2}}}{\pi i}\int d\alpha\prod_{j=1}^{2n}\varphi(\alpha-\beta_j)\phi(\alpha)\exp(n+1-2j)\alpha\,,$$

where $\phi(\alpha)$ is an arbitrary polynomial of degree $n-1$. A_i and ϕ are polynomials, which is why the exponential behaviour of the integrands is determined by $\varphi(\alpha-\beta_j)$ and $\exp(n+1-2j)\alpha$. The function $\varphi(\alpha-\beta_p)$ behaves for $\alpha \to \pm\infty$ as follows:

$$\varphi(\alpha-\beta_p) = 2^{\frac{1}{2}} \exp\left(-\frac{1}{2}|\alpha-\beta_p|\right) \alpha^{-\frac{1}{2}}(1+\mathrm{O}(\alpha^{-1})) .$$

We can estimate this roughly by the principal term of its asymptotics. Consider three regions on the real axis: $\alpha < \beta_j, j = 1,\ldots n$; $\beta_i < \alpha < \beta_j, i = 1,\ldots,n, j = n+1,\ldots 2n$; $\alpha > \beta_j, j = n+1,\ldots,2n$. The exponential behaviour of the integrands in these regions is described by $\exp((n+1-2j)\alpha + 2n\alpha - \sum_{j=1}^{2n}\beta_j)$, $\exp((n+1-2j)\alpha - \sum_{j=n+1}^{2n}\beta_j + \sum_{j=1}^{n}\beta_j)$, $\exp((n+1-2j)\alpha - 2n\alpha + \sum_{j=1}^{2n}\beta_j)$ respectively. The first exponent is always an increasing one, i.e. it decreases for $\alpha \to -\infty$; the third exponent is always a decreasing one, i.e. it decreases for $\alpha \to \infty$. The second exponent is an increasing one for $n+1-2j > 0$ and a decreasing one for $n+1-2j < 0$; for $n+1-2j = 0$ the exponential behaviour in the intermediate region disappears. These estimations show that for $n+1-2j > 0$ only the vicinity of $\alpha = \sigma$ gives significant contribution to the integral; $\prod_{p=1}^{n}\varphi(\alpha-\beta_p)$ can then be replaced by its asymptotics

$$\begin{aligned}
\mathcal{F}_{ij} &= 2^{\frac{1}{2}}\sigma^{-\frac{n}{2}} \exp\left(\left(\frac{n}{2}-2j+1\right)\sigma\right) \frac{1}{2\pi i}\int_{-\infty}^{\infty} d\tilde{\alpha} \\
&\times \prod_{p=n+1}^{2n} \varphi(\tilde{\alpha}-\tilde{\beta}_p) \exp\left(\left(\frac{n}{2}+1-2j\right)\tilde{\alpha}\right) \\
&\times A_{i-1}(\tilde{\alpha}+\sigma|\beta_{i_1},\ldots,\beta_{i_n}|\beta_{j_1},\ldots,\beta_{j_n})(1+\mathrm{O}(\sigma^{-1})), \quad n+1-2j>0 .
\end{aligned}$$

We have denoted $\phi(\alpha)$ by $A_0(\alpha)$. For $n+1-2j < 0$ only the vicinity of zero gives an essential contribution to the integral

$$\begin{aligned}
\mathcal{F}_{ij} &= 2^{\frac{1}{2}}\sigma^{-\frac{n}{2}} \\
&\times \exp\left(-\frac{n\sigma}{2}\right) \frac{1}{2\pi i}\int_{-\infty}^{\infty} d\alpha \prod_{p=1}^{n} \varphi(\alpha-\beta_p) \exp\left(\left(\frac{3n+1}{2}+1-2j\right)\alpha\right) \\
&\times A_{i-1}(\alpha|\beta_{i_1},\ldots,\beta_{i_n}|\beta_{j_1},\ldots,\beta_{j_n})(1+\mathrm{O}(\sigma^{-1})) , \quad n+1-2j<0 .
\end{aligned}$$

Finally, for odd n the case $n+1-2j = 0$ is possible. For this j we have for $\mathcal{F}_{ij}$ a plateau between the points 0 and α, which is estimated as

$$\mathcal{F}_{ij} = 2^{\frac{1}{2}} \frac{1}{2\pi i} \sigma^{-n} \exp\left(-\frac{n\sigma}{2}\right) \int_0^{\sigma} A_{i-1}(\alpha|\beta_{i_1}, \ldots, \beta_{i_n}|\beta_{j_1}, \ldots \beta_{j_n}) d\alpha$$
$$\times (1 + \mathrm{O}(\sigma^{-1})) , \quad n + 1 - 2j = 0 .$$

Hence, for $n = 2k$ the full exponential contribution to the asymptotics is $\exp((k^2 + \sum_{j=1}^{k}(1-2j))\sigma) = 1$, for $n = 2k+1$ the full exponential contribution is $\exp((\frac{1}{4}(2k+1)(2k+1) - \frac{1}{2}(2k+1) + \sum_{j=1}^{k}(1-2j))\sigma) = \exp(-\frac{\sigma}{4})$. These results are in agreement with Eq. (187).

Let us now estimate the power contribution. We consider only the case $n = 2k$ which is the most interesting. For $n = 2k+1$ the form factor decreases exponentially.

The above reasoning implies

$$\det \mathcal{F} \underset{\sigma \to \infty}{\sim} e^{-k^2\sigma} \sigma^{-2k^2} \left(\frac{1}{\sqrt{2}\pi i}\right)^{2k} \int d\alpha_1 \ldots d\alpha_k d\tilde{\alpha}_1 \ldots d\tilde{\alpha}_k$$
$$\times \prod_{i=1}^{k} \prod_{p=1}^{2k} \varphi(\alpha_i - \beta_p) \prod_{p=2k+1}^{4k} \varphi(\tilde{\alpha}_i - \beta_p) \exp\left(\sum_{j=1}^{k} (k+1-2j)(\alpha_j + \tilde{\alpha}_j)\right)$$
$$\times \Pi^0(\alpha_1, \ldots, \alpha_k, \tilde{\alpha}_1 + \sigma, \ldots, \tilde{\alpha} + \sigma|\beta_{i_1}, \ldots, \beta_{i_n}|\beta_{j_1}, \ldots, \beta_{j_n}) , \tag{191}$$

where Π^0 is given by Eq. (183). Our main purpose is the estimation of $\Pi^0(\alpha_1, \ldots, \alpha_{2k}|\lambda_1, \ldots, \lambda_{2k}|\mu_1, \ldots, \mu_{2k})$ for the case $\alpha_1, \ldots, \alpha_k \sim 0$, $\alpha_{k+1}, \ldots, \alpha_{2k} \sim \sigma$; the set $\lambda_1, \ldots, \lambda_{2k}, \mu_1, \ldots \mu_{2k}$ is divided into two subsets both containing $2k$ elements, elements of the first subset being ~ 0, elements of the second subset being $\sim \sigma$. More concretely, let $\lambda_1, \ldots, \lambda_{k+p} \sim 0, \lambda_{k+p+1}, \ldots, \lambda_{2k} \sim \sigma$, $\mu_1, \ldots, \mu_{k-p} \sim 0, \mu_{k-p+1}, \ldots, \mu_{2k} \sim \sigma$, p being an integer such that $|p| \leq k$. For the sake of definiteness we consider $p \geq 0$ also.

Let us use for Π^0 the presentation (183), taking $-\prod_{q=1}^{2k-1}(\alpha_j - \mu_q + \frac{\pi i}{2})$ for $\phi(\alpha_j)$ and $\prod_{q=2}^{2k}(\alpha_i - \lambda_q - \frac{\pi i}{2})$ for $\psi(\alpha_i)$. The matrix appears to be divided into four blocks:

$$\mathcal{D}^0 = \begin{array}{|cc|} \mathcal{D}_1 & \mathcal{D}_2 \\ & \\ \mathcal{D}_3 & \mathcal{D}_4 \end{array}$$

The matrix elements of $\mathcal{D}_1$ and $\mathcal{D}_4$ have the following asymptotics for $\sigma \to \infty$:

$$\mathcal{D}^0_{ij} = \sigma^{2k}\left\{\prod_{q=1}^{k+p}\left(\alpha_j - \lambda_q - \frac{\pi i}{2}\right)\sum_{m=1}\prod_{q=m+2}^{k-p}(\alpha_i - \mu_q)\sum_{s=0}\left(\frac{\pi i}{2}\right)^{2s+1}\right.$$
$$\times\frac{1}{(2s+1)!}\left(\frac{d}{d\alpha_j}\right)^{2s+1}\prod_{q=1}^{m}(\alpha_i - \mu_q) + \prod_{q=1}^{k-p}\left(\alpha_j - \mu_q + \frac{\pi i}{2}\right)$$
$$\times\sum_{m=1}\prod_{q=m+2}^{k+p}(\alpha_i - \lambda_q)\sum_{s=0}\left(\frac{\pi i}{2}\right)^{2s+1}\frac{1}{(2s+1)!}\left(\frac{d}{d\alpha_j}\right)^{2s+1}$$
$$\times\prod_{q=1}^{m}(\alpha_j - \lambda_q - \pi i)(1 + \mathrm{O}(\sigma^{-1}))\,, \quad i \le k, j \le k\,,$$

$$\mathcal{D}^0_{ij} = \sigma^{2k}\left\{\prod_{q=k+p+1}^{2k}(\alpha_j - \lambda_q - \pi i)\sum_{m=1}\prod_{q=m+2}^{2k}(\alpha_i - \mu_q)\sum_{s=0}\left(\frac{\pi i}{2}\right)^{2s+1}\right.$$
$$\times\frac{1}{(2s+1)!}\left(\frac{d}{d\alpha_j}\right)^{2s+1}\prod_{q=k-p+1}(\alpha_j - \mu_q) + \prod_{q=k-p+1}^{2k}\left(\alpha_j - \mu_q + \frac{\pi i}{2}\right)$$
$$\times\sum_{m=1}\prod_{q=m+2}^{2k}(\alpha_i - \lambda_q)\sum_{s=0}\left(\frac{\pi i}{2}\right)^{2s+1}\frac{1}{(2s+1)!}\left(\frac{d}{d\alpha_j}\right)^{2s+1}$$
$$\left.\to\prod_{q=k+p+1}^{m}(\alpha_j - \lambda_q - \pi i)\right\}(1 + \mathrm{O}(\sigma^{-1}))\,, \quad i > k, j > k\,.$$

From the matrix elements of the blocks $\mathcal{D}_2$ and $\mathcal{D}_3$ we extract only those terms which increase faster than σ^{2k-2}. They originate from the terms

$$\mathcal{D}^0_{ij} = \sum_{m=1}^{2p-2}\prod_{q=1}^{2k}\left(\alpha_j - \mu_q + \frac{\pi i}{2}\right)\sum_{s=0}\left(\frac{\pi i}{2}\right)^{2s+1}\frac{1}{(2s+1)!}$$
$$\times\left(\frac{d}{d\alpha_j}\right)^{2s+1}\prod_{q=1}^{m}(\alpha_j - \lambda_q)\prod_{q=m+2}^{2k}(\alpha_i - \lambda_q - \pi i) - \prod_{q=1}^{2k-1}\left(\alpha_j - \mu_q + \frac{\pi i}{2}\right)$$
$$\times\prod_{q=2}^{2k}(\alpha_i - \lambda_q - \pi i) + \mathrm{O}(\sigma^{2k-2})\,, \quad j \le k, i > k\,,$$

$$\mathcal{D}^0_{ij} = \sum_{m=1}^{2p-2} \prod_{q=1}^{2k} \left(\alpha_i - \lambda_q - \frac{\pi i}{2}\right) \sum_{s=0} \left(\frac{\pi}{2}\right)^{2s+1} \frac{1}{(2s+1)!}$$

$$\times \left(\frac{d}{d\alpha_j}\right)^{2s+1} \prod_{q=1}^{m} (\alpha_j - \mu_q) \prod_{q=m+2}^{2k} (\alpha_i - \mu_q) - \prod_{q=1}^{2k-1} \left(\alpha_j - \mu_q + \frac{\pi i}{2}\right)$$

$$\times \prod_{q=2}^{2k} (\alpha_i - \lambda_q - \pi i) + \mathrm{O}(\sigma^{2k-2}) \,, \quad j > k, i \leq k \,.$$

Notice that $\mathcal{D}^0_{ij}$ is in this case of the form $\sum_{m=1}^{2p-1} f^m_j g^m_i + \mathrm{O}(\sigma^{2k-2})$ where $f^m_j \sim \sigma^{k+p}$. One can transform the matrix $\mathcal{D}^0$, taking the linear combinations of columns in order to obtain in the changed blocks $\tilde{\mathcal{D}}_2$ and $\tilde{\mathcal{D}}_3$:

$$(\tilde{\mathcal{D}}_2)_{ij} = \sum_{m=j-k}^{2p-1} f^m_j g^m_i$$

$$(\tilde{\mathcal{D}}_3)_{ij} = \sum_{m=j}^{2p-1} \tilde{f}^m_j \tilde{g}^m_i \,.$$

In the course of this transformation the estimation $\mathrm{O}(\sigma^{2k})$ for the matrix elements of $\tilde{\mathcal{D}}_1$ and $\tilde{\mathcal{D}}_4$ is preserved. Now the dependence on α_i can be extracted precisely in $\tilde{\mathcal{D}}_2, \tilde{\mathcal{D}}_3$:

$$(\tilde{\mathcal{D}}_2)_{ij} = \sum_{m=0} \alpha_i^m P^{(j)}_m(\sigma) \,,$$

$$(\tilde{\mathcal{D}}_3)_{ij} = \sum_{m=0} \alpha_i^m \tilde{P}^{(j)}_m(\sigma) \,,$$

where $\tilde{P}^{(j)}_m, P^{(j)}_m$ are polynomials which depend, besides σ, on α_j, λ, μ, though this dependence is not essential for our purposes. The polynomials $P^{(j)}_m(\sigma)$ and $\tilde{P}^{(j)}_m(\sigma)$ have the degrees $2k+2p-(j-k)-1$ and $2k+2p-j-1$ respectively. Let us add to the second column of the determinant the first column multiplied by $-\tilde{P}^{(2)}_0(\sigma)[\tilde{P}^{(1)}_0(\sigma)]^{-1}$ to the third column – the second column multiplied by $\tilde{P}^{(3)}_1(\sigma)[\tilde{P}^{(2)}_1 - \tilde{P}^{(1)}_1(\tilde{P}^{(1)}_0)]^{-1}$ and the first column multiplied by $\tilde{P}^{(3)}_0(\sigma)[\tilde{P}^{(1)}_0(\sigma)]^{-1}$, etc., up to the k-th column. In the course of this process the asymptotic behaviour of $\mathcal{D}_1$ remains unchanged while the first $(p-1)$ columns of $\mathcal{D}_3$ change their asymptotic behaviour to

$\sigma^{2(k+p-j)}, j=1,\dots,p-1$; other columns of $\mathcal{D}_3$ still behave as $O(\sigma^{2k})$. A similar operation can be carried out with the task k columns of the determinant, whose result is: the first $p-1$ columns of $\mathcal{D}_2$ are $O(\sigma^{2(2k+p-j)})$, other columns of $\mathcal{D}_2$ and all the matrix elements of $\mathcal{D}_4$ are $O(\sigma^{2k})$. Hence

$$\det \mathcal{D}^0 = O(\sigma^{4k^2+2p(p-1)}) .$$

Consider in more detail the cases $p=0, p=1$. For $p=0$ the block $\mathcal{D}_3$ is $O(\sigma^{2k-2})$, hence

$$\det \mathcal{D}^0 = \sigma^{4k^2}[(\det \hat{\mathcal{D}}_1 + O(\sigma^{-1}))(\det \hat{\mathcal{D}}_4 + O(\sigma^{-1})) + O(\sigma^{-2})] ,$$

where $\hat{\mathcal{D}}_1$ is a $k\times k$ matrix with matrix elements

$$\begin{aligned}(\hat{\mathcal{D}}_1)_{ij} = &\prod_{q=1}^{k}\left(\alpha_j - \lambda_q - \frac{\pi i}{2}\right)\sum_{m=1}\prod_{q=m+2}^{k}(\alpha_i - \mu_q)\sum_{s=0}\left(\frac{\pi i}{2}\right)^{2s+1}\frac{1}{(2s+1)!}\\ &\times\left(\frac{d}{d\alpha_j}\right)^{2s+1}\prod_{q=1}^{m}(\alpha_j-\mu_q) + \prod_{q=1}^{k}\left(\alpha_j - \mu_q + \frac{\pi i}{2}\right)\\ &\times\sum_{s=0}\left(\frac{\pi i}{2}\right)^{2s+1}\frac{1}{(2s+1)!}\left(\frac{d}{d\alpha_j}\right)^{2s+1}\prod_{q=1}^{m}(\alpha_j-\mu_q) .\end{aligned}$$

The matrix $\hat{\mathcal{D}}_4$ differs from $\hat{\mathcal{D}}_1$ by the replacement

$$\alpha_l \to \alpha_{k+l}, \lambda_q \to \lambda_{k+q}, \mu_q \to \mu_{k+q} .$$

For $p=1$ the block $\mathcal{D}_3$ is $O(\sigma^{2k-1})$; the block $\mathcal{D}_2$ is $O(\sigma^{2k})$. Consequently

$$\begin{aligned}\det \mathcal{D}^0 = &\sigma^{4k^2}\det \mathcal{D}^1(\alpha_1,\dots,\alpha_k|\lambda_1,\dots,\lambda_{k+1}|\mu_1,\dots,\mu_{k-1})\\ &\times \det \mathcal{D}^{-1}(\alpha_{k+1},\dots,\alpha_{2k}|\lambda_{k+2},\dots,\lambda_{2k}|\mu_k,\dots,\mu_{2k})(1+O(\sigma^{-1})) .\end{aligned}$$

Matrices $\mathcal{D}^c$ are given by Eq. (184).

The case $p\le -1$ can be treated in a similar way. One has

$$\det \mathcal{D}^0 = O(\sigma^{4k^2+2p(p+1)}) .$$

For $p=-1$ this formula can be more precise:

$$\begin{aligned}\det \mathcal{D}^0 = &\sigma^{4k^2}\det \mathcal{D}^{-1}(\alpha_1,\dots,\alpha_k|\lambda_1,\dots,\lambda_{k-1}|\mu_1,\dots,\mu_{k+1})\\ &\times \det \mathcal{D}^1(\alpha_{k+1},\dots,\alpha_{2k}|\lambda_k,\dots,\lambda_{2k}|\mu_{k+2},\dots,\mu_{2k})(1+O(\sigma^{-1})) .\end{aligned}$$

Substituting these results into Eq. (191) one obtains

$$\det \mathcal{F} = O(e^{-k^2\sigma}\sigma^{k^2+2p^2-2}) \,, \quad (|p|>1) \,. \tag{192}$$

$$\det \mathcal{F} = 2^{k^2} e^{-k^2\sigma}\sigma^{k^2} \det \mathcal{F}^+(\beta_{i_1},\dots,\beta_{i_{k+1}}|\beta_{j_1},\dots,\beta_{j_{k-1}}) \times \det \mathcal{F}^-(\tilde\beta_{i_{k+2}},\dots,\tilde\beta_{i_{2k}}|\tilde\beta_{j_k},\dots,\tilde\beta_{j_{2k}})(1+O(\sigma^{-1})) \,, \quad p=1 \tag{193}$$

where $\beta_{i_1},\dots,\beta_{i_{k+1}},\beta_{j_1},\dots,\beta_{j_{k-1}} \sim 0, \beta_{i_{k+2}},\dots,\beta_{i_{2k}},\beta_{j_k},\dots,\beta_{j_{2k}} \sim \sigma$,

$$\det \mathcal{F} \sim 2^{k^2} e^{-k^2\sigma}\sigma^{k^2} \det \mathcal{F}^-(\beta_{i_1},\dots,\beta_{i_{k-1}}|\beta_{j_1},\dots,\beta_{j_{k+1}}) \times \det \mathcal{F}^+(\tilde\beta_{i_k},\dots,\tilde\beta_{i_{2k}}|\tilde\beta_{j_{k+2}},\dots,\tilde\beta_{j_{2k}})(1+O(\sigma^{-1})) \,, \quad p=-1 \,. \tag{194}$$

The case $p=0$ is to be considered in more detail. We have obtained the formula

$$\Pi^0(\alpha_1,\dots,\alpha_{2k}|\lambda_1,\dots,\lambda_{2k}|\mu_1,\dots,\mu_{2k}) \sim \sigma^{3k^2}\left\{\prod_{i<j}^{k}\frac{1}{(\alpha_i-\alpha_j)}(\det\hat{\mathcal{D}}_1 +O(\sigma^{-1}))\prod_{k<i<j}\frac{1}{(\alpha_i-\alpha_j)}(\det\hat{\mathcal{D}}_2+O(\sigma^{-1}))+O(\sigma^{-2})\right\} \,.$$

Let us apply to $\det\hat{\mathcal{D}}_1$ and $\det\hat{\mathcal{D}}_2$ a transformation inverse to that described in Lemma 2, dividing them by $\prod_{i<j<k}(\alpha_i-\alpha_j)$ and $\prod_{k<i<j}(\alpha_i-\alpha_j)$ respectively. The determinant of $\hat{\mathcal{D}}_1$ for example, is transformed to the determinant of a $k\times k$ matrix with the following elements:

$$\prod_{q=1}^{k}\left(\alpha_j-\lambda_q-\frac{\pi i}{2}\right)Q_i(\alpha_j-\pi i|\mu_1-\pi i,\dots,\mu_k-\pi i) + \prod_{q=1}^{k}\left(\alpha_j-\mu_q+\frac{\pi i}{2}\right)Q_i(\alpha_j|\lambda_1,\dots,\lambda_k), i,j=1,\dots,k \,.$$

Recall that

$$Q_k(\alpha|\nu_1,\dots,\nu_k) = \prod_q\left(\alpha-\nu_q-\frac{\pi i}{2}\right) - \prod_q\left(\alpha-\nu_q+\frac{\pi i}{2}\right) \,,$$

hence in the last row of the determinant one has

$$\prod_q \left(\alpha_j - \lambda_q - \frac{\pi i}{2}\right)\left(\alpha_j - \mu_q - \frac{\pi i}{2}\right)$$
$$- \prod_q \left(\alpha_j - \lambda_q + \frac{\pi}{2}\right)\left(\alpha_j - \mu_q + \frac{\pi i}{2}\right) .$$

Usual arguments which make use of the identity

$$\varphi(\alpha - 2\pi i) = \varphi(\alpha)\left(\frac{\alpha - \frac{\pi i}{2}}{\alpha - \frac{3\pi i}{2}}\right) ,$$

show that the last row vanishes after the integrations in (191) over $\alpha_1, \ldots,$ α_k are fulfilled. The same happens to the last row of $\det \hat{D}_2$ after the integrations over $\alpha_{k+1}, \ldots, \alpha_{2k}$ are fulfilled. Hence

$$\det \mathcal{F} = \mathrm{O}(e^{-k^2\sigma}\sigma^{k^2-2}) , \quad p = 0 . \tag{195}$$

Now we have to collect all the above facts together. Consider Eq. (190). The operator $B(\tau)$ can be written as follows:

$$B(\sigma) = B_1(\tau)D_2(\tau) + A_1(\tau)B_2(\tau) ,$$

where A_1, B_1 and D_2, B_2 are constructed via Eq. (111) through the first and the last $2k$ particles respectively. Let $\tau \sim 0$, then $D_2(\tau) \sim 1, B_2(\tau) = \mathrm{O}(\sigma^{-1})$, hence

$$B(\tau) = B_1(\tau) + \mathrm{O}(\sigma^{-1}) . \tag{196}$$

Let $\tau \sim \sigma$, then

$$B(\tau) = B_2(\tilde{\tau}) + \mathrm{O}(\sigma^{-1}) , \tag{197}$$

where $\tilde{\tau} = \tau - \sigma, B_2(\tilde{\tau}) = B_2(\tilde{\tau}|\tilde{\beta}_{2k+1}, \ldots, \tilde{\beta}_{4k})$.

Consider a fixed term in Eq. (190). Let $\beta_{i_1}, \ldots, \beta_{i_{k+p}}, \beta_{j_1}, \ldots, \beta_{j_{k-p}} \sim 0$ $\beta_{i_{k+p+1}}, \ldots, \beta_{i_{2k}}, \beta_{j_{k-p+1}}, \ldots, \beta_{j_{2k}} \sim \sigma$, then

$$\prod_{s,t=1}^{2k} \frac{1}{\beta_{i_s} - \beta_{j_t}} = \sigma^{-2k^2-2p^2} \prod_{s=1}^{k+p}\prod_{t=1}^{k-p} \frac{1}{\beta_{i_s} - \beta_{j_t}} \prod_{q=k+p+1}^{2k} \prod_{s=k-p+1}^{2k} \frac{1}{\beta_{i_q} - \beta_{j_s}} . \tag{198}$$

Combining Eqs. (192, 195, 193, 194, 198, 197) one gets

$$\langle F^3\rangle_{2k}(\beta_1,\dots,\beta_{4k}) = 2^{k^2}\sigma^{-1}e^{-k^2\sigma}\sigma^{-k^2}\pi\{\langle F^+\rangle_k(\beta_1,\dots,\beta_{2k}) \\ \times \langle F^-\rangle_k(\tilde\beta_{2k+1},\dots,\tilde\beta_{4k}) - \langle F^-\rangle_k(\beta_1,\dots,\beta_{2k}) \\ \times \langle F^+\rangle_k(\beta_{2k+1},\dots,\beta_{4k})\}(1+\mathrm{O}(\sigma^{-1}))\ .$$

Substituting this in (188) and using the asymptotics (189) one obtains finally

$$g^3(\beta_1,\dots,\beta_{2n}) = \frac{2\pi}{\sigma}\varepsilon^{3bc}g^b(\beta_1,\dots,\beta_n) \\ \times g^c(\tilde\beta_{n+1},\dots,\tilde\beta_{2n})(1+\mathrm{O}(\sigma^{-1}))\ .$$

The power contribution for $n = 2k+1$ can be obtained in a similar way. Consideration of the functions g^1, g^2, g involves no additional difficulties. Q.E.D.

Let us present without proof a similar theorem for SG model.

Theorem 2. The functions g_+, g_- in SG model have the following asymptotics for $\beta_1,\dots,\beta_k \sim 0, \beta_{k+1},\dots,\beta_{2n}\sim\sigma, \sigma\to\infty$:

$$\begin{aligned} g_+(\beta_1,\dots,\beta_{2n}) &= \mathrm{O}\left(\exp\left(-\frac{\pi}{4\xi}\sigma-\frac{1}{4}\sigma\right)\right)\ , \\ g_-(\beta_1,\dots,\beta_{2n}) &= \mathrm{O}\left(\exp\left(-\frac{\pi}{4\xi}\sigma-\frac{1}{4}\sigma\right)\right)\ , \qquad k\equiv 1(\mathrm{mod}\ 2)\ , \\ g_+(\beta_1,\dots,\beta_{2n}) &= \mathrm{O}\left(\exp\left(-2\min\left(1,\frac{\pi}{\xi}\right)\sigma\right)\right)\ , \\ g_-(\beta_1,\dots,\beta_{2n}) &= \mathrm{O}\left(\exp\left(-2\min\left(1,\frac{\pi}{\xi}\right)\sigma\right)\right)\ , \quad k\equiv 0(\mathrm{mod}\ 2)\ . \end{aligned} \tag{199}$$

The asymptotics in NLS can be obtained as a corollary of Theorem 1.

Theorem 3 (Corollary of Theorem 1). Consider form factors $f, f_{\mu\nu}, f^a_\mu, f^a$ in NLS. Introduce the functions

$$\begin{aligned} g^a(\beta_1,\dots,\beta_{2n}) &= \left(\sum \mathrm{ch}\beta_j\right)^{-1} f^a_1(\beta_1,\dots,\beta_{2n})\ , \\ g(\beta_1,\dots,\beta_{2n}) &= \left(\sum \mathrm{ch}\beta_j\right)^{-2} f_{11}(\beta_1,\dots,\beta_{2n})\ , \\ g(\beta_1,\dots,\beta_{2n+1}) &= \left(\left(\sum \mathrm{ch}\beta_j\right)^2 - \left(\sum \mathrm{sh}\beta_j\right)^2 - 1\right)^{-1} \\ &\quad \times f(\beta_1,\dots,\beta_{2n+1})\ . \end{aligned}$$

Let $\beta_1, \dots, \beta_k \sim 0, \beta_{k+1}, \dots, \beta_{2n} \sim \sigma$. The following asymptotics hold:

$$
\begin{aligned}
g^a(\beta_1, \dots, \beta_{2n}) &\sim \frac{2\pi}{\sigma}\varepsilon^{abc} g^b(\beta_1, \dots, \beta_k) g^c(\tilde{\beta}_{k+1}, \dots, \tilde{\beta}_{2n}) , \\
&\qquad\qquad\qquad\qquad k \equiv 0 (\mathrm{mod}\ 2) \ ; \\
g^a(\beta_1, \dots, \beta_{2n}) &\sim \frac{2\pi}{\sigma}\varepsilon^{abc} f^b(\beta_1, \dots, \beta_n) f^c(\tilde{\beta}_{k+1}, \dots, \tilde{\beta}_{2n}) , \\
&\qquad\qquad\qquad\qquad k \equiv 1 (\mathrm{mod}\ 2) \ ; \\
f^a(\beta_1, \dots, \beta_{2n+1}) &\sim \frac{2\pi}{\sigma}\varepsilon^{abc} f^b(\beta_1, \dots, \beta_k) g^c(\tilde{\beta}_{k+1}, \dots, \tilde{\beta}_{2n+1}) , \\
&\qquad\qquad\qquad\qquad k \equiv 1 (\mathrm{mod}\ 2) \ ; \\
f^a(\beta_1, \dots, \beta_{2n+1}) &\sim \frac{2\pi}{\sigma}\varepsilon^{abc} g^b(\beta_1, \dots, \beta_k) f^c(\tilde{\beta}_{k+1}, \dots, \tilde{\beta}_{2n+1}) , \\
&\qquad\qquad\qquad\qquad k \equiv 0 (\mathrm{mod}\ 2) \ ; \\
g(\beta_1, \dots, \beta_{2n}) &= \mathrm{O}(\sigma^{-2}) , \qquad\qquad (200)
\end{aligned}
$$

where $\tilde{\beta}_j = \beta_j - \sigma$.

The form factors in NLS are expressed in terms of elementary functions; that is why Eqs. (200) can be nicely illustrated by explicit consideration of the few-particle form factors. One can get from the general formulae of Sec. 9 the following results:

$$
f^3(\beta_1) = \frac{1}{4}, \ g^3(\beta_1, \beta_2) = \frac{1}{8}\mathrm{th}^2\frac{1}{2}\beta_{12}\frac{\pi i(\beta_{12} + \pi i)}{\beta_{12}(\beta_{12} + 2\pi i)}(e_{1,-1} - e_{-1,1})
$$

$$
\begin{aligned}
f^3(\beta_1, \beta_2, \beta_3) = \frac{1}{16}\prod_{i<j}\mathrm{th}^2\frac{1}{2}\beta_j\frac{(\beta_{ij} + \pi i)}{\beta_{ij}(\beta_{ij} + 2\pi i)}\pi^2\{-(\beta_{13} + 2\pi i) \\
\times (e_{0,0,0} - e_{1,0,1}, -e_{1,0-1}) + \beta_{12}(e_{0,0,0} - e_{1,-1,0} - e_{-1,1,0}) \\
+ \beta_{23}(e_{0,0,0} - e_{0,1,-1} - e_{0,-1,1})\} ,
\end{aligned}
$$

where $\beta_{ij} = \beta_i - \beta_j$.

In the next section we shall need one more asymptotic formula. Consider the form factor $f^a_\mu(\beta_1, \dots, \beta_{2n})$. Let $\beta_1 < \dots < \beta_{2n}$ and denote $\beta_{i+1} - \beta_i$ by Δ_i. We have shown that $f^a_\mu(\beta_1, \dots, \beta_{2n}) = \Sigma(e^{\beta_j} - (-1)^\mu e^{-\beta_j})$ $g^a(\beta_1, \dots, \beta_{2n})$, where $g^a_\cdot$ is invariant under simultaneous shift of rapidities, i.e., g^a depends only on $\Delta_1, \Delta_2, \dots, \Delta_{2n-1}$. We are interested in the behaviour of $\|g^a(\beta_1, \dots, \beta_{2n})\|^2$ when all Δ_i's are large.

Theorem 4. The function $||g^a(\beta_1, \dots, \beta_{2n})||^2$ in ITM is characterized by the following behaviour for

$$||g^a(\beta_1, \dots, \beta_{2n}||^2 \sim \prod_{p=1}^{n} \Delta_{2p-1}^{-\frac{1}{2}} \exp\left(-\frac{1}{2}\Delta_{2p+1}\right) \prod_{p=1}^{n-1} \Delta_{2p}^{-2} . \tag{201}$$

Sketch of the Proof. According to the remark made the end of Sec. 7 one can present $||g^3(\beta_1, \dots, \beta_{2n})||^2$ as follows:

$$\begin{aligned} ||g^3(\beta_1, \dots, \beta_{2n})||^2 = & \sum_{\{1,\dots,2n\}=\{i_1,\dots,i_n\}\cup\{j_1,\dots,j_n\}} \\ & \times (F^3(\beta_{i_1}, \dots, \beta_{i_n}, |\beta_{j_1}, \dots, \beta_{j_n}))^2 \\ & \times \prod_{p,q} \frac{1}{(\beta_{j_p} - \beta_{i_q})(\beta_{j_p} - \beta_{i_q} \quad \pi i)} , \end{aligned}$$

where the function F^3 is presented in Lemma 1. The main difficulty in obtaining on estimation for (201) lies in the estimation of $F^3(\beta_{i_1}, \dots, |\beta_{j_1}, \dots)$. The method of estimating F^3 is similar to that used in Theorem 1. One should separate, at first, the exponential contributions. Having done that, one should make sure that the integral over $\alpha_j (j = 1, \dots, n)$ is mainly concentrated on the segment $\beta_{2j-1} < \alpha_j < \beta_{2j}$ (we supported the β's to be sufficiently separated). Then the power contribution should be estimated. Corresponding calculations are very cumbersome and do not differ radically from those made in Theorem 1.

We present without proof a similar Theorem for SG model.

Theorem 5. The function $||g_\pm(\beta_1, \dots, \beta_{2n})||^2$ in SG model behaves for $\Delta_i \equiv \beta_{i+1} \gg 1$ as follows:

$$\begin{aligned} & ||g_\pm(\beta_1, \dots, \beta_{2n})||^2 \\ & \sim \prod_{p=1}^{n} \exp\left(-\frac{1}{2}\Delta_{2p-1} - \frac{\pi}{2\xi}\Delta_{2p-1}\right) \prod_{p=1}^{n-1} \exp\left(-\min\left(2, \frac{2\pi}{\xi}\right)\Delta_{2p}\right) . \end{aligned} \tag{202}$$

Finally, for NLS the following estimations of $||f^a||^2$ and $||g^a||^2$ hold.

Theorem 6. The functions $\|f^a\|^2$ and $\|g^a\|^2$ in NLS behave for $\Delta_i \equiv \beta_{i+1} - \beta_i \gg 1$ as follows:

$$\|g^a(\beta_1, \dots, \beta_{2n})\|^2 = \prod_{p=1}^{2n-1} \Delta_p^{-2} \,,$$

$$\|f^a(\beta_1, \dots, \beta_{2n+1}\|^2 = 2 \prod_{p=1}^{2n} \Delta_p^{-2} \,. \tag{203}$$

Sketch of the Proof. According to Eq. (181), $\|f^3(\beta_1, \dots, \beta_n)\|^2$ can be presented as follows:

$$\|f^3(\beta_1, \dots, \beta_n)\|^2 = \prod_{i<j} \mathrm{th}^4 \frac{1}{2}(\beta_i - \beta_j) \frac{(\beta_i - \beta_j)^2 + \pi^2}{(\beta_i - \beta_j)^2}$$

$$\times \sum_{B = B^- \cup B^0 \cup B^+} (\tilde{\phi}^0(B^-|B^0|B^+))^2 \prod \frac{(\beta^+ - \beta^- - \pi i)}{(\beta^+ - \beta^-)(\beta^+ - \beta^0)(\beta^0 - \beta^-)(\beta^0 - \beta^0 - \pi i)}$$

$$\times \prod \frac{1}{(\beta^+ - \beta^- + \pi i)(\beta^0 - \beta^0 - 2\pi i)(\beta^0 - \beta^- - 2\pi i)(\beta^+ - \beta^- - 2\pi i)(\beta^+ - \beta^0 - 2\pi i)} .$$

The main difficulty lies in the estimation of $\tilde{\phi}^0(B^-|B^0|B^+)$. For these determinants one can write formulae similar to (183), multiplying and dividing them by the Vandermond determinant $\prod_{i<j}(\beta_i - \beta_j)$. When this is done the estimation is reduced to complicated but rather straightforward calculations.

10

CURRENT ALGEBRAS

The present section is devoted to the calculation of the singularities of commutators at the origin of coordinates. We consider the following operators: j_μ in SG, j_μ^a in ITM, j_μ^a in NLS.

Let us recall several formulae from Sec. 2. Suppose the operators $O(x)$ $O(y)$ are defined by the form factors f_1, f_2. Then there are two equivalent forms of matrix element of

$$\langle \overrightarrow{A} | O(x)O(y) | \overleftarrow{B} \rangle_+ e^{i\kappa(A)x - i\kappa(B)y} = \sum_{\substack{A=A_1\cup A_2\cup A_3 \\ B=B_1\cup B_2\cup B_3}} S(\overrightarrow{A} | \overrightarrow{A}_1)$$

$$\times\, S(\overrightarrow{A_2 \cup A_3} | \overrightarrow{A}_3) \sum_{n(C)=0}^{\infty} \int d\overleftarrow{C}\, f_1(\overleftarrow{A}_1 + i0 | \overrightarrow{C} \cup \overrightarrow{B}_2) S(\overleftarrow{C} | \overleftarrow{A}_3 \cup \overleftarrow{C})$$

$$\times\, f_2((\overleftarrow{C} \cup \overleftarrow{A}_2) - i0 | \overrightarrow{B}_1) \exp(i(x-y)(\kappa(C) + \kappa(B_2) + \kappa(A_2) + \kappa(A_3)))$$

$$\times\, \Delta(A_3, B_3) S(\overleftarrow{B}_3 | \overleftarrow{B_2 \cup B_2}) S(\overleftarrow{B}_1 | \overleftarrow{B}) \,, \tag{204}$$

$$\langle \overrightarrow{A} | O(x)O(y) | \overleftarrow{B} \rangle_- e^{i\kappa(A)x - i\kappa(B)y} = \sum_{\substack{A=A_1\cup A_2\cup A_3 \\ B=B_1\cup B_2\cup B_3}} S(\overrightarrow{A} | \overrightarrow{A}_2)$$

$$\times\, S(\overrightarrow{A_1 \cup A_3} | \overrightarrow{A}_3) \sum_{n(C)=0}^{\infty} \int d\overleftarrow{C}\, f_1(\overleftarrow{A}_1 - i0 | \overrightarrow{C} \cup \overrightarrow{B}_2) S(\overrightarrow{C} \cup \overrightarrow{A}_3 | \overrightarrow{C})$$

$$\times\, f_2((\overleftarrow{C} \cup \overleftarrow{A}_2) + i0 | \overrightarrow{B}_1) \exp(i(x-y)(\kappa(C) + \kappa(B_1) + \kappa(A_3) + \kappa(A)))$$

$$\Delta(A_3, B_3) S(\overleftarrow{B}_3 | \overleftarrow{B_1 \cup B_3}) S(\overleftarrow{B}_2 | \overleftarrow{B}) \,. \tag{205}$$

Lemma 2 of Sec. 2 states that

$$\langle \overrightarrow{A}|O_1(x)O_2(y)|\overleftarrow{B}\rangle_t = \langle \overrightarrow{A}|O_1(x)O_2(y)|\overleftarrow{B}\rangle_- \,.$$

To prove the local commutativity theorem we have written the convolution $\int [O(x), O(0)]dx$ in the form

$$\begin{aligned}
&\langle \overrightarrow{A}| \int [O_1(x), O_2(0)] \exp(i\kappa(A)x)\varphi(x)dx|\overleftarrow{B}\rangle \\
&= \int \left(\langle \overrightarrow{A}|O_1(x)O_2(0)|\overleftarrow{B}\rangle_- - \langle \overrightarrow{A}|O_2(0)O_1(x)|\overrightarrow{B}\rangle_+\right) e^{i\kappa(A)x}\varphi_-(x)dx \\
&+ \int \left(\langle \overrightarrow{A}|O_1(x)O_2(0)|\overleftarrow{B}\rangle_+ - \langle \overrightarrow{A}|O_2(0)O_1(x)|\overrightarrow{B}\rangle_-\right) e^{i\kappa(A)x}\varphi_+(x)dx \,.
\end{aligned} \tag{206}$$

Vanishing of the matrix element follows from the possibility to deform the contour of integration to the line $\text{Im}\sigma = -\pi + 0$ in the integral

$$\begin{aligned}
&\int f_1(\overleftarrow{A}_1 + i0|\overrightarrow{C} \cup \overrightarrow{B}_2)S(\overleftarrow{C}|\overleftarrow{A}_3 \cup \overleftarrow{C})f_2((C \cup A_2) - i0|B_1) \\
&\quad \times \hat{\varphi}_-(\kappa(C) + \kappa(B_2) + \kappa(A_3) + \kappa(A_4))d\sigma \,,
\end{aligned}$$

where $\sigma \in \frac{1}{n(C)}\Sigma_{\gamma\in C}\gamma, \hat{\varphi}_-$ is the Fourier transform of φ_-. Function $\hat{\varphi}_-(\kappa(C) + \dots)$ decreases faster than an arbitrary power of $e^{-\sigma}$ for $\sigma \to \infty$, if $\varphi(x)$ vanishes with all its derivaties at the point $x = 0$. This is why it is sufficient to require that f_1, f_2 do not increase too fast for $\sigma \to \infty$. Evidently the asymptotics (187), (199), (200) satisfy this requirement. But as we intend to study the singularity of the commutator at the origin of coordinates we have to use the functions φ with a different behaviour at the point $x = 0$. Here some new possibilities arise.

Let us discuss one important circumstance which has been ignored until now. Dealing with the expressions (204), (205) one has to make sure that the integrals and the series in Eqs. (204, 205) are convergent. Actually, very rough estimation of the form factors guarantees the convergence. Consider, for example, the operators $T_{\mu\nu}, j_\mu$ in any model discussed in this book. The form factors of these operators can be presented as follows:

$$f(B) = (p_0(B))^\kappa (p_1(B))^l g(B) \,, \tag{207}$$

where κ, l are certain integers (for example, $\kappa = l = 1$ for T_{01}), the function $g(B)$ satisfies the condition

$$g(B + \sigma) = g(B) \,. \tag{208}$$

Let us suppose that the function g allows the following estimation:

$$\|g(A-\pi i,\ C-\frac{\pi i}{2},B)\| \le C^{n(A)+n(B)+n(C)} \prod_{\substack{\alpha\in A\\ \beta\in B}} \mathrm{cth}\frac{1}{2}(\alpha-\beta)\ , \qquad (209)$$

where C is some construct. Consider now the expression (205) for $x < y$. It is possible to move the contour of integration with respect to $\sigma = \sum_{\gamma\in C}\gamma$ to the line $\mathrm{Im}\sigma = \frac{\pi}{2}$. Then, using the estimation (209), one can easily get the following estimation for the matrix element:

$$\begin{aligned} |\langle \overrightarrow{A}|O(x)O(y)|\overleftarrow{B}\rangle| \ge & \sum_{\substack{A=A_1\cup A_2\cup A_3\\ B=B_1\cup B_2\cup B_3}} \Delta(A_3,B_3) \\ & \times \prod_{\substack{\alpha\in A_1\cup A_2\\ \beta\in B_1\cup B_2}} \mathrm{cth}\frac{1}{2}(\alpha-\beta) C^{n(A_1)+n(A_2)+n(B_1)+n(B_2)} \\ & \times \left(|p_0(A_1)-p_0(B_2)| + \frac{d}{dy}\right)^{\kappa_1} \left(|p_1(A_1)-p_1(B_1)| + \frac{d}{dy}\right)^{l_1} \\ & \times \left(|p_0(A_2)-p_0(B_1)| + \frac{d}{dy}\right)^{\kappa_2} \left(|p_1(A_2)-p_1(B_1)| + \frac{d}{dy}\right)^{l_2} \\ & \times \exp(C_1C_2K_0(m(x-y)))\ , \end{aligned} \qquad (210)$$

where K_0 is the following function:

$$K_0(x) = \int_0^\infty \exp(-x\ ch\ \beta)d\beta\ .$$

Thus, if the estimation (209) holds the expression (205) for the matrix element is finite for $x > y$ (similarly (204) is finite for $x < y$). When x approaches y, the matrix element can be singular but the estimation (210) proves that the singularity is of power character because the function $K_0(x)$ behaves for $x \to 0$ as follows:

$$K_0(x) = -\frac{1}{2}\ln x + \dots\ , \qquad x \to 0\ .$$

As to the estimation (209) it looks quite natural. The author proved it for SG in the reflectionless case. The investigation of the estimation (209) in the other cases encounters many technical problems but the author believes that they can be overcome.

For a detailed study of the singularities of the commutators at the origin of coordinates one needs some refined information. Some experiments and general reasonings have convinced us of the validity of the following statements which are formulated as hypotheses, since their rigorous proof is still lacking.

Hypothesis 1. In SG model and ITM the series

$$\sum_{n(C)=0}^{\infty} \int \|g(\overrightarrow{C})\|^2 d\overrightarrow{C}_0 , \quad \sum_{n(C)=0}^{\infty} \int \|g^a(\overrightarrow{C})\|^2 d\overrightarrow{C}_0 \tag{211}$$

converge (the convergence of the integrals $\int \|g(\overrightarrow{C})\|^2 d\overrightarrow{C}_0$ follows from Eqs. (201, 202)).

For what follows Hypothesis 1 will be sufficient in the form it is formulated. We have however more precise conjecture about the character of the convergence of the series (211). We expect the existence of the estimation

$$\int \|g(\overrightarrow{C})\|^2 d\overrightarrow{C}_0 \sim q^{n(C)}(\xi)$$

for large $n(C)$. The function $q(\xi) < 1$ for $0 < \xi \leq \infty$.

This estimation implies the existence of a similar estimation for ITM with $q = q(\infty)$. What is the sum of series (211)? This question is very important. It is natural to suppose that according to Coleman's equivalence Eq. (2)

$$\sum_{n(C)} \int \|g(\overrightarrow{C})\|^2 d\overrightarrow{C}_0 = \left(\frac{\xi}{\pi + \xi}\right) \frac{1}{2\pi^2} ,$$

and consequently for $\xi = \infty$,

$$\sum_{n(C)} \int \|g^a(\overrightarrow{C})\|^2 d\overrightarrow{C}_0 = \frac{1}{2\pi^2} .$$

To prove these formulae directly is an extremely difficult but tempting problem.

Let us pass on to NLS. For this model we suppose the following hypothesis to be valid

Hypothesis 2. In NLS model the series

$$\sum_{n(C)} \int \|g^a(\overrightarrow{C})\|^2 d\overrightarrow{C}_0 , \quad \sum_{n(C)} \int \|f^a(\overrightarrow{C})\ |^2 d\overrightarrow{C}_0$$

diverge (the convergence of the integrals $\int \|g^a\|^2 dC_0, \int \|f^a\|^2 dC_0$ follows from Eq. (203)). Asymptotically for $n \to \infty$ the n-th term of these series approach as some constant. When the ultraviolet cutoff is introduced, $\Delta_i \leq \Lambda$, and the series become convergent:

$$\sum_n \int_0^\Lambda d\Delta_1 \ldots \int_0^\Lambda d\Delta_{n-1} \|g^a(\vec{C})\|^2 < \infty \,,$$
$$\sum_n \int_0^\Lambda d\Delta_1 \ldots \int_0^\Lambda d\Delta_{n-1} \|f^a(\vec{C})\|^2 < \infty \,. \tag{212}$$

Assuming this hypothesis to be valid one can prove the following statement.

Lemma 2. The series (212) are characterized by the following behaviour for $\Lambda \to \infty$:

$$\sum_n \int_0^\Lambda d\Delta_1 \ldots \int_0^\Lambda d\Delta_{n-1} \|g^a(\vec{C})\|^2 \sim \frac{\Lambda}{(4\pi)^2} \,,$$
$$\sum_n \int_0^\Lambda d\Delta_1 \ldots \int_0^\Lambda d\Delta_{n-1} \|f^a(\vec{C})\|^2 \sim \frac{\Lambda}{(4\pi)^2} \,.$$

Proof. Denote the sums of the series

$$\sum_n \int_0^\Lambda d\Delta_1 \ldots \int_0^\Lambda d\Delta_{n-1} \|g^a(\vec{C})\|^2 \,, \quad \sum_n \int_0^\Lambda d\Delta_1 \ldots \int_0^\Lambda d\Delta_{n-1} \|f^a(\vec{C})\|^2$$

by $\eta_1(\Lambda), \eta_2(\Lambda)$ respectively. According to Hypothesis 2, $\eta_{1,2}(\Lambda) < \infty$. Differentiate $\eta_1(\Lambda)$ with respect to Λ:

$$\frac{d}{d\Lambda}\eta(\Lambda) = \sum_{n=0}^{\infty} \sum_{\kappa=1}^{n+1} \int_0^\Lambda d\Delta_1 \overset{\kappa}{\underset{\ldots}{\wedge}} \int_0^\Lambda d\Delta_{n-1} \|g^1(\gamma_1, \ldots \gamma_{\kappa-1}, \gamma_\kappa + \Lambda \,,$$
$$\ldots, \gamma_n + \Lambda)\|^2 \,.$$

For large Λ the following estimations (200) hold:

$$g^a(B_1, B_2 + \Lambda) \simeq \frac{2\pi}{\Lambda} \varepsilon^{abc} g^b(B_1) g^c(B_2) \,, \quad n(B_1) \equiv 0 (\mathrm{mod}\ 2) \,,$$
$$g^a(B_1, B_2 + \Lambda) \simeq \frac{2\pi}{\Lambda} \varepsilon^{abc} f^b(B_1) f^c(B_2) \,, \quad n(B_1) \equiv 1 (\mathrm{mod}\ 2) \,.$$

Hence for large Λ the formula is valid:

$$\frac{d}{d\Lambda}\eta_1(\Lambda) = 2(2\pi)^2\frac{1}{\Lambda^2}(\eta_1^2(\Lambda)+\eta_2^2(\Lambda)) .$$

Similarly,

$$\frac{d}{d\Lambda}\eta_2(\Lambda) = 2(2\pi)^2\frac{1}{\Lambda^2}2\eta_1(\Lambda)\eta_2(\Lambda) .$$

Hence, for the function $\eta(\Lambda) = \eta_1(\Lambda)+\eta_2(\Lambda)$ one has the equation

$$\frac{d}{d\Lambda}\eta(\Lambda) = 2(2\pi)^2\frac{1}{\Lambda^2}2\eta^2(\Lambda) .$$

The solution of this differential equation is

$$\eta(\Lambda) = \frac{\Lambda}{c\,\Lambda + 2(2\pi)^2} .$$

According to Hypothesis 2, $\eta(\Lambda)\to\infty$ for $\Lambda\to\infty$; hence $c=0$, which means that for large Λ the following asymptotic holds: $\eta(\Lambda)\sim\frac{1}{2(2\pi)^2}\Lambda$. Hence, $\eta_i(\Lambda) = c_i\Lambda$ for $\Lambda\to\infty$. Substituting these η_i into the above equations one gets

$$\eta_1(\Lambda)\simeq\eta_2(\Lambda)\simeq\frac{\Lambda}{(4\pi)^2} .$$

Q.E.D.

The possibility of indicating an explicit asymptotic behaviour of $\eta_{1,2}(\Lambda)$ in the framework of Hypothesis 2 is very important and, as it will be clear later, is a strong indirect argument in favour of this hypothesis.

Now, assuming the validity of Hypothesis 1, 2, let us pass on to calculation of the singularities of commutators at the origin of coordinates.

Theorem 1. The following commutation relations between currents on the space interval for ITM hold:

$$\begin{aligned}
[j_0^a(x),\ j_0^b(0)] &= i\varepsilon^{abc}j_0^c(0)\delta(x) ,\\
[j_1^a(x),\ j_1^b(0)] &= i\varepsilon^{abc}j_0^c(0)\delta(x) ,\\
[j_0^a(x),\ j_1^b(0)] &= i\varepsilon^{abc}j_1^c(x)\delta(x) + 4\delta^{ab}\eta\delta'(x)\pi i ,
\end{aligned} \qquad (213)$$

where

$$\eta = \sum_{n(C)=0}^{\infty}\int\|g^a(\overrightarrow{C})\|^2 d\overrightarrow{C}_0 .$$

Proof. Look at the formula (206),

$$\langle \overrightarrow{A}| \int [O_1(x), O_2(0)] e^{i\kappa(A)x} \varphi(x) dx | d\overleftarrow{B}\rangle$$
$$= \int \left\{ \langle \overrightarrow{A} | O_1(x) O_2(0) | \overleftarrow{B}\rangle_- - \langle \overrightarrow{A} | O_2(0) O_1(x) | \overleftarrow{B}\rangle_+ \right\} e^{i\kappa(A)x} \varphi_-(x) dx$$
$$+ \int \left\{ \langle \overrightarrow{A} | O_1(x) O_2(0) | \overleftarrow{B}\rangle_+ - \langle \overrightarrow{A} | O_2(0) O_1(x) | \overleftarrow{B}\rangle_- \right\} e^{i\kappa(A)x} \varphi_+(x) dx \ , \tag{214}$$

and consider the first term in RHS,

$$\int \left\{ \langle \overrightarrow{A} | O_1(x) O_2(0) | \overleftarrow{B}\rangle_- - \langle \overrightarrow{A} | O_2(0) O_1(x) | \overleftarrow{B}\rangle_+ \right\} e^{i\kappa(A)x} \varphi_-(x) dx$$
$$= \sum_{\substack{A = A_1 \cup A_2 \cup A_3 \\ B = B_1 \cup B_2 \cup B_3}} S(\overrightarrow{A} | \overrightarrow{A}_1) S(\overrightarrow{A_2 \cup A_3} | \overrightarrow{A}_3) \sum_{n(C)=0}^{\infty} \int dC \hat{G}_-(x | C, A_1, B_1, A_3)$$
$$\times \Delta(A_3, B_3) S(\overleftarrow{B}_2 | \overleftarrow{B_2 \cup B_3}) S(\overleftarrow{B_2 \cup B_3} | \overleftarrow{B}) \ , \tag{215}$$

where

$$\hat{G}_-(C, A_1, B_1, A_3) = (-1)^{n(B_1)+n(B_2)+n(C)} [f_1(\overleftarrow{A}_1 + i0 | \overrightarrow{C} \cup \overrightarrow{B}_2)$$
$$S(\overleftarrow{C} | \overleftarrow{A}_3 \cup \overleftarrow{C}) f_2((\overleftarrow{C} \cup \overleftarrow{A}_2) - i0 | \overrightarrow{B}_1) \hat{\varphi}_-(\kappa(C) + \kappa(B) + \kappa(A_3)$$
$$+ \kappa(A_2)) - f_2(\overleftarrow{A}_2 - i0 | \overrightarrow{B}_1 \cup \overrightarrow{C}) S(\overrightarrow{C} \cup \overrightarrow{A}_3 | \overrightarrow{C})$$
$$\times f_1((\overleftarrow{A}_1 \cup \overleftarrow{C}) + i0 | \overrightarrow{B}_2) \varphi_-(-\kappa(C) + \kappa(B_2) + \kappa(A_3) + \kappa(A_2)) \ . \tag{216}$$

The principal statement of Sec. 2 is that the matrix element (214) vanishes if $\varphi(x) \in \mathbf{C}_0^\infty$ and $\varphi(x)$ vanishes at the origin with all its derivatives. That means that the matrix element is a distribution whose support consists of one point $x = 0$. If $\varphi(x)$ vanishes with all its derivatives at $x = 0$ the Fourier transform $\hat{\varphi}(\kappa)$ decreases faster than an arbitrary power of κ^{-1}. Hence

$$\int d\sigma \hat{G}(C, A_1, B_1, A_3) = 0 \ , \tag{217}$$

because the first and second terms in (216) are the boundary values of a regular function in the strip $-\pi \le \text{Im}\sigma \le 0$ on the lines $\text{Im}\sigma = 0$ and $\text{Im}\sigma = -\pi + 0$, and this regular function decreases for $\sigma \to \pm\infty$.

The asymptotics (187) implies that f^a_μ and f^b_ν do not increase faster than $e^{|\sigma|}$ for $\sigma \to \pm\infty$. Hence the condition that $\hat{\varphi}_-(\kappa)$ decreases faster than an arbitrary power of κ^{-1} can be replaced by the following: $\hat{\varphi}_-(\kappa)$ decreases as κ^{-3} for $\kappa \to \infty$. Under this condition on $\hat{\varphi}_-(\kappa)$, Eq. (217) remains valid. The condition $\hat{\varphi}_-(\kappa) \sim \kappa^{-3}$ for $\kappa \to \infty$ is equivalent to $\varphi(0) = \varphi'(0) = 0$. There are no restrictions on other derivatives, hence the commutator $\langle \overrightarrow{A} | [j^a_\mu(x), j^b_\nu(0)] | \overleftarrow{B} \rangle$ does not contain other singularities but those proportional to δ and δ'.

We want to extract the δ and δ' singularities from the commutators. To this end we need to consider a convolution of commutators with the functions which do not vanish at the point $x = 0$. One should be careful at this point because in the cases we shall consider later the integral

$$\int d\sigma f_1(\overleftarrow{A}_1 + i0 | \overrightarrow{C} \cup \overrightarrow{B}_2) S(\overleftarrow{C} | \overleftarrow{A}_3 \cup \overleftarrow{C}) f_2(\overleftarrow{C} \cup \overleftarrow{A}_2 | \overrightarrow{B}_1 + i0) \\ \times \hat{\varphi}_-(\kappa(C) + \kappa(B_2) + \kappa(A_2) + \kappa(A_3)) \tag{218}$$

diverges, which means that the convolution

$$\int O_1(x) O_2(0) \varphi(x) dx$$

does not exist. However, the commutator is a less singular object than the product of operators. Let us introduce the regularization

$$\langle \overrightarrow{A} | [O_1(x), O_2(0)] | \overleftarrow{B} \rangle = \lim_{\Lambda \to \infty} \sum_{n(0)=0}^{\infty} \int (\langle \overrightarrow{A} | O_1(x) | \overleftarrow{D} \rangle \\ \times \langle \overrightarrow{D} | O_2(0) | \overleftarrow{B} \rangle_- - \langle \overrightarrow{A} | O_2(0) | \overleftarrow{D} \rangle \langle \overrightarrow{D} | O_1(x) | \overleftarrow{B} \rangle) \\ \times \prod_{j=1}^{n(0)} \theta(\Lambda - \delta_{j+1} + \delta_j) dD\,,$$

where $D = \{\delta_1, \ldots, \delta_{n(D)}\}, \delta_{n(D)} > \ldots > \delta_2 > \delta_1$. In other words, we demand that the difference between two nearest rapidities does not exceed Λ and then take the limit $\Lambda \to \infty$. Lemma 2 of Sec. 1 implies that $D = C \cup A_2 \cup B_2 \cup A_3$. Let us denote the minimal and maximal rapidities in $A_2 \cup B_2 \cup A_3$ by $\beta^{\max}$ and $\beta^{\min}$ respectively. Let $C = \{\gamma_1, \ldots, \gamma_{n(C)}\}, \Delta_i =$

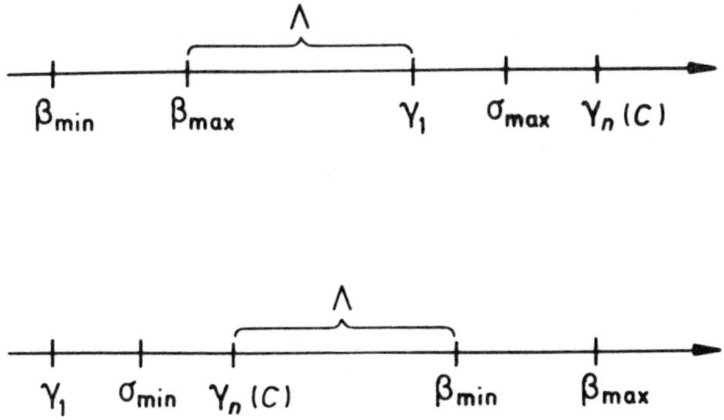

Fig. 6.

$\gamma_{i-1} - \gamma_i$, $\sigma = \frac{1}{n(C)}\sum_{\gamma\in C}\gamma$ and fix $\Delta_1, \dots, \Delta_{n(C)-1}$, then σ may vary in the range from $\sigma_{\min}$ to $\sigma_{\max}$ as indicated in Fig. 6.

The integral

$$\int d\sigma \hat{G}_-(C, A_1, B_1, A_3) \tag{219}$$

changes under the limit sign to

$$\int_{\sigma_{\min}}^{\sigma_{\max}} d\sigma \hat{G}_-(C, A_1, B_1, A_3) = \left(\int_{I_1} - \int_{I_2}\right) f_1\left(\overleftarrow{A} + i0|\overrightarrow{C} \cup \overrightarrow{B}_2\right) \times S(\overleftarrow{C}|\overleftarrow{A}_3 \cup \overleftarrow{C}) f_2(\overleftarrow{C} \cup \overleftarrow{A}_3|\overrightarrow{B}_1)\hat{\varphi}_-(\kappa(C) + \kappa(B_2) + \kappa(A_3) + \kappa(A_2)) \,, \tag{220}$$

where I_1, I_2 are segments parallel to the imaginary axes $I_1 = (\sigma_{\min}, \sigma_{\min} - \pi i), I_2 = (\sigma_{\max}, \sigma_{\max} - \pi i)$. Evidently the limit of the integral (220) can exist when the integral (219) diverges. It is desirable to work with such operators for which only the leading term of asymptotics contributes to (220) for $\sigma \to \pm\infty$.

The first commutation relation we intend to obtain is

$$[j_+^a(x)\ , j_-^b(0)] = 0 \,, \tag{221}$$

where $j_\pm^a = \frac{1}{2}(j_0^a \pm j_1^a)$. Form factors f_+^a, f_-^a are written through g^a as follows:

$$f_+^a(\beta_1, \dots, \beta_{2n}) = M(\sum e^{\beta_j}) g^a(\beta_1, \dots, \beta_{2n}) \,,$$
$$f_-^b(\beta_1, \dots, \beta_{2n}) = M(\sum e^{-\beta_j}) g^a(\beta_1, \dots, \beta_{2n}) \,.$$

Let $\varphi \in \mathbf{C}_0^\infty$ without any limitation on the behaviour of φ at the point $x = 0$. Then $\hat{\varphi}(\kappa) \sim \kappa^{-1}\varphi(0)$ for $|\kappa| \to \infty$. Fix $\Delta_1, \dots, \Delta_{n(C)-1}$, then from the asymptotics (187) and $S(\beta) \sim 1, \beta \to \infty$ it follows that

$$\begin{aligned}
&f_+^a(\overleftarrow{A}_1|\overrightarrow{C} \cup \overrightarrow{B}_2)S(\overleftarrow{C}|\overleftarrow{A}_3 \cup \overleftarrow{C})f_-^b(\overleftarrow{C} \cup \overleftarrow{A}_2|\overrightarrow{B}_2)\\
&\quad \times \varphi_-(\kappa(C) + \kappa(B_2) + \kappa(A_3) + \kappa(A_2))\\
&\sim \begin{cases} \varphi(0)\Lambda^{-2+\delta_{n(A_2)+n(B_2),0}+\delta_{n(A_2)+n(B_1),0}} \\ \times \exp(-\Lambda\, \delta_{n(A_2)+n(B_1),0} + \frac{\Lambda}{2}((-1)^{n(C)} - 1))\,, & \text{if } \sigma \sim \sigma_{\max} \\ \varphi(0)\Lambda^{-2+\delta_{n(A_1)+n(B_2),0}+\delta_{n(A_2)+n(B_1),0}} \\ \times \exp(-\Lambda\, \delta_{n(A_1)+n(B_2),0} + \frac{\Lambda}{2}((-1)^{n(C)} - 1))\,, & \text{if } \sigma \sim \sigma_{\min}\,. \end{cases}
\end{aligned} \tag{222}$$

Thus the function (222) decrease when $\Lambda \to \infty$ in the vicinities of $\sigma_{\min}$ and $\sigma_{\max}$. Hence, the first term in RHS of (214) disappears. The second term can be considered in a similar way. Eq. (222) is now proven.

Consider now the operators

$$Q_-^a(x) = \int_{-\infty}^{x} j_0^A(x)dx, \quad Q_+^a(x) = \int_{x}^{\infty} j_0^a(x)dx\,.$$

Matrix elements of Q_-^a, Q_+^a are constructed according to Eq. (28) (Eq. (29)) through the functions $g^a, -g^a$. According to the results of Sec. 7 the operator $Q^a = Q_-^a + Q_+^a$ is an operator of change. We want to calculate the commutator $[Q_-^a(x), j_\mu^b(0)]$. For its matrix element one has

$$\begin{aligned}
&\langle\overrightarrow{A}|\int [Q_-^a(x), j_\mu^b(0)]e^{i\kappa(A)x}\varphi(x)dx|\overleftarrow{B}\rangle\\
&\quad = \langle\overrightarrow{A}|\int [Q_-^a(x), j_\mu^b(0)]e^{i\kappa(A)x}\varphi_-(x)dx|\overleftarrow{B}\rangle\\
&\quad - \langle\overrightarrow{A}|\int [Q_+^a(x), j_\mu^b(0)]e^{i\kappa(A)x}\varphi_+(x)dx|\overleftarrow{B}\rangle\\
&\quad + \int \langle\overrightarrow{A}|[Q^a, j_\mu^b(0)]e^{i\kappa(A)x}\varphi_+(x)dx\,.
\end{aligned} \tag{223}$$

The last term in RHS can be easily evaluated using the formula

$$[Q^a, j_\mu^b(0)] = i\varepsilon^{abc} j_\mu^c(0)\,.$$

The first term in RHS is considered in the usual manner, and is calculated as the limit (220) for $\Lambda \to \infty$, when g^a, f_μ^b are taken as f_1, f_2. The asymptotics

(187) imply

$$g^a(\overleftarrow{A}|\overrightarrow{C}\cup\overrightarrow{B}_2)S(\overleftarrow{C}|\overleftarrow{A}_3\cup\overleftarrow{C})f^b_\mu(\overleftarrow{C}\cup\overleftarrow{A}_2|\overrightarrow{B}_1)$$
$$\times\,\hat{\varphi}(\kappa(C)+\kappa(B_2)+\kappa(A_3)+\kappa(A_2))$$
$$\sim\begin{cases}\varphi(0)\exp(\frac{\Lambda}{2}\,((-1)^{n(C)}-1))\,\Lambda^{-2+\delta_{n(A_1)+n(B_2),0}+\delta_{n(A_2)+n(B_1),0}} & \\ \qquad\text{if }\ \sigma\sim\sigma_{\max} & \\ \varphi(0)\exp(\frac{\Lambda}{2}\,((-1)^{n(C)}-1))\,\Lambda^{-2+\delta_{n(A_1)+n(B_2),0}+\delta_{n(A_2)+n(B_1),0}} & \\ \qquad\text{if }\ \sigma\sim\sigma_{\min}\,. & \end{cases} \tag{224}$$

Eq. (224) means that the integrand in (220) does not decrease when $\Lambda\to\infty$ in the vicinities of $\sigma_{\max}$ and $\sigma_{\min}$ in the one case, $n(A_1)=n(A_2)=n(B_1)=n(B_2)=0$. In that case the limit of the integral (220) is equal to $\pi i(1-(-1)^\mu)$. The condition $A_1=A_2=B_1=B_2=\emptyset$ is equivalent to $A_3=A, B_3=B$. Hence,

$$\int\langle\overrightarrow{A}|[Q^a_-(x),j^b_\mu(0)]|\overleftarrow{B}\rangle e^{i\kappa(A)x}\varphi_-(x)dx=\varphi(0)\Delta(A,B)\pi i(1-(-1)^\mu)\eta\,.$$

In a similar way,

$$\int\langle\overrightarrow{A}|[Q^a_+(x),j^b_\mu(0)]|\overleftarrow{B}\rangle e^{i\kappa(A)x}\varphi_+(x)dx=-\varphi(0)\Delta(A,B)\pi i(1-(-1)^\mu)\eta\,.$$

Hence

$$[Q^a_-(x),j^b_\mu(0)]=\theta(x)\varepsilon^{abc}j^c_\mu(0)+2\pi i(1-(-1)^\mu)\eta\delta(x)\,,$$

or, differentiating with respect to x

$$[j^a_0(x),j^b_\mu(0)]=i\varepsilon^{abc}j^c_\mu(0)\delta(x)+4\pi i\eta(1-(-1)^\mu)\delta'(x)\,. \tag{225}$$

Algebra (221), (225) is equivalent to (213).
Q.E.D.

A similar Theorem holds for SG model. We present it without proof.

Theorem 2. The currents in SG model satisfy the commutation relations

$$[j_0(x),j_0(0)]=[j_1(x),j_1(0)]=0\,,$$
$$[j_0(x),j_1(0)]=4\pi i\delta'(x)\eta(\xi)\,,$$

where

$$\eta(\xi) = \sum_{n(C)} \int \|g(\overrightarrow{C})\|^2 d\overrightarrow{C}_0 \,.$$

Let us pass on to NLS.

Theorem 3. The following commutation relations hold in NLS model:

$$\begin{aligned}
&[n^a(x), n^b(0)] = 0, [j_0^a(x), n^b(0)] = i\varepsilon^{abc} n^c(0)\delta(x) \,, \\
&[j_0^a(x), j_0^b(0)] = i\varepsilon^{abc} j_0^c(0)\delta(x) \,, \\
&[j_1^a(x), n^b(0)] = 0, [j_1^a(x), j_1^b(0)] = 0 \,, \\
&[j_0^a(x), j_1^b(0)] = i\varepsilon^{abc} j_1^c(0)\delta(x) + \infty\delta^{ab}\delta'(x) \,.
\end{aligned} \tag{226}$$

The last relation means that the convolution of $[j_0^a(x), j_1^b(0)]$ with an arbitrary function $\varphi(x)$ from $\mathbf{C}_0^\infty$ whose first derivative at $x = 0$ vanishes is finite, while the convolution with a function $\varphi(x)$ with nonvanishing first derivative at $x = 0$ is infinite.

Proof. The main idea of the proof is similar to that of Theorem 1. We introduce the cut-off and consider the limit (220) instead of (219). At this point, however, an essential difference arises. The fact is that the series

$$\sum \int \|g(\overrightarrow{C})\|^2 d\overrightarrow{C}_0 \tag{227}$$

is divergent. That is why even if the integrals (220) tend to zero for $\Lambda \to \infty$ the sum of the terms with fixed A_1, A_2, B_1, B_2 can give a non-zero contribution. So we are dealing with an indeterminacy of the kind $0 \cdot \infty$ and the cut-off should help us to uncover it.

As it follows from Lemma 2 the series (227) behaves as Λ for $\Lambda \to \infty$ after the cut-off is introduced. Consequently in the integrals

$$\begin{aligned}
&\left(\int_{I_1} - \int_{I_2}\right) d\sigma f_1(\overleftarrow{A}_1 + i0|\overrightarrow{C} \cup \overrightarrow{B}_2) S(\overleftarrow{C}|\overleftarrow{A}_3 \cup \overleftarrow{C}) f_2(\overleftarrow{C} \cup \overleftarrow{A}_2|\overrightarrow{B}_1) \\
&\quad \times \hat{\varphi}_-(\kappa(C) + \kappa(B_2) + \kappa(A_3) + \kappa(A_4)) d\sigma \,,
\end{aligned} \tag{228}$$

we are now interested not only in those terms which behave as $O(1)$ for $\Lambda \to \infty$ but also in those which behave as $O(\Lambda^{-1})$ for $\Lambda \to \infty$. But the terms which decrease as $o(\Lambda^{-1})$ disappear after the summation over $n(C)$ and the limit $\Lambda \to \infty$ are taken.

Let us check the commutation relations. We start with the commutator $[n^a(x), n^b(0)]$. The function $f^a(\overleftarrow{A}_1 + i0|\overrightarrow{C} \cup \overrightarrow{B}_2)\ S(\overleftarrow{C}|\overleftarrow{A}_3 \cup \overleftarrow{C})f^b(\overleftarrow{C} \cup \overleftarrow{A}_2|\overrightarrow{B}_1)$ behaves as $O(1)$ or decreases for $\Lambda \to 0$ while the function $\hat{\varphi}_-$ decreases as $\exp(-|\Lambda|)$. Hence the integral (228) decreases (at least as $O(e^{-|\Lambda|})$). Therefore $[n^a(x), n^b(0)] = 0$.

Now consider the commutator $[Q^a_-(x), O^b(0)]$, where O^b is n^b or j^b_μ. The following formula similar to (223), holds:

$$\begin{aligned}
&\int \langle \overrightarrow{A}|[Q^a_-(x), O^b(0)]|\overleftarrow{B}\rangle e^{i\kappa(A)x}\varphi(x)dx \\
&= \int \langle A|[Q^a_-(x), O^b(0)]|\overrightarrow{B}\rangle e^{i\kappa(A)x}\varphi_-(x)dx \\
&- \int \langle \overrightarrow{A}|[Q^a_+(x), O^b(0)]|\overleftarrow{B}\rangle e^{i\kappa(A)x}\varphi_+(x)dx \\
&+ \int \langle \overrightarrow{A}|[Q^a, O^b(0)]|\overleftarrow{B}\rangle e^{i\kappa(A)x}\varphi_+(x)dx\ . \qquad (229)
\end{aligned}$$

Evidently the last term in RHS is responsible for the appearance of $i\varepsilon^{abc}\theta(x)O^c(0)$ in the commutator. Consider the first term. Let $O^b = n^b$, then in the integral (228) the functions f_1, f_2 are taken as g^a, f^b, respectively. Obviously the integral goes to zero as $\exp(-|\Lambda|)$ for $\Lambda \to \infty$. The second term in RHS of (228) is treated in a similar way. Hence $[Q^a_-(x), n^b(0)] = i\varepsilon^{abc}\theta(x)n^c(0)$.

Now let $O^b = j^b_\mu$ and consider the function

$$\begin{aligned}
&(-1)^{n(C)+n(B_1)+n(B_2)}g^a(\overleftarrow{A}_1|\overrightarrow{C} \cup \overrightarrow{B}_2)S(\overleftarrow{C}|\overleftarrow{A}_3 \cup \overleftarrow{C}) \\
&\quad \times f^b_\mu(\overleftarrow{C} \cup \overleftarrow{A}_2|\overrightarrow{B}_1)\hat{\varphi}_-(\kappa(C) + \kappa(B_2) + \kappa(A_3) + \kappa(A_4))\ . \qquad (230)
\end{aligned}$$

The form factor f^b_μ is equal to

$$f^b_\mu(\overleftarrow{C} \cup \overleftarrow{A}_2|\overrightarrow{B}_1) = \varepsilon_{\mu\mu'}\left(p^{\mu'}(C) + p^{\mu'}(A_2) - p^{\mu'}(B_1)\right) g^b(\overleftarrow{C} \cup \overleftarrow{A}_2|\overrightarrow{B}_1)\ .$$

Hence, in the vicinity of $\sigma_{\max}$ the function (230) can behave both as $O(1)$ and as $O(\Lambda^{-1})$. It is easy to understand that the function (23) behaves as $O(1)$ only if $A_1 = A_2 = B_1 = B_2 = \emptyset$. In that case it is equal to

$$\varphi(0)\|g^a(\overrightarrow{C})\|^2\delta^{ab}(1 + O(\Lambda^{-1}))\ . \qquad (231)$$

When does the function (230) behave as $O(\Lambda^{-1})$? For $A_1 = B_2 = \emptyset$ one has

$$(-1)^{n(B_1)}\|g^a(\overrightarrow{C})\|^2\varepsilon^{bda}g^d(\overleftarrow{A}_2 - i0|\overrightarrow{B}_1)\frac{2\pi}{\Lambda}(1+O(\Lambda^{-1})) \ . \tag{232}$$

For $A_2 = B_1 = \emptyset$ one has

$$(-1)^{n(B_2)}\|g^a(\overrightarrow{C})\|^2\varepsilon^{adb}g^d(\overleftarrow{A}_2 + i0|\overrightarrow{B}_1)\frac{2\pi}{\Lambda}(1+O(\Lambda^{-1})) \ . \tag{233}$$

Finally, for $A_1 = A_2 = B_1 = B_2 = \emptyset$, the next term of asymptotics after (231) is

$$\varepsilon^{abd}\|g^a(C)\|^2\frac{2\pi}{\Lambda}\Sigma^d_{A_3} \ , \tag{234}$$

where $\sum_A^d$ is the sum $\sum_{\alpha\in a} S^d_{H\alpha}$, H_α is the isotopic space connected with the rapidity α. Having derived Eq. (234) we have to use the formula

$$S_{12}(\alpha) = 1 + 2\pi i\alpha^{-1}S_1^aS_2^a + O(\alpha^{-2}) \ , \alpha \to \infty \ .$$

For all A_1, A_2, B_1, B_2 which do not satisfy the requirement listed above the function (230) behaves as $O(\Lambda^{-2})$ for $\Lambda \to \infty$.

Fix C and compute the sum (218) of the functions (230) for $\sigma \sim \sigma_{\max}$. The term of order Λ^{-1} is equal, as it can be understood from (232), (233), (234), to

$$\varphi(0)\langle\overrightarrow{A}|Q^d_-(x) + Q^d_+(x) - Q^d|\overleftarrow{B}\rangle\varepsilon^{abd}\Lambda^{-1} = 0 \ .$$

Hence, the contribution $O(\Lambda^{-1})$ to the sum disappears. The point $\sigma_{\min}$ also does not give an $O(\Lambda^{-1})$ contribution.

Thus we have only to calculate the $O(1)$ contributions. Such a contribution from the point $\sigma_{\max}$ is defined by Eq. (231) while the contribution from the point $\sigma_{\min}$ differs by the multiplier $(-1)^\kappa$. Hence

$$\int\langle\overrightarrow{A}|[Q^a_-(x),\ j^b_\mu(0)]|\overleftarrow{B}\rangle\varphi_-(x)dx \sim i\varphi(0)(1-(-1)^\mu)\delta^{ab}\frac{\Lambda}{2\pi} \ .$$

Similarly,

$$-\int\langle\overrightarrow{A}|[Q^a_+(x),\ j^b_\mu(0)]|\overleftarrow{B}\rangle\varphi_+(x)dx \sim i\varphi(0)(1-(-1)^\mu)\delta^{ab}\frac{\Lambda}{2\pi} \ .$$

Hence

$$[Q^a_-(x),\ j^b_\mu(0)] = i\varepsilon^{abc}\theta(x)j^c_\mu(0) + i\delta^{ab}\delta(x)(1-(-1)^\mu)\infty\ , \tag{235}$$

where ∞ means a constant diverging as Λ for $\Lambda \to \infty$.

Let us now calculate the commutator $[n^a, j^b_\mu]$. Notice that this commutator has been calculated above, but here we give another method. The coincidence of the results provides us with a strong argument in favour of Hypothesis 2.

Consider the function

$$\begin{aligned}&(-1)^{n(C)+n(B_1)+n(B_2)} f^a(\overleftarrow{A}_1|\overrightarrow{C} \cup \overrightarrow{B}_2) S(\overleftarrow{C}\,|\overleftarrow{A}_3 \cup \overleftarrow{C})\\ &\quad \times f^b_\mu(\overleftarrow{C} \cup \overleftarrow{A}_2|\overrightarrow{B}_1)\hat{\varphi}_-(\kappa(C)+\kappa(B_2)+\kappa(A_3)+\kappa(A_4))\end{aligned}$$

and estimate it for $\sigma \sim \sigma_{\max}$. Everything is quite similar to the case considered above, the only difference being that now $n(A_1)+n(C)+n(B_2) \equiv 1 (\mathrm{mod}\ 2), n(C)+n(A_2)+n(B_1) \equiv 0(\mathrm{mod}\ 2)$. This fact implies that the terms with $A_1 = A_2 = B_1 = B_2 = \emptyset$ are absent, that is why the O(1) contribution is absent. The $O(\Lambda^{-1})$ contributions are produced by the terms with $A_1 = B_2 = \emptyset$ or $A_2 = B_1 = \emptyset$.

$$\begin{aligned}&(-1)^{n(B_2)}\varepsilon^{bda} f^d(\overleftarrow{A}_2 - i0|\overrightarrow{B}_1)\|g^b(\overrightarrow{C})\|^2\,\Lambda^{-1}\,2\pi\varphi(0)\ ,\\ &\quad -(-1)^{n(B_1)}\varepsilon^{adb} f^d(\overleftarrow{A}_1 + i0|\overrightarrow{B}_2)\|f^a(\overrightarrow{C'})\|^2\,\Lambda^{-1}\,2\pi\varphi(0)\ ,\end{aligned}$$

where $n(C) \equiv 0(\mathrm{mod}\ 2), n(C') \equiv 1(\mathrm{mod}\ 2)$, which fact stipulates the presence of the minus sign in the second formula. Summing over A_1, B_1, A_2, B_2, A_3 one obtains the contribution from $\sigma \sim \sigma_{\max}$:

$$\varphi(0)\frac{2\pi}{\Lambda}\varepsilon^{abc}\langle\overrightarrow{A}|n^c(0)|\overleftarrow{B}\rangle\left(\|f^d(C')\|^2 + \|g^d(C)\|^2\right)\ .$$

Now summing over C one obtains

$$\frac{1}{2\pi}\varphi(0)\varepsilon^{abc}\langle\overrightarrow{A}|n^c(0)|\overleftarrow{B}\rangle\ .$$

The contribution from $\sigma \sim \sigma_{\min}$ can be calculated in a similar way:

$$\frac{1}{2\pi}\varphi(0)\varepsilon^{abc}\langle\overrightarrow{A}|n^c(0)|\overleftarrow{B}\rangle(-1)^{\mu-1}\ .$$

Hence

$$\int \langle \overrightarrow{A} | [n^a(x),\ j_\mu^b(0)] | \overleftarrow{B} \rangle \varphi_-(x) e^{i\kappa(A)x} = \frac{i}{2}\varepsilon^{abc}$$
$$\times\ \varphi(0)(1+(-1)^\mu)\langle \overrightarrow{A} | n^c(0) | \overleftarrow{B} \rangle\ .$$

The same formula can be proven for the convolution with φ_+, hence

$$[n^a(x),\ j_\mu^b(0)] = i\varepsilon^{abc}(1+(-1)^\mu)\frac{1}{2}n^c(0)\delta(0)\ .$$

We have obtained all the commutation relations but $[j_1^a(x), j_1^b(0)] = 0$. Evidently, the commutators $[j_0^a, j_1^b], [j_0^a, j_0^b]$ having been calculated, it is sufficient to calculate $[j_+^a, j_-^b]$, which is what we are going to do now.

We have to elucidate the behaviour of the function

$$(-1)^{n(C)+n(B_1)+n(B_2)} f_+^a(\overleftarrow{A}_1 + i0 | \overrightarrow{C} \cup \overrightarrow{B}_2) S(\overleftarrow{C} | \overleftarrow{A}_3 \cup \overleftarrow{C})$$
$$\times\ f_-^b(\overleftarrow{C} \cup \overleftarrow{A}_2 | \overrightarrow{B}_1)\hat{\varphi}_-\left(\kappa(C)+\kappa(B_2)+\kappa(A_3)+\kappa(A_4)\right)$$

for $\sigma \sim \sigma_{\max}, \sigma \sim \sigma_{\min}$. Let us begin with $\sigma \sim \sigma_{\max}$. Form factors f_+^a, f_-^b can be written in terms of g^a, g^b as follows:

$$f_+^a(\overleftarrow{A}_1 | \overrightarrow{C} \cup \overrightarrow{B}_2) = \left(p^+(A_1) - p^+(C) - p^+(B_2)\right) g^a(\overleftarrow{A}_1 | \overrightarrow{C} \cup \overrightarrow{B}_2)\ ,$$
$$f_-^a(\overleftarrow{C} \cup \overleftarrow{A}_2 | \overrightarrow{B}_1) = \left(p^-(A_2) + p^-(C) - p^-(B_1)\right) g^b(\overleftarrow{C} \cup \overleftarrow{A}_2 | \overrightarrow{B}_1)\ ,$$

where

$$p^\pm(A) = M\sum_{\alpha\in A} e^{\pm\alpha}\ .$$

It is easy to understand that there is no O(1) contribution from $\sigma \sim \sigma_{\max}$. The $\mathrm{O}(\Lambda^{-1})$ contribution is given by the term with $A_1 = B_2 = \emptyset$:

$$\left(p^-(A_2) - p^-(B_1)\right)\varepsilon^{abc} g^c(\overleftarrow{A}_2 - i0 | \overrightarrow{B}_1) \| g^c(\overrightarrow{C}) \|^2 2\pi\, \Lambda^{-1}\, \varphi(0)\ .$$

Summing over A_2, B_1, A_3 and C one obtains for the O(1) contribution from $\sigma \sim \sigma_{\max}$:

$$\varepsilon^{abc}\varphi(0)\frac{1}{2\pi}\langle \overrightarrow{A} | j_-^c(0) | \overleftarrow{B} \rangle\ .$$

The contribution from $\sigma \sim \sigma_{\min}$ can be calculated similarly and it appears to be equal to

$$\varepsilon^{abc}\varphi(0)\frac{1}{2\pi}\langle\overrightarrow{A}|j_+^c(0)|\overleftarrow{B}\rangle \ .$$

Hence

$$\int\langle\overrightarrow{A}|[j_+^a(x),j_-^b(0)]|\overleftarrow{B}\rangle\varphi_-(x)dx = \frac{i}{2}\varepsilon^{abc}\varphi(0)\langle\overrightarrow{A}|j_0^c(0)|\overleftarrow{B}\rangle \ .$$

The same formula can be obtained for the convolution with φ_+, hence

$$[j_+^a(x),\ j_-^b(0)] = i\varepsilon^{abc}j_0^c(0)\delta(x) \ .$$

Combining this with Eq. (235) one gets

$$[j_1^a(x),\ j_1^b(0)] = 0 \ .$$

Q.E.D.

The commutation relations (226) coincide in general with corresponding Poisson brackets in a classical model, the only difference being in the coefficient before the δ' function (Schwinger term) in the commutator $[j_0^a(x), j_1^b(0)]$. In a classical model this coefficient is $(\delta^{ab} - n^a n^b)$. The infiniteness of the Schwinger term in the quantum case is quite logical: usually the Schwinger term is the result of quantum anomaly; it disappears in the classic case, being proportional to $\hbar$; in NLS the Schwinger term is finite in the classic case, which is why it is natural to suppose that tin the quantum case it is infinite, being proportional to $\hbar^{-1}$. This argument is in good agreement with the fact that the coefficient before δ' in (226) can be estimated as Λ (Λ is the ultraviolet cut-off of the rapidities which is connected with the ultraviolet cut-off of the moments M_0 via $\Lambda = \log\frac{M_0}{M}$). We have the formula

$$M = M_0 e^{-\frac{1}{g}} \ ,$$

which follows from asymptotic freedom (M is the mass of the physical particle, g is the coupling constant, M_0 is the ultraviolet cut-off). This formula is very general: the origin of M_0 is of no importance. Putting these facts together one realizes that the coefficient before δ' is proportional to g^{-1}. The coupling constant in NLS can be identified with $\hbar$. The absence of the nondiagonal term $n^a n^b$ in the quantum Schwinger term needs further understanding.

APPENDIX A

FORM FACTORS IN SU(N)-INVARIANT THIRRING MODEL (SU(N) CHIRAL GROSS-NEVEU MODEL)

In Sec. 7 formulae have been presented for currents and the energy-momentum tensor form factors in SU(2)-invariant Thirring model. This Appendix extends these results to the SU(N)-invariant model (chiral Gross-Neveu model) with Lagrangian

$$\mathcal{L} = \int (i\overline{\psi}\gamma_\mu\partial_\mu\psi - gK_{ab}(\overline{\psi}t^a\gamma_\mu\psi)(\overline{\psi}t^b\gamma_\mu\psi))dx \ ,$$

where $\{\psi_i^\alpha\} = \psi, i = 1, 2, \alpha = 1, 2, \ldots, N$, are Fermi fields, i, α being spinor and isotopic indices, t^a are the generators of SU(N), K_{ab} is the Killing form. It can be shown that the model leads to the same physical results as the chiral SU(N) Gross-Neveu model. The spectrum of physical excitations involves $N-1$ massive particles which transform with respect to $N-1$ fundamental representations of the group SU(N) [53, 54, 55]. The particle which transforms with respect to the l-fundamental representation will be referred to as rank-l particle. All the particles are the bound states of rank-1 particles (kinks). The rank-l particle is the antiparticle of the rank-($N-l$) particle. The mass of rank-l particle is

$$M_l = M_1 \frac{\sin\frac{\pi l}{N}}{\sin\frac{\pi}{N}} \ ,$$

where M_1 is the mass of the kink.

The rank-l particle will be parametrized by the rapidity β and the isotopic multi-index $\varepsilon = \{k^{(1)}, \ldots, k^{(l)}\}(k^{(1)} < \ldots < k^{(l)}, k^{(j)} = 1, \ldots, N)$. The number of k's in ε is denoted by $l(\varepsilon)$.

The S-matrix of two rank-1 particles can be presented as an operator in $V_1^{(1)} \otimes V_2^{(1)}$ ($V^{(1)}$ is the space of the vector representation of SU(N)):

$$S_{1,2}(\beta) = S_0(\beta)\hat{S}_{1,2}(\beta) ,$$

where

$$S_0(\beta) = \frac{\Gamma(\frac{N-1}{N} + \frac{\beta}{2\pi i})\Gamma(-\frac{\beta}{2\pi i})}{\Gamma(\frac{N-1}{N} - \frac{\beta}{2\pi i})\Gamma(\frac{\beta}{2\pi i})}, \quad \hat{S}_{1,2}(\beta) = \frac{\beta - \frac{2\pi i}{N}P_{1,2}}{\beta - \frac{2\pi i}{N}} .$$

$P_{1,2}$ is the permutation operator of spaces $V_1^{(1)}$ and $V_2^{(1)}$.

The S-matrix of rank-l_1 and rank-l_2 can be presented as an operator in $V_1^{(l_1)} \otimes V_2^{(l_2)}$ ($V^{(l)}$ is the space of l-th fundamental representation of SU(N)). The spaces $V_1^{(l_1)}$ and $V_2^{(l_2)}$ can be realized as completely skew-symmetric subspaces in $V_{a_1}^{(1)} \otimes \ldots \otimes V_{a_{l_1}}^{(l)}$ and $V_{b_1}^{(1)} \otimes \ldots \otimes V_{b_{l_2}}^{(1)}$ respectively. The S-matrix can be found in accordance with fusion procedure [50, 51]:

$$\begin{aligned}
S_{1,2}(\beta) &= P^-_{a_1,\ldots,a_{l_1}} P^-_{b_1,\ldots b_{l_2}} S_{a_1,b_1}\left(\beta + \frac{\pi i}{N}(l_1 + l_2 - 2)\right)\ldots \\
&\times S_{a_1,b_{l_2}}\left(\beta - \frac{\pi i}{N}(l_1 - l_2)\right)\ldots S_{a_{l_1}b_1}\left(\beta + \frac{\pi i}{N}(l_1 - l_2)\right)\ldots \\
&\times S_{a_{l_1},b_{l_2}}\left(\beta - \frac{\pi i}{N}(l_1 + l_2 - 2)\right) P^-_{a_1,\ldots,a_{l_1}} P^-_{b_1,\ldots,b_{l_2}} ,
\end{aligned}$$

where $P^-_{a_1,\ldots,a_{l_1}}, P^-_{b_1,\ldots,b_{l_2}}$ are the projection operators onto skew-symmetric subspaces.

To make the notations more close to those used in the basic part of this book we write the matrix element of all S-matrices as $S^{\varepsilon'_1,\varepsilon'_2}_{\varepsilon_1,\varepsilon_2}$. Evidently, only those $S^{\varepsilon'_1,\varepsilon'_2}_{\varepsilon_1,\varepsilon_2}$ can differ from zero for which $l(\varepsilon_1) = l(\varepsilon'_1), l(\varepsilon_2) = l(\varepsilon'_2)$.

The Zamolodchikov-Faddeev operators satisfy the usual algebra of commutation relations. "In" and "out" states are constructed in the usual way. The matrix C involved in the crossing-symmetry relation is

$$C_{\varepsilon_1,\varepsilon_2} = \begin{cases} (-1)^\pi , & \varepsilon_1 \cup \varepsilon_2 = \pi(1, \ldots, N) \\ 0 , & \varepsilon_1 \cup \varepsilon_2 \neq \pi(1, \ldots, N) \end{cases} .$$

We denote by ε^* such multi-index that $\varepsilon \cup \varepsilon^* = \pi(1, \ldots, N), \pi$ being some permutation.

The general matrix elements of the local operator $O(x)$ can be expressed through

$$f(\beta_1, \dots, \beta_n)_{\varepsilon_1, \dots, \varepsilon_n} = \langle \mathrm{ph}|O(0)|Z^*_{\varepsilon_n}(\beta_n), \dots, Z^*_{\varepsilon_1}(\beta_1)|\mathrm{ph}\rangle$$

in the usual way.

We consider form factors of currents $j^a_\mu = \overline{\psi}\gamma_\mu t^a \psi$, where t^a are the generators of SU(N) in vector representation, and the energy-momentum tensor $T_{\mu\nu}$. We use the following choice of generators: $H^i = \mathrm{diag}(0, \dots, 1^i, -1^{i+1}, \dots, 0)$ and E^{ij} with one nonzero element which is unity in the position ij. So a can be made equal to one of $N^2 - 1$ symbols: $\{i\}, i = 1, \dots, n-1; \{i,j\}, i \neq j$.

The fundamental requirements on form factors remain the following:

1. $$f(\beta_1, \dots, \beta_i, \beta_{i+1}, \dots, \beta_n)_{\varepsilon_1, \dots, \varepsilon_i, \varepsilon_{i+1}, \dots, \varepsilon_n} S^{\varepsilon_i, \varepsilon_{i+1}}_{\varepsilon'_i, \varepsilon'_{i+1}}(\beta_i - \beta_{i+1}) = f(\beta_1, \dots, \beta_{i+1}, \beta_i, \dots, \beta_n)_{\varepsilon_1, \dots, \varepsilon'_{i+1}\varepsilon'_i, \dots, \varepsilon_n}, \tag{A1}$$

2. $$f(\beta_1, \dots, \beta_{n-1}, \beta_n + 2\pi i)_{\varepsilon_1, \dots, \varepsilon_{n-1}, \varepsilon_n} = (-1)^{\frac{1}{N}\Sigma l(\varepsilon_j)} f(\beta_n, \beta_1, \dots, \beta_{n-1})_{\varepsilon_n, \varepsilon_1, \dots, \varepsilon_{n-1}}, \tag{A2}$$

3. $$\begin{aligned} \mathrm{res}\, f(\beta_1, \dots, \beta_{n-1}, \beta_n)_{\varepsilon_1, \dots, \varepsilon_{n-1}, \varepsilon_n} &= f(\beta_1, \dots, \beta_{n-2})_{\varepsilon'_1, \dots, \varepsilon'_{n-2}} \\ &\times C_{\varepsilon'_{n-1}, \varepsilon_n}\{\delta^{\varepsilon'_1}_{\varepsilon_1}, \dots, \delta^{\varepsilon'_{n-1}}_{\varepsilon_{n-1}} - (-1)^{\frac{1}{N}\Sigma l(\varepsilon_j)} S^{\varepsilon'_{n-1}, \varepsilon'_1}_{\tau_1, \varepsilon_1}(\beta_{n-1} - \beta_1) \dots \\ &\times S^{\tau_{n-3}\varepsilon'_{n-2}}_{\varepsilon_{n-1}, \varepsilon_{n-2}}(\beta_{n-1} - \beta_{n-2})\}, \quad \beta_n = \beta_{n-1} + \pi i . \end{aligned} \tag{A3}$$

Besides these identities there is, as in SG model, a number of formulae for residues corresponding to bound states. We would not list all these formulae, noting only one of the most fundamental. Suppose ε_{n-1} and ε_n do not contain coinciding k. Then $\varepsilon_{n-1} \cup \varepsilon_n$ can be reordered into a new multi-index ε by some permutation. There we require $f(\beta_1, \dots, \varepsilon_{n-1}, \varepsilon_n)_{\varepsilon_1, \dots, \varepsilon_{n-1}, \varepsilon_n}$ to have a simple pole at the point $\beta_n = \beta_{n-1} + \frac{\pi i}{2N}(l(\varepsilon_n) + l(\varepsilon_{n-1}))$ with the following residue:

$$\begin{aligned} &\mathrm{res}\, f(\beta_1, \dots, \beta_{n-1}, \beta_n)_{\varepsilon_1, \dots, \varepsilon_{n-1}, \varepsilon_n} \\ &\quad = (-1)^\pi a_{l(\varepsilon_n)l(\varepsilon_{n-1})} f\left(\beta_1, \dots, \varepsilon_{n-2}, \beta_{n-1} + \frac{\pi i}{N} l(\varepsilon_{n-1})\right)_{\varepsilon_1, \dots, \varepsilon_{n-2}, \varepsilon}, \end{aligned} \tag{A4}$$

where $\varepsilon = \pi(\varepsilon_1 \cup \varepsilon_2)$, and $a_{l(\varepsilon_n)l(\varepsilon_{n-1})}$ is certain constant connected with $S^{\varepsilon'_n, \varepsilon'_{n-1}}_{\varepsilon_n, \varepsilon_{n-1}}$.

It is convenient to present, first, the formulae for the form factors in the case $l(\varepsilon_1) = \ldots = l(\varepsilon_n) = 1$ (in the sector containing only kinks) and then to consider the general situation.

Form factors $f(\beta_1, \ldots, \beta_n)_{\varepsilon_1, \ldots, \varepsilon_n}, l(\varepsilon_1) = \ldots = l(\varepsilon_n) = 1$, can be combined into a vector $f(\beta_1, \ldots, \beta_n)$ in the space $(V_1^{(1)} \otimes \ldots \otimes V_n^{(1)})^*$. Let us introduce some orthonormal bases $e_{j,k}$ in the spaces $V_j^{(1)}$. The natural basis of the tensor product consists of the vectors $e_{k_1, \ldots, k_n} = \Pi \otimes e_{j,k_j}$. The form factor $f(\beta_1, \ldots, \beta_n)_{\varepsilon_1, \ldots, \varepsilon_n}$ is the component of $f(\beta_1, \ldots, \beta_n)$ in the base $e_{\varepsilon_1, \ldots, \varepsilon_n}(\varepsilon_j = \{k_j\})$. As usual, it is convenient to consider also another base $w_{k_1, \ldots, k_n}(\beta_1, \ldots, \beta_n)$, which we are now going to describe. Unfortunately, we have not succeeded in the generalization of the definition given in Sec. 7 for the SU(N) case which is why we shall give only an indirect definition of the base $w_{k_1, \ldots, k_n}(\beta_1, \ldots, \beta_n)$. To this end, let us first divide the space $(V_1^{(1)} \otimes \ldots, \otimes V_n^{(1)})^*$ into a direct sum of special subspaces. The algebra $sl(N)$ acts on $V_k^{(1)}$ with generators e_k^{ij}, h_k^i, which are represented as matrices E^{ij}, H_i in the base $e_{k,1}, \ldots, e_{k,N}$. This extends the action of $sl(N)$ to the space $(V_1^{(1)} \otimes \ldots \otimes V_n^{(1)})$ with the generators $\varepsilon^{ij} = \sum_k e_k^{ij}, \mathcal{H}^i = \sum_k h_k^i$. Obviously this is a restriction of the action of $N^2 - 1$ charges to the n-kinks sector. The space $(V_1^{(1)} \otimes \ldots V_n^{(1)})^*$ appears to be divided into a direct sum of subspaces $V^*_{l_1, \ldots, l_n}$, which are the eigensubspaces for all the $\mathcal{H}^i$ with the eigenvalues $l_i - l_{i+1}(\Sigma l_j = n)$. The basis $\omega_{k_1, \ldots, k_n}(\beta_1, \ldots, \beta_n)$ in the space $V^*_{l_1, \ldots, l_n}(k_1, \ldots, k_n$ is some permutation of the set $\underbrace{1, \ldots, 1}_{l_1}, \underbrace{2, \ldots, 2}_{l_2}, \ldots, \underbrace{N, \ldots, N}_{l_N})$ is defined by two relations:

$$\begin{aligned} &w(\beta_1, \ldots, \beta_n)_{\underbrace{N, \ldots, N}_{l_n}, \ldots, \underbrace{2, \ldots, 2}_{l_2}\underbrace{1, \ldots, 1}_{l_1}} \\ &\quad = e_{1,N} \otimes \ldots \otimes e_{l_N,N} \otimes \ldots \otimes e_{l_1 + \ldots + l_{N-1}+1,1} \otimes \ldots \otimes e_{n,1} \,, \end{aligned} \tag{A5}$$

$$\begin{aligned} &w(\beta_1, \ldots, \beta_i, \beta_{i+1}, \ldots, \beta_n)_{k_1, \ldots, k_i, k_{i+1}, \ldots, k_n} \hat{S}_{i,i+1}(\beta_i - \beta_{i+1}) \\ &\quad = w(\beta_1, \ldots, \beta_{i+1}, \beta_i, \ldots, \beta_n)_{k_1, \ldots, k_{i+1} k_i, \ldots, k_n} \,. \end{aligned} \tag{A6}$$

As usual, it is implied in the last formula that the rapidity β_j is always connected with the space $V_j^{(1)}$, i.e., when rapidities change places the spaces also change places.

Let us explain how to construct $w_{k_1,\dots,k_n}(\beta_1,\dots,\beta_n)$ using Eqs. (A5, A6). It is quite evident that using Eq. (A6) one can present $w_{k_1,\dots,k_n}$ $(\beta_1,\dots,\beta_n)$ in the form $w_{\underbrace{N,\dots,N}_{l_n},\dots,\underbrace{1,\dots,1}_{l_1}}(\beta_{\pi(1)},\dots,\beta_{\pi(n)})$ (some product of operator S), where π is some permutation. From Eq. (A5) it follows that

$$w_{\underbrace{N,\dots,N}_{l_N},\underbrace{N-1,\dots,N-1}_{l_{N-1}},\dots,\underbrace{1,\dots,1}_{l_1}}(\beta_{\pi(1)},\dots,\beta_{\pi(n)})$$
$$= e_{\pi(1),N}\otimes\dots\otimes e_{\pi(l_N)}\otimes\dots\otimes e_{\pi(n),1}\ .$$

This definition cannot lead to any contradiction because of the validity of the Yang-Baxter equation for $\hat{S}$ and of the equation

$$e_{i,k}\otimes e_{j,k}S_{ij}(\beta_i-\beta_j)=e_{i,k}\otimes e_{j,k}\ .$$

It is clear that the logic of Sec. 7 is borne out: Eqs. (A5, A6) which are considered as properties of $w_{k_1,\dots,k_n}(\beta_1,\dots,\beta_n)$ in Sec. 7 are now considered as definitions.

It can be shown that, as well as in the SU(2) case, the transformation $\{e_{k_1,\dots,k_n}\}\to\{w_{k_1,\dots,k_n}\}$ is triangular. Vector $w_{k_1,\dots,k_n}(\beta_1,\dots,\beta_n)$ has simple poles at the points $\beta_j=\beta_i+\frac{2\pi i}{N}$ for $j>i$, $k_j>k_i$.

Let us introduce the usual compact notations for the sets of rapidities. We are going to define the polynomials $\Delta(A|B^{(1)}|\dots|B^{(N)})$, $\Delta^i(A|B^{(1)}|\dots|B^{(N)})$, $(n(B^{(i)})=m, n(A)=(N-1)m-1)$, $\Delta^{ij}(A|B^{(1)}|\dots|B^{(N)})$, $(n(B^{(k)})=m-\delta_{ki}+\delta_{kj}, n(A)=(N-1)m-1)$.
The role of these polynomials is similar to that of the polynomials Δ,Δ^3, $\Delta^\pm$ from Sec. 7. The polynomials $\Delta,\Delta^i,\Delta^{ij}$ are invariant under the permutation of rapidities within the sets $B^{(j)}$. They are antisymmetric with respect to the rapidities composing A.

Polynomial Δ is the determinant of an $((N-1)m-1)((N-1)m-1)$ matrix with the following matrix elements:

$$A_{ij}=A_i(\alpha_j|B^{(1)}|\dots|B^{(N)})\ ,$$
$$A_i(\alpha|B^{(1)}|\dots|B^{(N)})=\sum_{s=1}^{N}\prod_{\beta\in B^{(s)}}\left(\alpha-\beta-\frac{\pi i}{N}(N+1-2s)\right)$$
$$\times\, Q_i^{(N)}\left(\left(\alpha-\frac{\pi i}{N}(N-2s+1)\right)\right)|\bigcup_{\substack{p=1\\p\neq s}}^{N}\left(B^{(p)}+\frac{\pi i}{N}\mathrm{sgn}(s-p)\right)\ ,$$

where

$$Q_k^{(N)}(\alpha|B)=\sum_{l=0}^{k}\left(\left(\alpha+\frac{\pi i}{N}\right)^l-\left(\alpha-\frac{\pi i}{N}\right)^l\right)(-1)^{k-l}\sigma_{k-l}(B)\,,$$

and σ_i is an elementary symmetrical polynomial of degree i.

The polynomial Δ^i is

$$\Delta^i(A|B^{(1)}|\dots|B^{(N)})=\left(\sum_{\beta\in B^{(i+1)}}\beta-\sum_{\beta\in B^{(i)}}\beta-\frac{2\pi im}{N}\right)$$
$$\times\,\Delta(A|B^{(1)}|\dots|B^{(N)})\,.$$

The polynomial Δ^{ij} are the determinants of $((N-1)m-1)\times((N-1)m-1)$ matrices A^{ij} with the following matrix elements:

$$A_{kl}^{ij}=A_k^{ij}(\alpha_l|B^{(1)}|\dots|B^{(N)})\,,$$
$$A_k^{ij}(\alpha|B^{(1)}|\dots|B^{(N)})=\sum_{s=1}^{N}\prod_{\beta\in B^{(s)}}\left(\alpha-\beta+\frac{\pi i}{N}(N+1-2s)\right)$$
$$\times\left\{Q_{k+\delta_{si}-\delta_{sj}}\left(\alpha-\frac{\pi i}{N}(N-2s+1)|\bigcup_{p\neq s}\left(B^{(p)}+\frac{\pi i}{N}\mathrm{sgn}(s-p)\right)\right)\right.$$
$$\left.-\,\delta_{si}\sigma_k\left(\bigcup_{p\neq j}(B^{(p)}+\frac{\pi i}{N}\mathrm{sgn}(s-j))\right)\right\}\,.$$

Let us define also the antisymmetric functions, which are analogs of $\exp(\pm\Sigma\alpha_i)\exp(\mp\frac{1}{2}\Sigma\beta_j)\prod_{i<j}\mathrm{sh}(\alpha_i-\alpha_j)$ involved in the integral transformation $\phi_{n,\sigma}$ from Sec. 7. The functions $w^{\pm}(A)$ are defined by

$$w^{\pm}(A)=2^{-\frac{((N-1)m-2)(m-1)}{2}}\exp\left(\mp\frac{N}{2}\sum_{\alpha\in A}\alpha\pm\frac{1}{2}\sum_{\beta\in B}\beta\right)b^{\pm}(A)\,,$$

where $B=\bigcup_{p=1}^{N}B^{(p)}$. $b^{\pm}$ are the determinants of the $((N-1)m-1)\times((N-1)m-1)$ matrices $\mathcal{B}^{\pm}$ with the following matrix elements:

$$\mathcal{B}_{kl}^{\pm}=\exp\left(\pm\left(k+\left[\frac{k-1}{N-1}\right]\right)\alpha_l\right)\,.$$

Otherwise, $b^{\pm}$ are the Vandermond determinants of the variables $\exp(\pm\alpha_l)$ with all the rows whose numbers are integer multiples of N omitted. Evidently, for $n=2$ we deal with the Vandermond determinants of $\exp(\pm 2\alpha_l)$.

Now we are in a position to define the functions

$$F_{\mu\nu}(B^{(1)}|\dots|B^{(N)}) = \sum_{\beta\in B}(e^{\beta}-(-1)^{\mu}e^{-\beta})\left(\frac{1}{2\pi i}\right)^{n(A)}\int dA$$
$$\to \prod_{\substack{\beta\in B\\ \alpha\in A}}\varphi_1^{(N)}(\alpha-\beta)\Delta(A|B^{(1)}|\dots|B^{(N)})(w^{+}(A)-(-1)^{\nu}w^{-}(A)),$$
$$F_{\mu}^{a}(B^{(1)}|\dots|B^{(N)}) = \int dA\left(\frac{1}{2\pi i}\right)^{n(A)}\prod_{\alpha\in A,\beta\in B}\varphi_1^{(N)}(\alpha-\beta)$$
$$\times\,\Delta^{a}(A|B^{(1)}|\dots|B^{(N)})(w^{+}(A)-(-1)^{\mu}w^{-}(A)),$$
$$\varphi_1^{(N)}(\alpha) = 4^{-\frac{1}{N}}\pi^{-\frac{N-1}{2N}}\Gamma\left(\frac{N-1}{2N}-\frac{\alpha}{2\pi i}\right)\Gamma\left(\frac{N-1}{2N}+\frac{\alpha}{2\pi i}\right), \qquad \text{(A7)}$$

where $\int dA = \int_{-\infty}^{\infty}d\alpha_1\dots\int_{-\infty}^{\infty}d\alpha_{n(A)}$. Evidently, we are dealing with a generalization of the integral transformation $\phi_{n,\sigma}$.

It is quite clear that the operators $T_{\mu\nu}, j_{\mu}^{a}$ can only transform the vacuum into a state containing Nm kinks ($m\in\mathbb{Z}_+$), which means that the corresponding form factors depend on Nm rapidities $\{\beta_1,\dots,\beta_{Nm}\}=B$. The form factors of $T_{\mu\nu}$ are given by the formula

$$f_{\mu\nu}(\beta_1,\dots,\beta_{Nm}) = \prod_{i<j}\zeta_{1,1}(\beta_i-\beta_j)\sum_{\substack{B=\cup B^{(p)}\\ n(B^{(p)})=m,\forall p}}F_{\mu\nu}(B^{(1)}|\dots|B^{(N)})$$
$$\times\prod_{p<q}\prod_{\substack{\beta\in B^{(p)}\\ \beta'\in B^{(q)}}}\frac{1}{(\beta-\beta')}w_{k_1,\dots,k_{mN}}(\beta_1,\dots,\beta_{Nm}), \qquad \text{(A8)}$$

where $k_j=k, \beta_j\in B^{(k)}$

$$\zeta_{1,1}(\beta) = c\,\mathrm{sh}\frac{\beta}{2}\exp\left\{-2\int_0^{\infty}\frac{\sin^2\frac{1}{2}(\beta+\pi i)k\exp(-\frac{N-1}{N}\pi k)\mathrm{sh}\frac{\pi k}{N}}{k\,\mathrm{sh}^2\pi k}dk\right\},$$
$$c = 2\pi^{\frac{N+1}{2N}}\Gamma^{-1}\left(\frac{1}{2N}\right)\exp\left(-2\int_0^{\infty}\frac{\mathrm{sh}^2\frac{\pi k}{4}\exp\left(-\frac{N-1}{N}\pi k\right)\mathrm{sh}\frac{\pi k}{N}}{k(\mathrm{sh}\,\pi k)^2}dk\right).$$

Similar formulae can be written for $f_\mu^{(a)}$. The only difference is that in the formula for $f_\mu^{(ij)}$ one has to take the sum over all the partitions of B into $B^{(p)}$ which satisfy the condition $n(B^{(p)}) = m - \delta_{pi} + \delta_{pj}$.

The validity of (A1) for the form factors is contained, as usual, in the very defiition and follows from Eq. (A6) and the formulae

$$\zeta_{1,1}(\beta)S_0(\beta) = \zeta_{1,1}(-\beta), \quad \zeta(\beta - 2\pi i) = \zeta(-\beta) .$$

We would not discuss the properties (A2), (A3) at this stage, but pass on to the problem of bound states.

It is easy to understand that the problem of calculations of form factors corresponding to bound states (particles of higher ranks) can be reduced essentially to the calculation of $F(B^{(1)}|\dots|B^{(N)})$ in the case $\beta_1^{(k_j)} = \tau - \frac{\pi i}{N}(l - 2j + 1), l \le N - 1, j = 1, \dots, l, \beta_1^{(k_j)} \in B^{(k_j)}$. Let us restrict our consideration to one form factor, namely to the form factor $f_{++}(\beta_1, \dots, \beta_n)$. The polynomial $Q_i^{(N)}$ satisfies the relations

$$Q_i^{(N)}(\alpha|B) = Q_i^{(N)}(\alpha|B\backslash\beta) - \beta Q_{i-1}^{(N)}(\alpha|B\backslash\beta) ,$$

$$Q_i^{(N)}(\alpha|B) = \left(\alpha + \frac{\pi i}{N}\right)^{i-n(B)} \prod_{\beta\in B}\left(\alpha - \beta + \frac{\pi i}{N}\right) - \left(\alpha - \frac{\pi i}{N}\right)^{i-n(B)} \prod_{\beta\in B}\left(\alpha - \beta - \frac{\pi i}{N}\right) , \quad i \ge n(B) .$$

Using these relations one can show that

$$\Delta(A|B^{(1)}|\dots|B^{(N)})|_{\substack{\beta^{(k_j)} = \tau - \frac{\pi i}{N}(l-2j+1) \\ j=1\dots l}} = \sum_{p_1<\dots<p_{l-1}\le(N-1)m-1} \Delta_{k_1,\dots,k_l}(A\backslash(\alpha_{p_1},\dots,\alpha_{p_{l-1}})|\tilde{B}^{(1)}|\dots|\tilde{B}^{(N)}|\tau) \times M(\alpha_{p_1},\dots,\alpha_{p_{l-1}})(-1)^{p_1+\dots+p_{l-1}} ,] \tag{A9}$$

where $\tilde{B}^{(q)} = B^{(q)}$ if $q \ne k_j, \tilde{B}^{k_j} = B^{(k_j)}\backslash\beta_1^{(k_j)}$, $M(\alpha_1, \dots, \alpha_{l-1})$ is the determinant of an $(l-1)\times(l-1)$ matrix $\mathcal{M}$ with the following matrix elements:

$$\mathcal{M}_{ij} = \mathcal{M}_i(\alpha_j) ,$$

$$\mathcal{M}(\alpha) = \left(\alpha - \tau + \frac{\pi i}{N}(N-l)\right)(\alpha + \pi i)^{i-1}\prod_{\beta\in\tilde{B}}\left(\alpha - \beta + \frac{\pi i}{N}(N-1)\right) - \left(\alpha - \tau - \frac{\pi i}{N}(N-l)\right)(\alpha - \pi i)^{i-1}\prod_{\beta\in\tilde{B}}\left(\alpha - \beta - \frac{\pi i}{N}(N-1)\right) ,$$

where $\tilde{B} = \cup \tilde{B}^{(p)}$. The polynomial $\Delta_{k_1,\ldots,k_l}(A|B^{(1)}|\ldots|B^{(N)}|\tau)$, $n(B^{(p)}) = m - \sum_j \delta_{p,k_j}$, $n(A) = (N-1)m - l$, is the determinant of an $((N-1)m - l) \times ((N-1)m - l)$ matrix with the following elements:

$$
\begin{aligned}
A_{ij}^{(k_1,\ldots,k_l)} &= A_i^{(k_1,\ldots,k_l)}(\alpha_j|B^{(1)}|\ldots|B^{(N)}|\tau) , \\
A_i^{(k_1,\ldots,k_l)}(\alpha|B^{(1)}|\ldots|B^{(N)}|\tau) &= \sum_{\substack{s=1 \\ s\neq k_1,\ldots,k_l}}^{N} \prod_{\beta\in B^{(s)}} \left(\alpha - \beta - \frac{\pi i}{N}(N+1-2s)\right) \\
&\times Q_i^{(N)}\left(\alpha - \pi i \frac{N-2s+1}{N} \Big| \bigcup_{\substack{p=1 \\ p\neq s}}^{N} \left(B^{(p)} + \frac{\pi i}{N}\mathrm{sgn}(s-k)\right) \cup \left(\tau + \frac{\pi i}{N}\right.\right. \\
&\times \left.\left.(-l + q(s,k_1,\ldots,k_l))\right)\right) + \sum_{j=1}^{l} \prod_{\beta\in B^{(k_j)}} \left(\alpha - \beta - \frac{\pi i}{N}(N+1-2k_j)\right) \\
&\times \left(\alpha - \tau - \frac{\pi i}{N}(N - l - 2k_j + 2j)\right) \\
&\times Q_i^{(N)}\left(\alpha - \frac{\pi i}{N}(N - 2k_j + 1) \Big| \cup_{p\neq k_j} \left(B^{(p)} + \frac{\pi i}{N}\mathrm{sgn}(s-p)\right)\right) ,
\end{aligned}
$$

where $q(s, k_1, \ldots, k_l)$ is an integer defined by the relation $k_q < s < k_{q+1}$. Relation (A9) is valid also for the case $l = N$. In this particular case, $\Delta_{1,\ldots,N}(A|B^{(1)}|\ldots|B^{(N)}|\tau) = \prod_\alpha(\alpha - \tau)\Delta(A|B^{(1)}|\ldots|B^{(N)})$, and we obtain a direct analog Eq. (120).

Consider $F_{++}(B^{(1)}|\ldots|B^{(N)})$. The determinant of matrix $\mathcal{M}$ can be transformed to the determinant of matrix $\tilde{\mathcal{M}}$ whose elements are

$$
\begin{aligned}
\tilde{\mathcal{M}}_{ij} &= \tilde{\mathcal{M}}_i(\alpha_j) , \\
\mathcal{M}_i(\alpha) &= \left(\alpha - \tau + \frac{\pi i}{N}(N-l)\right) \prod_{p=0}^{i-1} \left(\alpha - \frac{\pi i}{N}(N - l + 2p - 2) + 2\pi i\right) \\
&\times \prod \left(\alpha - \beta + \frac{\pi i}{N}(N-1)\right) - \left(\alpha - \tau - \frac{\pi i}{N}(N-l)\right) \\
&\times \prod_{p=0}^{i-1} \left(\alpha - \tau - \frac{\pi i}{N}(N - l + 2p - 2)\right) \prod \left(\alpha - \beta - \frac{\pi i}{N}(N-1)\right) .
\end{aligned}
$$

Notice now that

$$\prod_{j=1}^{l}\varphi_1^{(N)}\left(\alpha-\frac{\pi i}{N}(l+1-2j)\right)=\varphi_l^{(N)}(\alpha)$$
$$\times\prod_{j=1}^{l-1}\mathrm{sh}^{-1}\frac{1}{2}\left(\alpha-\frac{\pi i}{N}(N-l+2j-2)\right),$$
$$\varphi_l(\alpha)=4^{-\frac{l}{N}}\pi^{\frac{N+l}{N}}\Gamma\left(\frac{N-l}{2N}-\frac{\alpha}{2\pi i}\right)\Gamma\left(\frac{N-l}{2N}+\frac{\alpha}{2\pi i}\right).$$

Substituting these into Eqs. (A7) and taking into account the antisymmetry with respect to variables α_j one obtains

$$F_{++}(B^{(1)}|\dots|B^{(N)})|_{\beta_1^{(k_j)}=\tau-\frac{\pi i}{N}(l-2j+1)}$$
$$=\left(\frac{1}{2\pi i}\right)^{n(\tilde{A})}\int d\tilde{A}\int d\alpha_1\dots\int d\alpha_{l-1}\left(\frac{1}{2\pi i}\right)^{l-1}\prod_{\alpha\in A}\prod_{\beta\in\tilde{B}}\varphi_1^{(N)}(\alpha-\beta)$$
$$\times\prod_{\alpha\in A}\varphi_l(\alpha-\tau)\prod_{j=1}^{l-1}\mathrm{sh}^{-1}\left(\alpha-\tau-\frac{\pi i}{N}(N-l+2j-2)\right)$$
$$\times\Delta_{k_1,\dots,k_l}(A|\tilde{B}^{(1)}|\dots|\tilde{B}^{(N)}|\tau)\left\{\prod_{i=1}^{l-1}\left(\alpha_i-\tau+\frac{\pi i}{N}(N-l)\right)\right.$$
$$\times\prod_{p=0}^{i-1}\left(\alpha_i-\tau-\frac{\pi i}{N}(N-l+2p-2)+2\pi i\right)\prod_{\beta\in\tilde{B}}\left(\alpha_i-\beta+\frac{\pi i}{N}(N-1)\right)$$
$$-\left(\alpha_i-\tau-\frac{\pi i}{N}(N-l)\right)\prod_{p=0}^{i-1}\left(\alpha_i-\tau-\frac{\pi i}{N}(N-l+2p-2)\right)$$
$$\left.\times\prod_{\beta\in\tilde{B}}\left(\alpha-\beta-\frac{\pi i}{N}(N-1)\right)\right\}w_+(A),$$

where $A=\tilde{A}\cup\{\alpha_1,\dots,\alpha_{l-1}\}$. The integrals over $\alpha_1,\dots,\alpha_{l-1}$ can be calculated using the usual trick. The fact is that the integrals over $\alpha_1,\dots,$

α_{l-1} can be rewritten in the form

$$\frac{1}{2\pi i}\left(\int_{\mathbb{R}+2\pi i}-\int_{\mathbb{R}}\right)\prod_{\beta\in\tilde{B}}\varphi_1^{(N)}(\alpha_i-\beta)\varphi_l^{(N)}(\alpha_i-\tau)$$
$$\times\prod_{j=0}^{l-1}\frac{1}{\mathrm{sh}\frac{1}{2}(\alpha_i-\tau-\frac{\pi i}{N}(N-l+2j-2)}\left(\alpha_i-\tau-\frac{\pi i}{N}(N-l)\right)$$
$$\times\prod_{\beta\in\tilde{B}}\left(\alpha_i-\beta-\frac{\pi i}{N}(N-1)\right)\prod_{p=0}^{i-1}\left(\alpha_i-\tau-\frac{\pi i}{N}(N-l+2p-2)\right)$$
$$\times w_+(A)\dots,$$

where the dots denote a function not depending on α_i. Thus these integrals are equal to the sum of residues of the integrand in the strip $0<\mathrm{Im}\alpha_i<2\pi i$. Consider first the integral over α_{l-1}. The integrand has only one simple pole at the point $\alpha_{l-1}=\tau+\frac{\pi i}{N}(N+l-4)$. Now consider the integral over α_{l-2}. The integrand has two simple poles at the points $\alpha_{l-2}=\tau+\frac{\pi i}{N}(N+l-4)$ and $\alpha_{l-2}=\tau+\frac{\pi i}{N}(N+l-6)$. However, the first pole is cancelled by the zero of w_+, whose arguments after calculating the integral over α_{l-1} become $\tilde{A},\alpha_1,\dots,\alpha_{l-2},\tau+\frac{\pi i}{N}(N+l-4)$. Continuing these calculations one makes sure that the integrals over $\alpha_i(i=1,\dots,l-1)$ are equal to the residues at the points $\alpha_i=\tau+\frac{\pi i}{N}(N-l+2i-2)$. Finally, one obtains for F_{++}

$$F_{++}(B^{(1)}|\dots|B^{(N)})|_{\beta_1^{(k_j)}=\tau-\frac{\pi i}{N}(l-2j+1)}$$
$$=\prod_{j=1}^{l-1}\prod_{\beta\in\tilde{B}}\varphi_1^{(N)}\left(\tau-\beta+\frac{\pi i}{N}(N-l+2j-2)\right)$$
$$\times\left(\frac{1}{2\pi i}\right)^{n(\tilde{A})}\int d\tilde{A}\prod_{\alpha\in\tilde{A}}\varphi_l^{(N)}(\alpha-\tau)\prod_{\beta\in\tilde{B}}\varphi_1^{(N)}(\alpha-\beta)$$
$$\times\Delta_{k_1,\dots,k_l}(\tilde{A}|B^{(1)}|\dots|B^{(N)})w_+(\tilde{A}\cup\cup_{j=1}^{l-1}\left(\tau+\frac{\pi i}{N}(N-l+2j-2))\right)$$
$$\times\frac{1}{\prod_{j=1}^{l-1}\mathrm{sh}\frac{1}{2}(\alpha-\tau-\frac{\pi i}{N}(N-l+2j-2)))}. \qquad \text{(A10)}$$

Using the above procedure one can obtain the form factors which correspond to an arbitrary set of particles of all possible ranks. We present the final result, restricting our consideration to f_{++}. Consider the set of rapidities $B = \{\beta_1, \dots, \beta_n\}$. Let the rapidity β_j correspond to the rank-l_j particle. The set B appears to be divided into the sum $\bigcup_{l=1}^{N-1} B_l$, the subsets B_l containing rapidities which correspond to particles of fixed ranks. Form factors corresponding to this situation create the vector $f_{++}(\beta_1, \dots, \beta_n)$ belonging to the space $(V_1^{(l_1)} \otimes \dots \otimes V_n^{(l_n)})^*$. This vector is given by the formula

$$f_{++}(\beta_1, \dots, \beta_n) = \prod_{i<j} \zeta_{l_1, l_j}(\beta_i - \beta_j) \frac{1}{(\beta_i - \beta_j + \frac{\pi i}{N}(l_i - l_j))}$$

$$\times \sum_{B_l = \cup B^{(\varepsilon)}} \Big\{ \prod_{\varepsilon_1 \neq \varepsilon_2, \varepsilon_{1,2} = \{k_{1,2}^{(j)}\}} \prod_{\substack{p,q: \\ k_1^{(p)} = k_2^{(q)}}} \prod_{\substack{\beta \in B^{(\varepsilon_1)} \\ \beta' \in B^{(\varepsilon_2)}}} (\beta - \beta'$$

$$- \frac{\pi i}{N}(l_2 - l_1 - 2p - 2q)) \Big\} F_{++}(B^{(1)}| \dots |B^{(N)}|B^{(1,2)}| \dots$$

$$|B^{(N-1,N)}| \dots |B^{(1,\dots,N-1)}| \dots |B^{(2,\dots,N)})$$

$$\times w_{\varepsilon_1, \dots, \varepsilon_n}(\beta_1, \dots, \beta_n) , \quad l(\varepsilon) = l, \Sigma n(B^{(\varepsilon)})\varepsilon = m\{1 \dots N\} ,$$

where $\varepsilon_j = \varepsilon$ if $\beta_j \in B^{(\varepsilon)}$, and

$$w_{\{k_1^{(1)}, \dots, k_1^{(l_1)}\}, \{k_2^{(1)}, \dots, k_2^{(l_2)}\} \dots \{k_n^{(1)}, \dots, k_n^{(l_n)}\}}(\beta_1, \dots, \beta_n)$$

$$= \lim_{\varepsilon \to 0} \left(\frac{i\varepsilon}{N}\right)^{\Sigma(l_i - 1)} w_{k_1^{(1)}, \dots, k_1^{(l_1)}, k_2^{(l_1)}, \dots, k_2^{(l_2)}, \dots, k_n^{(1)}, \dots, k_n^{(l_n)}}$$

$$\times \Big(\beta_1 - \frac{\pi + \varepsilon}{N} i(l_1 - 1), \beta_1 - \frac{\pi + \varepsilon}{N} i(l_1 - 3) \dots$$

$$\beta_1 + \frac{\pi + \varepsilon}{N} i(l_1 - 1), \beta_2 - \frac{\pi + \varepsilon}{N} i(l_2 - 1), \dots$$

$$\beta_2 + \frac{\pi + \varepsilon}{N} i(l_2 - 1), \dots, \beta_n - \frac{\pi + \varepsilon}{N} i(l_n - 1), \dots, \beta_n + \frac{\pi + \varepsilon}{N} i(l_n - 1)\Big)$$

$$\times P^-_{1, \dots, l_1} P^-_{l_1+1, \dots, l_1+l_2} \cdots P^-_{l_1 + \dots + l_{n-1}+1, \dots, l_1 + \dots + l_n} ;$$

$$\zeta_{l_i,l_j}(\beta) = c_{l_i,l_j} \mathrm{sh}\frac{\beta}{2} \prod_{j_1=1}^{l_i} \prod_{j_2=1}^{l_j} \mathrm{sh}\frac{1}{2}\left(\beta + \frac{\pi i}{N}(j_1 - j_2)\right)$$

$$\times \prod_{p=1}^{l_i-1} \Gamma\left(\frac{l_i - l_j - 2p}{2N} - \frac{\beta}{2\pi i}\right) \Gamma\left(\frac{2N - l_i - l_j + 2p}{2N} + \frac{\beta}{2\pi i}\right)$$

$$\times \prod_{p=1}^{l_j-1} \Gamma\left(\frac{2N - l_i - l_j - 2p}{2N} - \frac{\beta}{2\pi i}\right) \Gamma\left(\frac{l_j - l_i - 2p}{2N} + \frac{\beta}{2\pi i}\right)$$

$$\times \exp\left\{-\int_0^\infty \frac{\sin^2 \frac{1}{2}(\beta + \pi i)k \exp(-\frac{N-1}{N}\pi k)\mathrm{sh}(\frac{\pi l_j}{N}k)\mathrm{sh}(\frac{\pi l_i}{N}k)}{k\mathrm{sh}^2\pi k \mathrm{sh}\frac{\pi k}{N}} dk\right\} .$$

The function F_{++} is given by the following integral:

$$F_{++}(B^{(1)}|\dots|B^{(N)}|B^{(1,2)}|\dots|B^{(2,3,\dots,N)})$$

$$= \left(\frac{1}{2\pi i}\right)^{n(A)} \int dA \prod_{\alpha\in A} \prod_{\beta\in B_l} \varphi_l^{(N)}(\alpha - \beta)$$

$$\times \frac{w_+(A \,|\, \bigcup_{l>1} \bigcup_{j=0}^{l-1} \bigcup_{\beta\in B_l} (\beta - \frac{\pi i}{N}(N - l + 2j - 2)))}{\prod_{\alpha\in A} \prod_{\beta\in B_l} \mathrm{sh}\frac{1}{2}(\alpha - \beta + \frac{\pi i}{N}(N - l + 2j - 2))}$$

$$\times \Delta(A|B^{(1)}|\dots|B^{(N)}|B^{(1,2)}|\dots|B^{(N-1,N)}|\dots|B^{(2,3,\dots,N)}) ,$$

where $n(A) = \Sigma(N - l)n(B_l), A$ is the determinant of the $n(A) \times n(A)$ matrix A with the following elements:

$$A_{ij} = A_i(\alpha_j|B^{(1)}|\dots|B^{(N)}|B^{(1,2)}|\dots|B^{(N-1,N)}|\dots|B^{(2,3,\dots,N)}) ,$$

$$A_i(\alpha|B^{(1)}|\dots|B^{(N)}|B^{(1,2)}|\dots|B^{(N-1,N)}|\dots|B^{(2,3,\dots,N)})$$

$$= \sum_{s=1}^{n} \sum_{l=1}^{N-1} \prod_{\substack{\varepsilon=\{k^{(1)},\dots,k^{(l)}\}:\\ \exists k^{(j)}=s}} \prod_{\beta\in B^{(\varepsilon)}} \left(\alpha - \beta - \frac{\pi i}{N}(N - l - 2s + 2j)\right)$$

$$\times Q_i^{(N)}\left(\alpha - \frac{\pi i}{N}(N + 1 - 2s) \,|\right.$$

$$\left.\times \bigcup_{l=1}^{N-1} \bigcup_{\substack{\varepsilon=\{k^{(1)},\dots k^{(2)}\}:\\ \forall k^{(j)}\neq s}} \bigcup_{\beta\in B^{(\varepsilon)}} \left(\beta + \frac{\pi i}{N}(-l + 2q(s, k^{(1)}, \dots, k^{(l)})\right)\right) .$$

Relations (A9) imply an enormous number of recurrent relations for $\Delta(B^{(1)}|\ldots|B^{(2,3,\ldots,N)})$. Using these relations one can prove the validity of Eqs. (A2,A3) for $f_{\mu\nu}, f^a_\mu$. The method of proof differs only in detail from that proposed in Sec. 4. We would not like to repeat once more this cumbersome proof. We believe that having read the book a reader interested in techniques should have obtained enough experience to reproduce this proof himself.

Let us list several important properties of $f_{\mu\nu}, f^a_\mu$. The corresponding proofs are similar to those presented in Secs. 4, 7.

1. The operators $T_{\mu\nu}$ and j^a_μ defined by $f_{\mu\nu}$ and f^a_μ transform as a (1, 1) tensor and a vector respectively with respect to Lorentz transformations.

2. The operators $T_{\mu\nu}$ are isotopic scalar while j^a_μ form the base of adjoint representation of SU(N).

3. The following equations hold:

$$\sum_{\mu=0}^{1}\sum_{j=1}^{n}(e^{-\beta_j}+(-1)^\mu e^{\beta_j})M_{l(\varepsilon_j)}f^a_\mu(\beta_1,\ldots,\beta_n)_{\varepsilon_1,\ldots,\varepsilon_n}=0\ ,$$

$$\sum_{\mu=0}^{1}\sum_{j=1}^{n}(e^{-\beta_j}+(-1)^\mu e^{\beta_j})M_{l(\varepsilon_j)}f_{\mu\nu}(\beta_1,\ldots,\beta_n)_{\varepsilon_1,\ldots,\beta_n}=0\ ,$$

$$f_{\mu\nu}(\beta_1,\ldots,\beta_n)_{\varepsilon_1,\ldots,\varepsilon_n}=f_{\nu\mu}(\beta_1,\ldots,\beta_n)_{\varepsilon_1,\ldots,\varepsilon_n}\ ,$$

which imply that

$$\partial_\mu j^a_\mu=0,\ \partial_\mu T_{\mu\nu}=0,\ T_{\mu\nu}=T_{\nu\mu}\ .$$

4. The functions

$$g^a(\beta_1,\ldots\beta_n)_{\varepsilon_1,\ldots,\varepsilon_n}=f^a_0(\beta_1,\ldots,\beta_n)_{\varepsilon_1,\ldots,\varepsilon_n}\left(\sum M_{l(\varepsilon_j)}\mathrm{sh}\beta_j\right)^{-1}\ ,$$

$$g_\mu(\beta_1,\ldots,\beta_n)_{\varepsilon_1,\ldots,\varepsilon_n}=f_{\mu 0}(\beta_1,\ldots,\beta_n)_{\varepsilon_1,\ldots,\varepsilon_n}\left(\sum M_{l(\varepsilon_j)}\mathrm{sh}\beta_j\right)^{-1}$$

have no additional singularities in comparison with the functions $f^a_\mu, f_{\mu\nu}$ for $n>2$. For $n=2$ there are additional simple poles at the point $\beta_2=\beta_1+\pi i$ with the residues

$$\begin{aligned}\operatorname{res} g^a(\beta_1,\beta_2)_{\varepsilon_1,\varepsilon_2}&=c_{\varepsilon_1,\varepsilon'_1}(t^a)^{\varepsilon'_1}_{\varepsilon_2}\ ,\\ \operatorname{res} g_\mu(\beta_1,\beta_2)_{\varepsilon_1,\varepsilon_2}&=c_{\varepsilon_1,\varepsilon_2}(e^{\beta_1}-(-1)^\mu e^{-\beta_1})\ ,\end{aligned}$$

where $(t^a)^{\varepsilon_2}_{\varepsilon_1}$ is the matrix element of t^a in the $l(\varepsilon_1)$-th fundamental representation for $l(\varepsilon_1) = l(\varepsilon_2), (t^a)^{\varepsilon_2}_{\varepsilon_1} = 0$ for $l(\varepsilon_1) \neq l(\varepsilon_2)$. This property implies that the operators $Q^a = \int j^a_0(x_1)dx_1$ and $P_\mu = \int T_{\mu 0}(x_1)dx_1$ are the charges and energy-momentum respectively.

5. The following asymptotics hold:

$$\begin{aligned} &g^a(\beta_1, \dots, \beta_k, \beta_{k+1} + \Lambda, \dots, \beta_n + \Lambda)_{\varepsilon_1,\dots,\varepsilon_n} \\ &\quad = O\left(\Lambda^{-\varphi(m)} \exp\left(-\frac{m(N-m)}{N}\Lambda\right)\right) \\ &g_\mu(\beta_1, \dots, \beta_k, \beta_{k+1} + \Lambda, \dots, \beta_n + \Lambda)_{\varepsilon_1,\dots,\varepsilon_n} \\ &\quad = O\left(\Lambda^{-\varphi(m)-1} \exp\left(-\frac{m(N-m)}{N}\Lambda\right)\right), \\ &\qquad \Lambda \to \infty \end{aligned} \tag{A11}$$

where $m = \Sigma l(\varepsilon_j) (\text{mod } N)$, $0 \leq m \leq N-1$, and $\varphi(m)$ is some function, whose value for $m = 0$ when the exponentionally decreasing factor disappears is important: $\varphi(0) = 1$.

All the properties of form factors mentioned above have their analogs in the SU(2) case. That is why we have not dwelt on their proofs. There is, however, one interesting property, trivial in the SU(2) case, which is to be discussed individually. This property is concerned with C-invariance which requires the following equations to hold:

$$\begin{aligned} f_{\mu\nu}(\beta_1, \dots, \beta_n)_{\varepsilon_1,\dots,\varepsilon_n} &= (-1)^{\frac{1}{N}\Sigma l(\varepsilon_j)} f_{\mu\nu}(\beta_1, \dots, \beta_n)_{\varepsilon^*_1,\dots,\varepsilon^*_n}, \\ f^a_\mu(\beta_1, \dots, \beta_n)_{\varepsilon_1,\dots,\varepsilon_n} &= -(-1)^{\frac{1}{N}\Sigma l(\varepsilon_j)} f^a_\mu(\beta_1, \dots, \beta_n)_{\varepsilon^*_1,\dots,\varepsilon^*_n}. \end{aligned} \tag{A12}$$

It is not evident from Eqs. (A8) that these equations take place. To prove Eqs. (A12) let us, first, prove the following lemma, which is also of interest by itself.

Lemma 1. The form factors $f(\beta_1, \dots, \beta_n)_{\varepsilon_1,\dots,\varepsilon_n}$ are defined uniquely by Eqs. (A1, A2, A3), the analytical properties described at the beginning of the Appendix, the asymptotics (A11), the properties with respect to the Lorentz transformations and the two-particle form factors $f(\beta_1, \beta_2)_{\varepsilon_1,\varepsilon_2}$ for $l(\varepsilon_1) = 1, l(\varepsilon_2) = N-1$.

Proof. We do not concretize any particular form factor we deal with because all following reasonings are applicable to both $f_{\mu\nu}$ and f^a_μ.

Evidently, it is sufficient to prove the uniqueness of $f(\beta_1,\dots,\beta_n)_{\varepsilon_1,\dots,\varepsilon_n}$ for $l(\varepsilon_1)=\dots=l(\varepsilon_n)=1$, because all the form factors can be obtained as residues using (A4). Form factors $f(\beta_1,\dots,\beta_n)_{\varepsilon_1,\dots,\varepsilon_n}$ for $l(\varepsilon_1)=\dots=l(\varepsilon_n)=1$ form the vector $f(\beta_1,\dots,\beta_n)$ belonging to $(V_1^{(1)}\otimes\dots\otimes V_n^{(1)})^*$. Consider the vector $h(\beta_1,\dots,\beta_n)$ defined by

$$h(\beta_1,\dots,\beta_n)=\prod_{i<j}\zeta_{1,1}^{-1}(\beta_i-\beta_j)f(\beta_1,\dots,\beta_n)\ .$$

$h(\beta_1,\dots,\beta_n)$ possesses the properties

1. $h(\beta_1,\dots,\beta_n)\hat{S}_{i,i+1}(\beta_i-\beta_{i+1})$
$= h(\beta_1,\dots,\beta_{i+1},\beta_i,\dots,\beta_n)\,,$

2. $h(\beta_1,\dots,\beta_{n-1},\beta_n+2\pi i)=h(\beta_n,\beta_1,\dots,\beta_{n-1})$
$= h(\beta_1,\dots,\beta_n)\hat{S}_{n-1,n}(\beta_{n-1}-\beta_n)\dots\hat{S}_{1,n}(\beta_1-\beta_n)\ ,\qquad$ (A13)

3. $h(\beta_1,\dots,\beta_n)$ is an analytical function of β_n whose only singularities in the strip $0<\mathrm{Im}\beta_n<\pi$ are the simple poles at the points $\beta_n=\beta_j+\frac{2\pi}{N}$ with residues

$$\begin{aligned}\mathrm{res}h(\beta_1,\dots,\beta_n)&=h_2\left(\beta_1,\dots,\hat{\beta}_j,\dots,\beta_{n-1},\beta_j+\frac{\pi i}{N}\right)\\&\times\hat{S}_{n-1,j}(\beta_{n-1}-\beta_j)\dots\hat{S}_{j+1,j}(\beta_{j+1}-\beta_j)\\&\times\prod_{i=1}^{n-1}\Gamma\left(\frac{N-1}{2N}-\frac{\beta_j-\beta_i}{2\pi i}\right)\Gamma\left(\frac{N-1}{2N}+\frac{\beta_j-\beta_i}{2\pi i}\right)\ ,\end{aligned}$$

where

$$\begin{aligned}&h_2(\beta_1,\dots,\beta_{n-2},\tau)\\&=\prod_{i<j}^{n-2}\zeta_{1,1}^{-1}(\beta_i-\beta_j)\prod_{i=1}^{n-2}\zeta_{1,2}^{-1}(\beta_i-\tau)f(\beta_1,\dots,\beta_{n-2},\tau)\ ,\end{aligned}$$

where $f(\beta_1,\dots,\beta_{n-2},\tau)$ is a vector from $(V_1^{(1)}\otimes\dots\otimes V_{n-2}^{(1)}\otimes V^{(2)})^*$ with components $f(\beta_1,\dots,\beta_{n-2},\tau)_{\varepsilon_1,\dots,\varepsilon_{n-2},\varepsilon_{n-1}}$, $l(\varepsilon_1)=\dots=l(\varepsilon_{n-2})=1$, $l(\varepsilon_{n-1})=2$.

4. $h(\beta_1,\dots,\beta_n)$ decreases exponentially for $\alpha\to\pm\infty$. What singularities does $h(\beta_1,\dots,\beta_n)$ have as functions of β_1 in the strip $-\pi<\mathrm{Im}\beta_1<0$? From the properties with respect to Lorentz transformations it follows

that it makes no difference to move β_1 into lower half plane or to move $\beta_2, \ldots, \beta_n$ into upper half plane. That is why this function has only simple poles at the points $\beta_1 = \beta_j - \frac{2\pi i}{N}$ with residues expressed in terms of $h_2(\beta_2, \ldots, \hat{\beta}_j, \ldots, \beta_n, \tau)$.

What singularities does $h(\beta_1, \ldots, \beta_n)$ have as functions of β_n in the strip $\pi < \mathrm{Im}\beta_n < 2\pi$? Eq. (A13) implies that $h(\beta_1, \ldots, \beta_n) = h(\beta_n - 2\pi i, \beta_1, \ldots, \beta_{n-1})$, hence the singularities at the strip $+\pi < \mathrm{Im}\beta_n < 2\pi$ are connected with the singularities with respect to the first argument in the strip $-\pi < \mathrm{Im}\beta_1 < 0$. Namely, there are simple poles at the points $\beta_n = \beta_j + 2\pi i - \frac{2\pi}{N}$ the residues of which are expressed in terms of $h_2(\beta_1, \ldots, \hat{\beta}_j, \ldots, \beta_{n-1}, \tau)$.

What singularities does $h(\beta_1, \ldots, \beta_n)$ have as functions of β in the strip $2\pi k \leq \mathrm{Im}\beta_n \leq 2\pi(k+1), k > 0$? Let $\beta_n = \tau + 2\pi k, 0 \leq \mathrm{Im}\tau \leq 2\pi$, then, as it follows from (A13),

$$\begin{aligned}
h(\beta_1, \ldots, \beta_{n-1}, \tau + 2\pi ik) &= h(\beta_1, \ldots, \beta_{n-1}, \tau)\hat{S}(\beta_{n-1}, -\tau) \\
&\times \ldots \times \hat{S}(\beta_1 - \tau)\hat{S}(\beta_{n-1} - \tau - 2\pi i) \ldots \hat{S}(\beta_1 - \tau - 2\pi i) \\
&\times \hat{S}(\beta_{n-1} - \tau - 2\pi i(k-1))\hat{S}(\beta_1 - \tau - 2\pi i(k-1)) \,. \qquad \text{(A14)}
\end{aligned}$$

The only singularity of $\hat{S}(\beta)$ is the simple pole at the point $\beta = \frac{2\pi i}{N}$. Consequently, the singularities of (A14) are caused by the singularities of $h(\beta_1, \ldots, \beta_{n-1}, \tau)$, i.e., they are simple poles at the points $\beta_n = \tau + 2\pi i = \beta_j + 2\pi ik \pm (\pi - \frac{2\pi}{N})i$ whose residues can be expressed through $h_2(\beta_1, \ldots, \beta_n)$. What singularities does $h(\beta_1, \ldots, \beta_n)$ have in the strip $-2\pi(k+1) < \mathrm{Im}\beta_n < -2\pi k, k \geq 0$? Let $\beta_n = \tau - 2\pi i(k+1)$. Then

$$\begin{aligned}
h(\beta_1, \ldots, \beta_{n-1}, \tau - 2\pi i(k+1)) &= h(\tau - 2\pi i(k+1), \beta_1, \ldots, \beta_{n-1}) \\
&\times \hat{S}(\tau - 2\pi i(k+1) - \beta_1) \ldots \hat{S}(\tau - 2\pi i(k+1) - \beta_{n-1}) \\
&= h(\beta_1, \ldots, \beta_{n-1}, \tau - 2\pi ik)\hat{S}(\tau - 2\pi i(k+1) - \beta_1) \\
&\times \ldots \hat{S}(\tau - 2\pi i(k+1) - \beta_{n-1}) = h(\beta_1, \ldots, \beta_{n-1}, \tau) \\
&\times \hat{S}(\tau - \beta_1 - 2\pi i) \ldots \hat{S}(\tau - \beta_{n-1} - 2\pi i)\hat{S}(\tau - \beta_1 - 4\pi i) \\
&\times \ldots \hat{S}(\tau - \beta_{n-1} - 4\pi i)\hat{S}(\tau - \beta_1 - 2\pi(k+1)) \\
&\times \ldots \hat{S}(\tau - \beta_{n-1} - 2\pi i(k+1)) \,.
\end{aligned}$$

Since $\hat{S}(\beta)$ has only one simple pole at the point $\beta = \frac{2\pi i}{N}$, the singularities of $h(\beta_1, \ldots, \beta_n)$ in the strip are caused by the singularities of

$h(\beta_1,\dots,\beta_{n-1},\tau)$, i.e., they are simple poles with residues expressed through $h_2(\beta_1,\dots,\beta_{n-2},\tau)$.

Suppose we know the function $h_2(\beta_1,\dots,\beta_{n-2},\tau)$. Then if we construct two functions $h(\beta_1,\dots,\beta_n)$ which satisfy all the above requirements, the difference between them $\Delta h(\beta_1,\dots,\beta_n)$ is an analytical function of β_n which has no singularities in the finite part of the complex plane and decreases exponentially for $\beta_n\to\pm\infty, \beta_n\in\mathbb{R}$. Such a function is, evidently, equal to zero.

Similarly, it can be shown that $h_2(\beta_1,\dots,\beta_{n-2},\tau)$ can be reconstructed uniquely through $h_3(\beta_1,\dots,\beta_{n-3},\tau)\in(V_1^{(1)}\otimes\dots\otimes V_{n-3}^{(1)}\otimes V^{(3)})^*$:

$$h_3(\beta_1,\dots,\beta_{n-3},\tau)=f(\beta_1,\dots,\beta_{n-3},\tau)\prod_{i<j}\zeta_{1,1}^{-1}(\beta_i-\beta_j)$$
$$\times\prod_{i=1}^{n-3}\prod_{i=1}^{n-3}\zeta_{1,3}^{-1}(\beta_i-\tau)\ ,$$

where $f(\beta_1,\dots,\beta_{n-3},\tau)$ is form factor corresponding to $n-3$ kinks and one rank-3 particle.

Further application of this procedure shows that $f(\beta_1,\dots,\beta_n)_{\varepsilon_1,\dots,\varepsilon_n}$, $l(\varepsilon_1)=\dots l(\varepsilon_n)=1$, is defined uniquely through $f(\varepsilon_1,\dots,\beta_{n-N+1},\beta_{n-N+2})_{\varepsilon_1,\dots,\varepsilon_{n-N+1},\varepsilon_{n-N+2}}$, $l(\varepsilon_j)=1, j\le n-N+1, l(\varepsilon_{n-N+2)}=N-1$. The latter form the vectors $f(\beta_1,\dots,\beta_{n-N+2})$ belonging to $(V_1^{(1)}\otimes\dots\otimes V_{n-N+1}^{(1)}\otimes V_{n-N+2}^{(N-1)})^*$. Define the function

$$h_{N-1}(\beta_1,\dots,\beta_{n-N+1},\tau)=f(\beta_1,\dots,\beta_{n-N+1},\tau)\ ,$$
$$\times\prod_{i<j}^{n-N+1}\zeta_{1,1}^{-1}(\beta_i-\beta_j)\prod_{i=1}^{n-N+1}\zeta_{1,N-1}^{-1}(\beta_i-\tau)\ .$$

It can be shown that the only singularities of the function in the strip $0\le\mathrm{Im}\le2\pi$ are simple poles at the points $\tau=\beta_j+\pi i$ with residues

$$\mathrm{res}\,h_{N-1}(\beta_1,\dots,\beta_{n-N+1},\tau)=\prod_{i\ne j}\Gamma\left(\frac{1}{N}-\frac{\beta_i-\beta_j}{2\pi i}\right)\Gamma\left(\frac{1}{N}+\frac{\beta_i-\beta_j}{2\pi i}\right)$$
$$\times h(\beta_1,\dots,\hat\beta_j,\dots,\beta_{n-N+1})\mathrm{sing}_{j,n-N+2}(S(\beta_{n-N+1}-\beta_j)$$
$$\times S(\beta_{j+1}-\beta_j)-S(\beta_j-\beta_1)\dots S(\beta_j-\beta_{j-1}))\ .$$

It is easy to show that all the singularities of $h_{N-1}(\dots,\tau)$ in the τ plane are simple poles with residues expressed through $h(\beta_1,\dots,\beta_{n-N})$. For $\tau\to\infty$

the function $h_{N-1}(\beta_1,\dots,\beta_{n-N+1},\tau)$ decreases exponentially for $n\geq N$ and, hence, is defined uniquely through $h(\beta_1,\dots,\beta_{n-N})$.

Thus the recurrent procedure is obtained for proving the uniqueness of $h(\beta_1,\dots,\beta_n), l(\varepsilon_1)=\dots=l(\varepsilon_n)=1$.

Q.E.D.

Theorem 1. The form factors $f_{\mu\nu}, f^a_\mu$ satisfy the requirements of C-invariance:

$$\begin{aligned} f_{\mu\nu}(\beta_1,\dots,\beta_n)_{\varepsilon_1,\dots,\varepsilon_n} &= (-1)^{\frac{1}{N}\Sigma l(\varepsilon_j)} f_{\mu\nu}(\beta_1,\dots,\beta_n)_{\varepsilon_1^*,\dots,\varepsilon_n^*}, \\ f^a_\mu(\beta_1,\dots,\beta_n)_{\varepsilon_1,\dots,\varepsilon_n} &= -(-1)^{\frac{1}{N}\Sigma l(\varepsilon_j)} f^a_\mu(\beta_1,\dots,\beta_n)_{\varepsilon_1^*,\dots,\varepsilon_n^*}\,. \end{aligned} \tag{A15}$$

Proof. Functions $(-1)^{\frac{1}{N}\Sigma l(\varepsilon_j)} f_{\mu\nu}(\beta_1,\dots,\beta_n)_{\varepsilon_1^*,\dots,\varepsilon_n^*}$ and $-(-1)^{\frac{1}{N}\Sigma l(\varepsilon_j)}$ $f^a_\mu(\beta_1,\dots,\beta_n)_{\varepsilon_1^*,\dots,\varepsilon_n^*}$ possess properties similar to those of functions $f_{\mu\nu}(\beta_1,\dots,\beta_n)_{\varepsilon_1,\dots,\varepsilon_n}$ and $f^a_\mu(\beta_1,\dots,\beta_n)_{\varepsilon_1,\dots,\varepsilon_n}$; consequently they can be reconstructed uniquely through $f_{\mu\nu}(\beta_1,\beta_2)_{\varepsilon_1^*,\varepsilon_2^*}$ and $f^a_\mu(\beta_1,\beta_2)_{\varepsilon_1^*,\varepsilon_2^*}$ in the spirit of Lemma 1. Thus the proof of the theorems reduces to the checking of (A15) for $n=2, l(\varepsilon_1)=1, l(\varepsilon_2)=N-1$, which is trivial.

Q.E.D.

C-invariance leads to interesting identities. Consider, for example, form factors $f_{++}=f_{00}+f_{11}+2f_{10}$ which form the vector $f_{++}(\beta_1,\dots,\beta_n)$. Construct the operator $\mathrm{C}_1=C_{1,1}\otimes\dots\otimes C_{n,1}$, where $C_{j,1}$ is an operator acting in $V_j^{(1)}$ via $e_{j,k}C_{j,1}=e_{j,N-k}$. It can be shown that

$$f_{++}(\beta_1,\dots,\beta_n)\mathrm{C}_1 = f_{++}(\beta_1,\dots,\beta_n)\,. \tag{A16}$$

We construct also another operator $\mathrm{C}_2=C_{1,2}\otimes\dots\otimes C_{n,2}$, where $C_{j,2}$ acts from the space of the first fundamental representation $V_j^{(1)}$ to the space of the $(N-1)$-th fundamental representation $V_j^{(N-1)}: e_{j,k}C_{j,2}=e_{j,\{1,\dots,\hat{k},\dots N\}}$. C-invariance together with (A16) imply that $f_{++}(\beta_1,\dots,\beta_n)$ $\mathrm{C}_1\mathrm{C}_2$ coincides with the form factor corresponding to rank-$(N-1)$ particles with rapidities $\beta_1,\dots,\beta_n$. Using the identities

$$w_{k_1,\dots,k_n}(\beta_1,\dots,\beta_n)\mathrm{C}_1\mathrm{C}_2 = w_{\{k_1\}^*,\dots,\{k_n\}^*}(\beta_1,\dots,\beta_n)$$
$$\times \prod_{i<j}\prod_{p=0}^{N} \frac{1}{\beta_i-\beta_j-j\frac{2\pi i}{N}} \prod_{\substack{i<j\\ k_i<k_j}} \left(\beta_j-\beta_i-\frac{2\pi i}{N}\right)^{k_j-k_i-1}$$
$$\times (\beta_i-\beta_j)^{N-k_j-k_i} \prod_{\substack{i<j\\ k_i=k_j}} (\beta_i-\beta_j)^{N-1} \prod_{\substack{i<j\\ k_i>k_j}} \left(\beta_i-\beta_j+\frac{2\pi i}{N}\right)^{k_i-k_j-1},$$

$$\zeta_{N-1,N-1}(\beta) = \zeta_{1,1}(\beta)\Gamma\left(\frac{N-1}{N}-\frac{\beta}{2\pi i}\right)\Gamma\left(\frac{N-1}{N}+\frac{\beta}{2\pi i}\right)$$
$$\times \prod_{j=1}^{N-3} \frac{1}{\mathrm{sh}\frac{1}{2}(\beta+\frac{2\pi i}{N}(N-2j+1))} \prod_{j=1}^{N-2} \frac{1}{(\beta-\frac{2\pi i}{N}j)},$$

the definition of form factors (A8) and the linear independence of $w_{k_1,\dots,k_n}$ $(\beta_1,\dots,\beta_n)$, one obtains a nice identity

$$\int dA \prod_{\alpha\in A,\beta\in B} \varphi_1^{(N)}(\alpha-\beta)\Delta(A|B^{(1)}|\dots|B^{(N)})w^+(A)$$
$$= \prod_{i<j}\Gamma\left(\frac{N-1}{N}-\frac{\beta_i-\beta_j}{2\pi i}\right)\Gamma\left(\frac{N-1}{N}+\frac{\beta_i-\beta_j}{2\pi i}\right)$$
$$\times \prod_{p=1}^{N-3} \frac{1}{\mathrm{sh}\frac{1}{2}(\beta_i-\beta_j+\frac{2\pi i}{N}(N-2p-1)}$$
$$\times \int dA' \prod_{\alpha\in A',\beta\in B} \varphi_{N-1}^{(N)}(\alpha-\beta)\tilde{\Delta}(A'|B^{(1)}|\dots|B^{(N)})$$
$$\times \frac{w_+(A'\cup \bigcup_{j=0}^{N-2}\bigcup_{\beta\in B}(\beta-\frac{\pi i}{N}(2j-1))}{\prod_{p=0}^{N-2}\prod_{\alpha\in A}\prod_{\beta\in B}\mathrm{sh}\frac{1}{2}(\alpha-\beta+\frac{\pi i}{N}(2p-1))}, \tag{A17}$$

where $B=\cup B^{(j)}, n(B^{(j)})=m$ for $\forall i, n(A)=(N-1)m-1, n(A')=m-1$ and $\tilde{\Delta}$ is the determinant of an $(m-1)\times(m-1)$ matrix with the following matrix elements:

$$\tilde{A}_{ij} = \tilde{A}_i(\alpha_j|B^{(1)}|\dots|B^{(N)}) ,$$

$$\tilde{A}_i(\alpha|B^{(1)}|\dots|B^{(N)}) = \sum_{s=1}^{N}\prod_{j\neq s}\prod_{\beta\in B^{(j)}}\left(\alpha-\beta-\frac{\pi i}{N}(j-s)\right)$$

$$\times Q_i^{(N)}\left(\alpha-\frac{\pi i}{N}(N+1-2s)|B^{(s)}+\frac{\pi i}{N}(-N+1+2s)\right) .$$

Even in the simplest case $N=3, n=3$, the identity (A17) leads to non-trivial result:

$$\int_{-\infty}^{\infty} d\alpha \prod_{j=1}^{3}\Gamma\left(\frac{1}{3}-\frac{\alpha-\beta_j}{2\pi i}\right)\Gamma\left(\frac{1}{3}+\frac{\alpha-\beta_j}{2\pi i}\right)(3\alpha-\beta_1-\beta_2-\beta_3)e^{\alpha}$$

$$=\prod_{i<j}\Gamma\left(\frac{2}{3}-\frac{\beta_i-\beta_j}{2\pi i}\right)\Gamma\left(\frac{2}{3}+\frac{\beta_i-\beta_j}{2\pi i}\right)(e^{\beta_1}+e^{\beta_2}+e^{\beta_3}) .$$

We believe in the convergence of the series

$$\eta=\sum_{n=2}^{\infty}\sum_{l_1,\dots,l_n}\int_{\beta_n>\dots>\beta_1}||g^a(\beta_1,\dots,\beta_n)_{l_1,\dots,l_n}||^2 d\beta,\dots d\beta_n ,$$

where $a=\{i\}, i=1,\dots,N-1$, the sum being taken over all possible values of the ranks of particles with rapidities $\beta_1,\dots,\beta_n$. Assuming the convergence one can calculate the singularities of commutators of currents at the origin of coordinates:

$$[j_0^a(x),j_0^b(0)] = if^{abc}j_0^c(0)\delta(x) ,$$
$$[j_1^a(x),j_1^b(0)] = if^{abc}j_0^c(0)\delta(x) ,$$
$$[j_0^a(x),j_1^b(0)] = if^{abc}j_1^c(0)\delta(x)+4\pi i\eta\delta^{ab}\delta'(x) .$$

At this point we finish the consideration of the SU(N)-invariant Thirring model. The sketchiness of the exposition is caused by the fact that full proofs, while not very instructive, would occupy to much place.

APPENDIX B
PHENOMENOLOGICAL REASONINGS

This Appendix contains some heuristic speculations on the derivation of Axioms 2, 3 in the framework of the axiomatic field theory.

Consider for simplicity the form factor of a scalar operator calculated between the vacuum and a state which contains n one-component particles with rapidities $\beta_1, \dots, \beta_n$. Denote this form factor by

$$f(\beta_1, \dots, \beta_n) = \langle \mathrm{ph}|O(0)Z^*(\beta_n)\dots Z^*(\beta_1)|\mathrm{ph}\rangle\ .$$

It is natural to suppose that $f(\beta_1, \dots, \beta_n)$ allows analytical continuation with respect to all its arguments. Let us concentrate on analytical properties of $f(\beta_1, \dots, \beta_n)$ considered as a function of β_n. Suppose $\beta_1 < \dots < \beta_{n-1} \leq \beta_n$, then

$$f(\beta_1, \dots, \beta_n) = \langle \mathrm{ph}|O(0)|\beta_1, \dots, \beta_n\rangle_{\mathrm{in}}\ .$$

Let us use the usual trick (see for example [56, 57]) convoluting the particle β_n:

$$\begin{aligned}
f(\beta_1, \dots, \beta_n) &= \langle \mathrm{ph}|O(0)a_{\mathrm{in}}^*(\beta_n)|\beta_1, \dots, \beta_{n-1}\rangle_{\mathrm{in}} \\
&= \lim_{x_0 \to -\infty} \int e^{-ip^\mu(\beta_n)x_\mu} \overleftrightarrow{\partial}_0 \langle \mathrm{ph}|O(0)\varphi_{\mathrm{in}}(x)|\beta_1, \dots, \beta_{n-1}\rangle_{\mathrm{in}} dx_1 \\
&= \lim_{x_0 \to -\infty} \int e^{ip^\mu(\beta_n)x_\mu} \overleftrightarrow{\partial}_0 \langle \mathrm{ph}|[O(0), \varphi_{\mathrm{in}}(x)]|\beta_1, \dots, \beta_{n-1}\rangle_{\mathrm{in}} dx_1 \\
&\quad + \langle \mathrm{ph}|a_{\mathrm{in}}^*(\beta_n)O(0)|\beta_1, \dots, \beta_{n-1}\rangle_{\mathrm{in}}\ .
\end{aligned}$$

Evidently the last term is equal to zero. Proceeding in the usual calculation [57] one gets

$$f(\beta_1, \dots, \beta_n) = \int dx e^{-ip^\mu(\beta_n)x_\mu} \theta(-x_0) \langle \mathrm{ph} | [O(0), \eta(x)] | \beta_1, \dots, \beta_{n-1} \rangle_{\mathrm{in}} , \tag{B1}$$

where $\eta(x)$ is defined by $(\Box + M^2)\varphi(x) = \eta(x)$, $\varphi(x)$ is the interpolating field: $\varphi(x) \xrightarrow{w} \varphi_{\mathrm{in}}(x), \varphi(x) \xrightarrow{w} \varphi_{\mathrm{out}}(x)$. The integral in (B1) is taken over all the space-time. It is clear, however, that the real range of integration coincides with the lower light cone because of the presence of $\theta(-x_0)$ in (B1) and of the annihilation of $[O(0), \eta(x)]$ on the space-like interval. Roughly speaking, the possibility of analytical continuation is determined by the behaviour of the exponent $\exp(ip^0(\beta_n)x_0)$ for $x_0 < 0$. Clearly, this exponent decreases when $0 < \mathrm{Im}\beta_n < \pi, \mathrm{Re}\beta_n > 0$. The condition $\mathrm{Re}\beta_n > 0$ is to be understood as meaning that $\mathrm{Re}\beta_n$ is greater than all the remaining rapidities (evidently the initial definition of $f(\beta_1, \dots, \beta_n)$ allow us to shift simultaneously all the rapidities). More refined consideration needs all the particles to be convoluted. When this is done one makes sure that the "true" value of $\mathrm{Re}\beta_n$ is determined by its value taken relative to $\beta_1, \dots, \beta_{n-1}$. So we have presented some speculations in favour of the possibility to continue analytically the integral (B1) into the range $0 < \mathrm{Im}\beta_n < \pi$.

Similarly for

$$f(\beta_n, \beta_{n-1}, \dots, \beta_1) = \langle \mathrm{ph} | O(0) | \beta_1, \dots, \beta_n \rangle_{\mathrm{out}} ,$$

one gets

$$f(\beta_n, \beta_{n-1}, \dots, \beta_1) = \int dx e^{-ip^\mu(\beta_n)x_\mu} \theta(x_0) \langle \mathrm{ph} | [O(0), \eta(x)] | \beta_1, \dots, \beta_{n-1} \rangle_{\mathrm{out}} . \tag{B2}$$

It seems likely that the integral (B2) allows analytical continuation into the range $-\pi < \mathrm{Im}\beta_n < 0$.

Consider also two matrix elements

$$f(\beta_n | \beta_1, \dots, \beta_{n-1}) = \langle \beta_n | O(0) \rangle | \beta_1, \dots, \beta_{n-1} \rangle_{\mathrm{in}} ,$$
$$f(\beta_n | \beta_{n-1}, \dots, \beta_1) = \langle \beta_n | O(0) | \beta_1, \dots, \beta_{n-1} \rangle_{\mathrm{out}} .$$

One can obtain for them presentations similar to (B1), (B2). There is, however, one difference. When convoluting the particle β_n in the form factor $f(\beta_n|\beta_1,\dots,\beta_{n-1})$, one obtains the expression $\varphi_{\rm in}(x)O(0)$ and wants to replace it by the commutator $[\varphi_{\rm in}(x), O(0)]$. Here the following discrepancy appears:

$$\delta(\beta_n-\beta_{n-1})\langle {\rm ph}|O(0)|\beta_1,\dots,\beta_{n-1}\rangle_{\rm in}\ .$$

Keeping in mind that β_n can coincide only with β_{n-1} one makes sure that this discrepancy is equal to

$$\langle {\rm ph}|O(0)a_{\rm in}(\beta_n)|\beta_1,\dots,\beta_{n-1}\rangle_{\rm in}\ .$$

Taking into account this remark one gets

$$\begin{aligned} f(\beta_n|\beta_1,\dots,\beta_{n-1}) &= \delta(\beta_n-\beta_{n-1})f(\beta_1,\dots,\beta_{n-2}) \\ &+\int dx e^{ip^\mu(\beta_n)x_\mu}\theta(-x_0)\langle {\rm ph}|[O(0),\eta(x)]|\beta_1,\dots,\beta_{n-1}\rangle_{\rm in}\ , \\ f(\beta_n|\beta_{n-1},\dots,\beta_1) &= \delta(\beta_n-\beta_{n-1})f(\beta_{n-2},\dots,\beta_1) \\ &+\int dx e^{ip^\mu(\beta_n)x_\mu}\theta(x_0)\langle {\rm ph}|[O(0),\eta(x)]|\beta_1,\dots,\beta_{n-1}\rangle_{\rm out}\ . \end{aligned} \tag{B3}$$

Comparing (B1), (B2), (B3) and using the symmetry property

$$f(\beta_1,\dots,\beta_i,\beta_{i+1},\dots,\beta_n)S(\beta_i-\beta_{i+1}) = f(\beta_1,\dots,\beta_{i+1},\beta_i,\dots,\beta_n)\ , \tag{B4}$$

one gets

$$\begin{aligned} f(\beta_1,\dots,\beta_n+\pi i-i0) &= f(\beta_n-\pi i+i0,\beta_1,\dots,\beta_{n-1}) \\ &+\delta(\beta_n-\beta_{n-1})f(\beta_1,\dots,\beta_{n-2})\left(1-\prod_{j=1}^{n-2}S(\beta_{n-1}-\beta_j)\right)\ , \\ \beta_n \ge \beta_{n-1} &> \beta_{n-2} > \dots > \beta_1\ . \end{aligned}$$

This equation means that the function $f(\beta_1,\dots,\beta_n)$ can be continued analytically up to ${\rm Im}\beta_n = \pi$ and has on this line a simple pole at the point $\beta_n = \beta_{n-1}+\pi i$ with the residue

$$\frac{1}{2\pi i}f(\beta_1,\dots,\beta_{n-2})\left(1-\prod_{j=1}^{n-2}S(\beta_{n-1}-\beta_j)\right)\ ,$$

and it coincides with the analytical continuation of $f(\beta_n, \beta_1, \ldots, \beta_{n-1})$ to the line $\mathrm{Im}\beta_n = -\pi$. The symmetry (B4) implies that if no restriction of $\beta_1, \ldots, \beta_n$ is imposed $f(\beta_1, \ldots, \beta_n)$ has simple poles at the points $\beta_n = \beta_j + \pi i$ with the proper residues.

Although the above speculations cannot be considered as a rigorous proof of Axioms 2, 3 it seems that they can be refined in order to get a proof on the level of rigour acceptable in the axiomatic field theory.

REFERENCES

1. V. E. Korepin and L. D. Faddeev, *Teor. Mat. Fiz.* **25** (1975) 1039.
2. L. D. Faddeev and V. E. Korepin, *Physics Reports* **420** (1978) 1–87.
3. S. Coleman, *Phys. Rev.* **D11** (1975) 2088.
4. A. M. Polyakov, *Phys. Lett.* **B59** (1975) 79.
5. A. B. Zamolodchikov, *Pisma ZhETF* **25** (1977) 499.
6. A. B. Zamolodchikov and Al. B. Zamolodchikov, *Annals of Phys. (NY)* **120** (1979) 253–291.
7. B. Berg, M. Karowski, V. Kurak and P. Weisz, *Nucl. Phys.* **B134** (1978) 125.
8. M. Karowski, *Nucl. Phys.* **B153** (1980) 244–252.
9. M. Karowski and H. J. Thun, *Nucl. Phys.* **B190** (1981) 61–92.
10. L. D. Faddeev and E. K. Sklyanin, *Dokl. Acad. Nauk SSSR* **243** (1978) 1430.
11. E. K. Sklyanin, *Dokl. Acad. Nauk SSSR* **244** (1978) 1337.
12. L. D. Faddeev, E. K. Sklyanin and L. A. Takhtajan, *Teor. Mat. Fiz.* **40** (1979) 194.
13. V. E. Korepin, *Teor. Mat. Fiz.* **41** (1979) 169.
14. L. D. Faddeev, *Soviet Sci. Reviews, Math. Phys.* **C1** (1980) 107.
15. P. P. Kulish and E. K. Sklyanin, *Lect. Notes in Physics* **151** (1982) 61.
16. A. G. Izergin and V. E. Korepin, *Elementary Particles and Atomic Nuclei* **13** 207.
17. D. B. Creamer, H. B. Thacker and D. Wilkinson, *Phys. Rev.* **D21** (1980) 1523.
18. M. Göckeler, *Z. Phys.* **C11** (1981) 125.
19. F. A. Smirnov, *Dokl. Acad. Nauk SSSR* **262** (1982) 78.
20. P. P. Kulish and F. A. Smirnov, *Phys. Lett.* **A90** (1982) 74.
21. P. P. Kulish and F. A. Smirnov, *J. Phys.* **C18** (1985) 1037.
22. P. P. Kulish, N. Yu. Reshetikhin and E. K. Sklyanin, *Lett. Math. Phys.* **5** (1981) 393.
23. V. E. Korepin, *Comm. Math. Phys.* **94** (1984) 93.

24. N. M. Bogoliubov, A. G. Izergin and V. E. Korepin, *Lect. Notes in Physics* **242** (1985).
25. F. A. Smirnov, *Teor. Mat. Fiz.* **60** (1980) 356.
26. F. A. Smirnov, *J. Phys.* **A17** (1984) L873.
27. F. A. Smirnov, Preprint LOMI E-3-86, 1986.
28. F. A. Smirnov, *Teor. Mat. Fiz.* **67** (1986) 40.
29. F. A. Smirnov, *Teor. Mat. Fiz.* **71** (1987) 341.
30. F. A. Smirnov, *J. Phys.* **A19** (1986) L575.
31. F. A. Smirnov, *Zap. nauchn. semin. LOMI* **161** (1987) 98.
32. I. M. Khamitov, *Zap. nauchn. semin. LOMI* **146** (1985) 147.
33. A. N. Kirillov, *Zap. nauchn. semin. LOMI* **146** (1975) 9.
34. M. Karowski and P. Weisz, *Nucl. Phys.* **B139** (1978) 455.
35. A. N. Kirillov and F. A. Smirnov, *Phys. Lett.* **B198** (1987) 506.
36. A. N. Kirillov and F. A. Smirnov, *Zap. nauchn. semin. LOMI* **164** (1987) 80.
37. A. N. Kirillov and F. A. Smirnov, *Int. J. of Modern Phys.* **A3** (1988) 731.
38. F. A. Smirnov, Preprint LOMI E-1-88, 1988.
39. C. N. Yang, *Phys. Rev.* **168** (1968) 1920.
40. P. J. Baxter, *Exactly Solved Models in Statistical Mechanics*, Academic Press, London, 1982.
41. A. G. Izergin and V. E. Korepin, *Lett. Math. Phys.* **5** (1981) 199.
42. V. E. Korepin, *Pisma ZhETF* **30** (1979) 633.
43. N. M. Bogoliubov and A. G. Izergin, *Teor. Mat. Fiz.* **61** (1984) 364.
44. H. Bergkhnoff and H. B. Thacker, *Phys. Rev. Lett.* **42** (1979) 135.
45. G. I. Japaridze, A. A. Nersesyan and P. B. Wiegmann, *Phys. ser.* **27** (1983) 5.
46. A. A. Belavin, *Phys. Lett.* **B87** (1979) 117.
47. L. D. Faddeev and L. A. Takhtajan, Preprint LOMI E-4-83, 1983.
48. P. B. Wiegmann, *Pisma ZhETF* **1** (1985) 79.
49. L. D. Faddeev and N. Yu. Reshetikhin, *Annals of Phys.* **167** (1986) 227.
50. V. A. Fatteev, I. V. Frolov and A. S. Schwartz, *Nucl. Phys.* **B154** (1979) 1.
51. P. Berg and M. Lüsher, *Comm. Math. Phys.* **96** (1979) 57.
52. E. Brezin and J. Zinn-Justin, *Phys. Rev.* **B69** (1976) 3110.
53. A. M. Polyakov, *Phys. Lett.* **B72** (1977) 224.
54. I. Ya. Aref'eva, P. P. Kulish, E. R. Nissimov and S. J. Pacheva, Preprint LOMI E-1-79, 1979.
55. P. P. Kulish, *Zap. nauchn. semin. LOMI* **109** (1981) 83.
56. J. D. Bjorken and S. D. Drell, *Relativistic Quantum Fields*, McGraw-Hill Book Company.
57. G. Barton, *Introduction to Dispersion Techniques in Field Theory*, W. A. Benjamin Inc., 1965.
58. C. Destry and J. H. Lewenstein, *Nucl. Phys.* **B205** (1982) 239.
59. L. D. Faddeev and L. A. Takhtajan, *Phys. Lett.* **A85** (1981) 375.
60. N. Andrei and J. Lowenstein, *Phys. Lett.* **B90** (1980) 106.
61. P. P. Kulish and N. Yu. Reshetikhin, *ZhETF* **80** (1981) 214.
62. F. A. Berezin, *Method of Secondary Quantization*, Nauka, Moscow, 1965.